Ocean of Life

CALLUM ROBERTS

Ocean of Life

How Our Seas Are Changing

ALLEN LANE
an imprint of
PENGUIN BOOKS

To my daughters, who love the sea

ALLEN LANE

Published by the Penguin Group

Penguin Books Ltd, 80 Strand, London WC2R 0RL, England

Penguin Group (USA) Inc., 375 Hudson Street, New York, New York 10014, USA

Penguin Group (Canada), 90 Eglinton Avenue East, Suite 700, Toronto, Ontario, Canada M4P 2Y3
(a division of Pearson Penguin Canada Inc.)

Penguin Ireland, 25 St Stephen's Green, Dublin 2, Ireland (a division of Penguin Books Ltd)

Penguin Group (Australia), 250 Camberwell Road, Camberwell, Victoria 3124,
Australia (a division of Pearson Australia Group Pty Ltd)

Penguin Books India Pvt Ltd, 11 Community Centre,
Panchsheel Park, New Delhi – 110 017, India

Penguin Group (NZ), 67 Apollo Drive, Rosedale, Auckland 0632, New Zealand
(a division of Pearson New Zealand Ltd)

Penguin Books (South Africa) (Pty) Ltd, Block D, Rosebank Office Park,
181 Jan Smuts Avenue, Parktown North, Gauteng 2193, South Africa

Penguin Books Ltd, Registered Offices: 80 Strand, London WC2R 0RL, England

www.penguin.com

First published 2012
001

Set in 10.5/14 pt Sabon LT Std
Typeset by Jouve (UK), Milton Keynes
Printed in Great Britain by Clays Ltd, St Ives plc

978-1-846-14394-6

npenguin.co.uk

MIX
Paper from
responsible sources
FSC
www.fsc.org   FSC™ C018179

Penguin Books is committed to a sustainable
future for our business, our readers and our planet.
This book is made from Forest Stewardship
Council™ certified paper.

ALWAYS LEARNING   PEARSON

Contents

CONTENTS

List of Illustrations

'Roll on, thou deep and dark blue Ocean – roll!
Ten thousand fleets sweep over thee in vain;
Man marks the earth with ruin – his control
Stops with the shore'

Byron, *Childe Harold's Pilgrimage*

Prologue

The water felt chilly as I waded out to fetch the battered skiff from its mooring. It slid easily over the glassy lagoon to the beach where Julie waited with our diving gear. We had been married a month and, in lieu of a honeymoon, I had persuaded her to accept two months of fieldwork studying fish behaviour on this remote patch of Australia's Great Barrier Reef. It was June 1987, winter in Australia and summer back home. Two herons picked their way along the shoreline looking for breakfast. They flapped off as the outboard engine coughed to life and we set course for a spot a mile away across a maze of coral so complex that it would have baffled the most capable navigator.

We anchored on a rubble ridge that separated the lagoon from the open sea. This was our first ocean dive here and the thrill of expectation was tempered by a frisson of fear. Ahead of us, the homely greens and browns of the shallow reef gave way to the dark indigo of the deep. Huge buttresses of coral plunged down hundreds of feet in parallel walls separated by deep channels. Vivid purple finger corals vied for space with yellow lettuce-coral, while great mounds of blue and green polyps rose from the bottom.

The reef was a confusing swirl of movement, like Grand Central Station at rush hour. Stocky surgeonfish with electric-blue stripes raced around purposefully while loose groups of gaudy parrotfish dodged them. Clouds of damselfish and anthias hovered above, delicately plucking invisible food from the water. Along the edge of the reef I counted eight loggerhead turtles, one for each buttress in view. A grey reef shark headed our way, but barely acknowledged us as he passed by. The whole scene felt timeless and primordial. I was transfixed.

I wonder what I would have thought if, on our return to shore, we had been greeted by a curmudgeonly sage who prophesied that in one hundred years this magnificent reef would be a crumbling ruin, that its bright coral escarpments would be replaced by green fuzz, and the ranks of fish would have thinned and given way to swarms of jellyfish and gelatinous plankton. I would probably have thought him mad. Nothing could have seemed less likely. And yet, less than twenty-five years later, most serious marine scientists predict just such a fate. We can already see it happening.

Eleven years after our honeymoon dive, in a foretaste of a warmer world to come, the oceans heated so dramatically that a quarter of the world's coral died. In much of the Indian Ocean, between 70 and 90 per cent of all corals died, taking with them countless creatures that used coral for living space or food. If three-quarters of our forests had withered and died that year, people would have demanded to know why and aggressive plans would have been drawn up for their recovery. Yet outside the world of marine science, this global catastrophe passed largely unseen and unremarked.

For most of civilized history, the world's oceans have been very stable. Since the seas levelled off six or seven thousand years ago, after the last ice age, they have for the most part been predictable. Yes, coasts have retreated and advanced under the relentless influence of wave and tide, but the oceans themselves seemed changeless. Their constancy contrasted with the world above water, where landscapes underwent dramatic alterations as first pastoralists and agriculture spread, and then later cities and industry. Today it is the turn of the oceans.

This is a book about the sea change unfolding around our planet. In the last half century human dominion over nature has finally reached the oceans. The speed and extent of these changes has caught us unprepared. The sea is becoming more hostile to life, and not just for the creatures that swim, scuttle or crawl beneath the waves but for us too. Only in the last decade or so have we begun to recognize how our activities are reshaping the oceans, and what that means for our own well-being.

We have long known how humankind has changed the land, modifying landscapes to suit our needs and affecting its wildlife across thousands of years – tens of thousands of years if you consider how Australian Aboriginals and Native Americans used fire to clear vegetation and facilitate hunting and gathering. But we persist in believing, as Lord Byron put it, that humanity's dominion stops at the sea. Yet, even in Byron's time human impacts on the sea were significant. The great auk stood two decades from extinction, and the Atlantic grey whale had vanished for ever. Fishing had begun to deplete stocks close to coasts and to damage habitats with trawls and dredges. People were building coastal defences and large areas of marsh and estuary had been reclaimed for ports and agriculture. In populous regions, rivers filled coastal estuaries and bays with mud washed from land exposed by the plough.

Our influence has grown exponentially since then. The last two hundred years have seen marine habitats wiped out or transformed beyond recognition. And with an ever-accelerating tide of human impact, the oceans have changed more in the last thirty years than in all of human history before. In most places, the oceans have lost upwards of 75 per cent of their megafauna – large animals such as whales, dolphins, sharks, rays and turtles – as fishing and hunting has spread in waves across the face of the planet. For some species, numbers are down as much as 99 per cent, as is the case for the oceanic whitetip sharks of the high seas, American sawfishes and the 'common' skate of Northern Europe. By the end of the twentieth century, almost nowhere shallower than a kilometre remained untouched by commercial fishing, and some places are now fished to three kilometres down.

For thousands of years the oceans have been our conduit for commerce. Today they are the highways of a globalized world and the roar of engines can be heard in every corner of the sea, even beneath polar ice. Increasingly, they also provide us with oil and gas and growing scarcity has driven us to venture deeper and further offshore. We are within a decade or so of the onset of deep sea mining. Riches beckon in the blackness of the abyss, thousands of metres down. Dark nuggets of precious metals and rare earths lie scattered over the ocean bottom, seamounts are crusted in cobalt, and deposits of gold, silver and manganese spewed forth from superheated springs are nearly within reach of the miners.

Why, in the face of widespread evidence of human impact, do so many people persist in thinking that the oceans remain wild and beyond our influence? The answer lies in part in the creeping rate of change. Each generation forms its own view of the state of the environment. Younger people generally fail to perceive changes experienced by the elderly, and so knowledge of past conditions fades with time. Younger generations are often dismissive of the tales of old-timers, rejecting their stories in favour of things they have experienced themselves. The result is a phenomenon known as 'shifting baseline syndrome', as we take for granted things that two generations ago would have seemed inconceivable.

Loren McClenachan unearthed a telling example of shifting baselines in the archives of the Monroe County Library in Florida when she was doing research there as a graduate at Scripps Institution of Oceanography.[1] She found a series of photographs of catches of fish landed into Key West by one recreational fishing charter company between the 1950s and the 1980s. She extended the series into the twenty-first century with her own pictures taken at the same dock. In the 1950s, huge Goliath groupers and sharks dominated catches, many of them bigger and fatter than the anglers. As the years pass by, the fish shrink and groupers and sharks give way to smaller snappers and grunts. But the grins on the anglers' faces are just as broad today as they were in the 1950s. Modern-day tourists have no idea that anything has changed.

The oceans are changing faster than at almost any time in Earth's history and we are the agents of that transformation. Many of these changes will test the ability of its denizens to survive into the future. These alterations are also reshaping our own relationship with the sea and threaten many of the things that we most cherish and take for granted. Our failure to notice creeping environmental degradation has compromised more than our quality of life. In extreme circumstances it threatens human welfare. History offers many examples of civilizations that have been destroyed by environmental catastrophes that they have unwittingly brought on themselves. Easter Islanders cut down all their homeland's trees to raise statues to their gods and then starved when their soil dried up. Mesopotamians invented sophisticated irrigation agriculture, but the technique eventually left fields so salty they could no longer grow crops. Mayan hill-slope farming practices stripped the region of its topsoil, precipitating the collapse of this extraordinary civilization during prolonged drought. In these and many similar cases, adverse effects were local to an island or region. Today, our influences are global, and we will have to act globally to reverse the impact of what we have already done.

I began my career studying coral reef fish. From the moment I first dived a Red Sea reef, I was hooked (my wife, Julie, was similarly hooked by our honeymoon trip to the Great Barrier Reef and also became a marine biologist). Thirty years on, fish are still at the heart of my research but my outlook has expanded to a much wider interest in the relations between people and the sea over the course of history. Nonetheless, when I began the research for this book there were huge stretches of ocean science of which I was only dimly aware. Scientists are specialists and devote their lives to research within narrow fields that become further constricted as time passes. Each pores over a fragment of the world, turning it over and over in his or her mind like a chip of some mosaic. Management of pollution is segregated from that of fisheries, which in turn are rarely considered in the same place as shipping or climate change. This means that impacts are discussed in isolation at different meetings and by different people who never quite see the overall picture. I decided to write this book because I felt there was an urgent need to bring all these separate fiefdoms together. What I found along the way has been a revelation.

We fear change and resist it. Perhaps that resistance is hardwired into our genes: the familiar seems safer than the unknown. Many animals go to strenuous lengths to return to the place where they were born to breed, probably because past success there gives greater assurance of future success. This is a dynamic world and change sometimes brings good things, but some changes, notably those which erode our planet's resilience, are harmful, for when resilience is worn away all bets are off for the future. The path we are on today, as I will show, is pushing ocean ecosystems to the edge of viability. Heedless of consequences, we are emptying the seas of fish and filling them with pollution, and our unplanned experiment with greenhouse gases is gradually infiltrating the deep sea.

Although a few of the human influences that I will describe in this book have been underway for centuries, others have really only taken hold in the last fifty years. In this sense, our impacts on the oceans have been sudden – instantaneous almost – given that they have occurred in less than one-thousandth of the 150,000 years or so during which modern humans have existed. The response demanded to counter these impacts will also have to be sudden and global in scale. Few people yet grasp the gravity of our predicament.

In this book I will take you on a voyage beneath the waves to reveal the seas as few know them. I will show how human activities have for centuries been unpicking the fabric of marine life. Until recently we have been able to ignore much of the harm done by our heedless use of the sea. But, as the scale and intensity of human influence have grown, the rate of change has accelerated and we must now confront the consequences.

To understand the present we must first know the past. I begin at the beginning, when the world began, before picking up our own story when humankind first makes its appearance on Earth. For tens of thousands of years, our only real impact on the oceans was to remove fish and shellfish, so I start with a brief history of hunting and fishing and how they have evolved through time. The industrial revolution heralded the emergence of people as agents of planetary change, and I move on to describe how the use of fossil fuels, and their impact on currents and climate, is transforming the sea in ways not seen for hundreds of thousands or even millions of years. Sea levels are rising faster than the highest rates predicted only twenty years ago. They now threaten dozens of the world's great metropolises, and within fifty years could inundate vast tracts of our best agricultural land, imperilling food security. In one of the least known but potentially most damaging effects of greenhouse gas emissions, within the seas themselves acidity is going up in step with carbon dioxide. The outcome could be catastrophic. Shell-forming animals, including many that sustain ocean food webs, and thereby our own fisheries, will find their lives increasingly difficult. Not for 55 million years has there been oceanic disruption of comparable severity to the calamity that lies just a hundred years ahead if we fail to curtail emissions fast.

By absorbing heat the oceans have so far spared us from the worst of global warming. But warming seas have set life on the move, so in the years to come the fortunes of some fishing nations will wane as others rise. But warming will have far more grave effects on marine productivity, driving it to excess in some places while turning others into oceanic deserts. If climate change and fishing were our only threats to life in the sea, matters would be bad enough. But ocean life faces other pressures, like pollution. I survey pollution effects ranging from toxic chemicals and the now ubiquitous plastic, to sewage and fertilizers and less familiar pollutants like noise and invasive species. Pollution problems have grown in severity with time, so much so that in many regions we see the emergence and spread of dead zones as oxygen is sucked from the water by decaying plankton. The cocktail of human stresses is mixed differently from place to place, but the result is the same: their effects in combination are far worse than in isolation. We are transforming life in the sea, and with it we are undermining our own existence.

This book is not a catalogue of unavoidable disasters ahead. There is much we can do to change course, if we take the opportunity now. But time is of the essence. The longer we ignore problems or prevaricate, the less leeway we have to avoid the direst of our possible futures. I devote the rest of the book to how we can plot a new course to safeguard the oceans and ourselves. What we need, I will argue, is an ambitious plan to reverse long-term trends of depletion and degradation, to recapitalize the value of the oceans to people and wildlife, and to improve the quality of everyone's lives, especially of generations yet to come. We don't have to look on helpless as all that we love about the sea is sullied. Change for good is within our reach.

I

Four and a Half Billion Years

In a remote part of Western Australia there is a range of hills so smoothed by time that little more than undulations are left. Summer sun bakes the red earth and sparse trees dig deep to quench their thirst. Close up, jumbled slabs reveal beds of smoothed pebbles brought here by ancient floods – very ancient, for the Jack Hills rocks were laid down 3 billion years ago. This landlocked desert may seem like a strange place to begin a book about the oceans, but, as unremarkable as they appear, these stones have rewritten the earliest history of our planet.

Zircons are pyramid-shaped crystals that coalesce from cooling magma. They are incredibly resilient and survive through repeated re-melting and hardening as the crust of our restless planet is recycled by plate-tectonic movements. Zircons form gems of many hues. The name is thought to come from the Persian *zargun*, which means golden, probably in reference to the translucent yellow stones that were once traded through Persia from Sri Lanka to Europe and China. Brilliant sky-blue crystals were mined from river mud by the people who built the temple city of Angkor Wat in Cambodia.

Jack Hills' zircons are less showy. In fact they are so small they can hardly be seen with the naked eye. It took over 200 pounds of Jack Hills rock to yield less than a thimbleful of zircons.[1] But it was worth the effort. Each crystal was found to contain traces of uranium that decays over time into lead, making it a geological chronometer that ticks away the eons. By measuring the ratio of uranium to lead, you can date the time of crystal formation to within a few tens of millions of years. The most ancient Jack Hills crystal is 4.4 billion years old, which takes us to within a whisker of the formation of Earth (well, a

170-million-year whisker, but that is close!). It lets us glimpse a planet in its infancy.

We can never be sure of all the details of Earth's birth and childhood; no one was there to see it, and we can't rewind time to have a look. The best we can do is to search the world for ancient rocks and fossils and probe them for their secrets. We spin theories that fit the evidence and rework them when new data find them wanting. These days, new ways of interrogating the past are developed with breathtaking speed, so the picture is coming into clearer focus. Although many uncertainties remain in the story I will tell in this chapter, the broad pattern of events is well established. About 4.57 billion years ago, the Earth coalesced from the rotating disc of dust and debris that would become our solar system.[2] The sun is around the same age, so our planet formed when the match was struck to light the solar system. As debris pounded the growing planet into shape it grew hotter until the rocks melted. Seen from the darkness of space, our world would have glowed like a faint sun. Sometime around 4.53 billion years ago, a planet the size of Mars barrelled into this new world and smashed off pieces that became the moon. The impact was so enormous it vaporized rock, and the primeval world was shrouded by a thick atmosphere of this vaporized rock and other gases. As the planet cooled, a time came when minerals condensed, and for 2,000 years the skies rained molten rock onto an ocean of magma below. Even after this thousand degree deluge ended, the atmosphere remained thick with vapour and gases, and atmospheric pressure at the surface of the flaming sea would have been hundreds of times higher than today.[3]

We owe a lot to this collision. It knocked Earth's axis of rotation askew, which gives us seasons.[4] Over the vast plain of geological time, the moon has slowed and stabilized the Earth's rotation, giving us longer days. A billion years ago, a day was just 18 hours and the year lasted 480 days. Early on, the moon was much closer to Earth and would have loomed large in the sky. The moon's gravitational pull gives us much greater tides than the more distant sun alone and the flood and ebb of tides would have been violent as they rose higher and fell faster in the short days.[5]

The searing heat of fireball Earth is called the Hadean. For a long time, we thought these hellish conditions lasted for the first half billion years of Earth history, but a zircon from Jack Hills has recently transformed that view. The 4.4-billion-year-old crystal and tiny particles of other materials trapped within it formed at temperatures characteristic of coalescing granite in contact with liquid water.[6] The continents are built from granite, so we can see within this insignificant speck the imprint of a cooler world with land and water long before such a place was suspected to exist. Instead of seething oceans of fire, the world might have been more like a steamy sauna.

The Earth is unique within the solar system for its liquid oceans and seas. Where did all this water come from? There was water in the disc of dust and gas from which the solar system formed, but some scientists believe the inner region of this disc where the Earth formed was too hot for water.[7] They contend that icy comets and asteroids delivered water from the further reaches of the solar system long after Earth's formation. This barrage goes on today; every few seconds a 'snowball' the size of a large truck melts into the outer atmosphere. But measurements of the isotopic composition of these snowballs suggest they only contributed a few per cent of our water.[8] Meteorites were assumed to have delivered most of the rest, but recently a comet has been found with a very similar isotopic composition of water to Earth's oceans, so comets could have contributed a bigger share than once thought.[9] Another view is that water molecules drifting in space stuck to particles of dust, and that rock and water came together at the same time. Our world actually has enough water for five to ten oceans, perhaps more; most remains trapped within the bedrock. By weight, about half a per cent of basalt rock is water, trapped within the mineral lattice.[10] As the early world heated up and rocks melted, water boiled off into the atmosphere. For over a hundred million years, the oceans were in the sky as a dense shroud of vapour that churned above the glowing surface of our planet.

When the Earth cooled down a bit, a new rain began to fall, this time of scalding water. The downpour lasted for thousands of years and would be repeated several times over the next half billion years, as giant impacts from extraterrestrial debris boiled off the upper layers of the seas. For a half billion years more, the Earth continued to be struck by asteroids far bigger than the one that killed off the dinosaurs 65 million years ago. Few traces of these impacts remain, for the Earth's crust is continuously reworked into the planet's interior, but we can read their fury from the moon whose cratered surface records the bombardment and which has remained immobile since it cooled.

To begin with, oceans covered most of the world. Their volume could have been twice that of the present seas.[11] Islands cropped up where blocks of the Earth's crust collided and volcanoes built ash and lava mountains, but there were no continents. Those came later and formed slowly. The world today is broken up into continents and sea because the crust is made up of two kinds of rock. Below the sea it is made of basalt and is slightly denser than the granites that make up the continents. Repeated re-melting in the furnace of our young world separated off the lighter granitic rocks that form today's continents from the basalt. Both kinds of rock float on a sea of hot, viscous magma in the Earth's mantle below, but the continents float higher. Like icebergs, how much you see above water depends on how much lies unseen. Continents have deep roots so they float higher in the mantle than the oceanic crust. This is why the average height of land is 840 metres above sea level, while the average depth of the oceans is 3,680 metres, a consequence of the different densities of their rocks.[12]

To us, the oceans are immense. They cover 360 million square kilometres and fill a volume of 1.33 billion cubic kilometres. It is hard to imagine a cubic kilometre. If you flooded all of New York's Central Park with water to the height of a thirty-storey building, that would be about half a cubic kilometre: it would take 2.66 billion Central Parks of water to fill the oceans, an almost unthinkable volume. Yet at the scale of the planet, the oceans form a layer only as thick as the skin of an apple.[13]

The sticky magma of the Earth's mantle keeps the world's surface in constant motion. Hot magma rising from deep down creates new crust in some places. Since the world is not getting bigger, that creative force is matched by destruction somewhere else as crust slides back into the mantle. The Earth's surface is thus divided into blocks, called plates, which move in a slow-motion geological dance through time. Here again, the density difference between oceanic and continental crusts comes into play. Because oceanic crust is thin and dense (four to ten kilometres thick, compared to the thirty to forty kilometres of most continental granite), it is recycled quickly into the mantle, about ten times more quickly than the continents. (None of the oceanic crust is greater than 200 million years old, whereas about 7 per cent of land is older than 2.5 billion years.[14]) Think of the continents as the froth bobbing on a pool beneath a waterfall. The water pours on beneath, but the froth endures.

Continents began to form very early on. To begin with crust was recycled quickly on a hot Earth, but the rate slowed as the planet cooled and the intense meteorite bombardment came to an end four billion years ago. Continents grew over time and reached roughly their present landmass 2.5 billion years ago. From then on, they have been recycled at about the same rate as they are created. Today, the plates creep slowly; the Atlantic Ocean is opening at an inch a year, a little slower than fingernails grow.

Zircons aside, the oldest rock in the world is the 4-billion-year-old Acasta Gneiss outcrop in northern Canada. This formed deep underground so it tells us little about what was going on at the surface. The oldest surface rocks are the highly metamorphosed Isua sediments of southern Greenland, which formed underwater and give us the first direct evidence of oceans.[15] Remarkably, these deposits suggest that life had already evolved by the time they were formed. There are no fossils in the Isua sediments, but the chemical composition of carbon buried in these rocks is characteristic of the presence of life.

The first life forms evolved in the early Archean Eon – the billion and a half year eon that followed the Hadean. It was a very different world from the one we live in today. There was almost no free oxygen for a start, and the sun burned a quarter less brightly.[16] Sometime between 3.8 and 4.1 billion years ago, microbes evolved that produced methane, a greenhouse gas twenty-five times more powerful than carbon dioxide.[17] Methane levels rose and the planet warmed.[18] For over a billion years, until at least 2.5 billion years ago, the greenhouse shroud sustained the liquid seas as the sun's fire warmed the Earth. If it hadn't been for this dense blanket of gases in the atmosphere, the oceans would have frozen and life might never have kindled, or it could have started only to be snuffed out early on.

For hundreds of millions of years after the first spark of life, there is not a trace or shadow of its form in the rocks beyond chemical alterations.[19] But spectacular recent advances in genetics and computing enable us to grow the tree of life backwards, from its leaves to the tip of the root from which it sprang. Every living thing, from the humblest virus to the greatest of the whales, shares a common heritage that is written in their genes. That similarity tells us, just as Charles Darwin predicted, that everything living today is descended from life's primordial spark. The genes that code for different metabolic functions can be placed in position on the tree of life to show the sequence of their emergence, and something of their timing. They tell us when life passed critical evolutionary milestones and from them we can infer how the environment was changing. In some cases, early life forms had to respond to planetary upheavals, but often they were themselves responsible for changing the world around them.

In the deep sea one can find hot springs under intense pressure that spew forth water superheated to 300–400°C and so laden with minerals that their plumes are opaque black or white. They deposit metal-rich compounds nearby that might have catalysed important chemical reactions. Today these hot springs support rich communities based entirely on energy captured from chemical reactions undertaken by microbes. Here, then, is a vestige of how the earliest microbes might have made their living, perhaps even of the place where life itself was forged.

For more than 3 billion years, the planet was ruled by singleton cells and microbial slime. It is hard to grasp such a vast timescale. Think of it this way: if every one of those years lasted just a second, it would take ninety-five years for 3 billion to pass by. Microbes evolve fast because they have fleeting generations. Even taking into account the likelihood that Archean oceans were far less productive than our own, there was time for hundreds of billions of generations to come and go.[20] Every new generation offered the possibility of variations that were the raw material of evolutionary innovation. This was a time of extraordinary inventiveness, when the foundations were laid for almost everything that life does today.

Early on, microbes developed the ability to produce energy by converting hydrogen sulphide – the gas that gives rotten eggs their smell and that pours forth from deep sea hot springs – into sulphates.[21] This was an essential step. Some developed complex chemical machinery to draw energy from sunlight to accomplish this conversion in shallow waters, and photosynthesis was born. Microbes are not easily fossilized, but they do get petrified into flinty rocks called chert. The earliest fossils are microscopic microbe threads locked into 3.45-billion-year-old Australian cherts, although their interpretation is controversial.[22] They look like cyanobacteria, a group still common today. These were the creatures that would later on develop photosynthesis, the method of generating energy from sunshine that dominates all primary production today. They used the sun's energy to create carbon compounds – food in other words – from carbon dioxide and water. Their waste product was oxygen.

The first chemical traces of oxygen-producing photosynthesis were found in 2.7-billion-year-old shales rich in organic matter.[23] We must thank this innovation for the way our world works because this kind of photosynthesis has produced essentially all of the free oxygen around us today. It took hundreds of millions of years to make enough oxygen for us to detect its traces. Dark shales from Mount McRae in Western Australia, laid down 2.5 billion years ago, give us the first whiff of oxygen.[24]

Soon after (at least in geological terms, since it took another 50 million years), we begin to find evidence of oxygen in rocks all over the world. The next 150 million years is known to geologists as the 'Great Oxidation Event', because it heralds the first major step in the formation of the atmosphere we have today. Far from being a boon (that would come later), oxygen first plunged the Earth into a crisis. When oxygen and methane get together, the result is carbon dioxide and water. Carbon dioxide is twenty-five times less powerful a greenhouse gas than methane, so the Earth's comfort blanket thinned and the planet froze.

This ice age is thought by some scientists to have been so severe that the sea and continents iced over all the way to the tropics.[25] When ice forms at latitudes lower than about 30°, it reflects so much heat back into space that glaciation runs away with itself. Some simulations of such a world suggest that the oceans could freeze to a thousand metres deep. The ice would only have retreated after millions of years of volcanic activity had added enough carbon dioxide to the atmosphere to warm Earth enough to melt it. It is hard to see how life could have survived a 'Snowball Earth', so one suggestion is that the tilt of the planet in relation to the sun must have been different at this time, so that the poles were warmer than the tropics and ice-free conditions persisted through the great freeze in some places.[26] Another possibility is that weather systems kept some areas of ocean ice free.

The oceans of the past were not places in which you would have wanted to swim. In Hadean and Archean times they are thought to have been rich in dissolved iron and anoxic, or oxygen-free. The iron came from deep-sea hot springs and weathered rock. In the absence of oxygen, iron dissolves and is easily washed to sea, whereas if oxygen is present iron oxides tend to stay put. Early microbes put this iron to work. They developed ways to use the power of sunshine to oxidize free iron and make food from carbon dioxide and water. Iron was precipitated at the bottom of these seas to form thick deposits known today as banded iron formations.[27]

A slice of 2.5-billion-year-old seabed sits on my shelf at home. It is only a centimetre thick, but is surprisingly heavy. Wavy layers of rust brown, yellow and orange silica alternate through the slab with dark black and grey stripes of magnetite. Thicker layers of shimmering tiger eye fill gaps where the rock was later twisted and deformed under pressure. It is beautiful. I find it amazing to run my finger across this fragment of our primordial ocean. Most banded iron formations have long since been recycled back into the Earth's mantle, but bits of ancient seabed lie stranded in rock formations in Australia, Canada, Russia and elsewhere to form some of the richest sources of iron ore in the world. Banded iron formations are largely confined to rocks older than 2.4 billion years. They disappear from the record for the next 400 million years. For a long time it was thought that free oxygen produced by cyanobacteria had dissolved in the ocean and stripped it of iron, but another possibility now seems more likely, based on the chemistry of rock deposits. Donald Canfield, a polymath geobiologist based in Denmark, thinks that below a thin surface layer, the sea had no oxygen for hundreds of millions of years after oxygen first began to rise in the atmosphere. His idea is that oxygen reacted with sulphides in terrestrial rocks and washed into the sea as sulphate. It was the sulphate that stripped the oceans of their iron, not the oxygen. The Great Oxidation Event would not penetrate to the deep sea until a billion years later.

Why did it take so long to ventilate the deep? Oxygen levels in the atmosphere were still very low, no more than about 1 per cent of today's level. We would have suffocated in no time. Oxygen is transported to the deep sea when surface waters plunge down, but it is used up breaking down organic matter – dead microbes in other words – that sinks from the surface. It would not take much sinking organic matter to strip oxygen from the deep faster than it could be replaced. Still, it is hard to explain how deep water could remain oxygen-free for such an immense stretch of time after the rise of oxygen-producing photosynthesis. Early oxygen-producers lived in shallow sunlit waters. Just like life today, they needed nutrients to fuel their growth as well as sunshine. Since most nutrients sink with the dead bodies of microbes, they have to be recycled by the upward mixing of deep water. An intriguing new idea is that anoxic waters, rich in hydrogen sulphide, reached up into the lower sunlit layer[28] where they encountered photosynthesizers that used up hydrogen sulphide but didn't make oxygen, and who intercepted most of the nutrients before they could reach the oxygen-producing cells closer to the surface.[29] There is another twist in the nutrient tale. Very early on, life incorporated the trace metals iron and molybdenum into enzymes that fix nitrogen, thus securing one of life's essential nutrients. Sulphidic seas would have little of either of these metals, because both would have reacted with hydrogen sulphide and been deposited at the seabed. So it was that quirks of oceanography and nutrient supply appear to have held back planetary oxygenation for what seems like an eternity.

There are still places today where sulphur-based photosynthesis continues. The Black Sea is an almost completely enclosed basin of water which is capped by a warm sunlit surface layer that is less dense than the cool water underneath. Little of the oxygen in this warm surface layer mixes downward, so, below about 150 metres, the Black Sea has not had any oxygen for thousands of years. If you were to somehow bring a sample to the surface, it would stink of rotten eggs, just as the sulphidic ancient oceans would have. As this anoxic water nears the surface, green and purple sulphur bacteria use sunlight to produce food and energy, just as their predecessors did eons ago.

It is hard to know if all the world's oceans were once sulphidic like today's Black Sea. Most of the rocks that could tell us were long ago recycled back into the mantle, so there are few places to look for clues. What we do know is that 800–900 million years ago conditions began to change. What palaeontologists refer to as the 'boring billion years' ended. This period saw a shift in ocean chemistry towards the composition of modern oceans. A phase of intense mountain building around this time may have enhanced the delivery of trace nutrients to surface waters, enabling oxygen-producers to flourish. With rising oxygen in the atmosphere, the sea became ventilated beyond the lower limit of the sunlit layers, ending the hegemony of the sulphur bacteria and freeing more nutrients for oxygen-producers. This positive feedback loop led gradually to the irreversible oxygenation of our atmosphere. It set the scene for life's next great innovation: multicellularity.

When Charles Darwin wrote *On the Origin of Species*, the fossil record seemed to fizzle out at the onset of the great Cambrian explosion, when life forms with hard body parts which readily fossilized first came into being. Before that, nothing: the rocks appeared blank. It was as if some creator had conjured forth a menagerie of primitive life that would later diversify into all subsequent beings. It has taken 150 years of intensive searching to find the enigmatic shadows of earlier life, much of it microbial, but it is there. Darwin would have loved to know what we know today.

Oxygenation of the Earth's atmosphere was a key event in the history of life. A metabolism based on oxygen respiration produces sixteen times more energy than equivalent anoxic pathways. When free oxygen became available, it created tremendous opportunities that were soon realized. The first oxygen-using enzymes have been traced by following the evolution of proteins back in time using the latest genetic sequencing libraries and powerful computer techniques. Although there are precious few pre-Cambrian fossils, in a sense every one of us contains a fossil library of genes that goes all the way back to the Archean Eon, billions of years ago. The history of changing protein shapes gives us a molecular clock from which we can read the time when new innovations first emerged. The first hesitant steps towards the use of free oxygen seem to have been taken 2.9 billion years ago, 400 million years before the first clear evidence of free oxygen appears in the rocks.[30] Proteins involved in oxygen metabolism multiply quickly through the Great Oxidation Event. Then there is a second wave of evolutionary innovation in oxygen use that begins about 1.2 billion years ago and continues all the way to the Cambrian explosion of life. By the onset of the Cambrian Period, 542 million years ago, atmospheric oxygen had increased to about 12 per cent of today's level.

Oxygen is produced by photosynthesis when organic matter is created, and consumed both by respiration and by the decay of that organic matter. If the two sides of the equation were exactly equal, oxygen would not build up in the air. But some organic matter is buried at the bottom of the sea or in swamps and lake beds. During late pre-Cambrian times, when oxygen levels in the atmosphere rose, the rate of organic carbon burial must have increased. Free oxygen is also produced when sulphate combines with iron to produce iron sulphide (pyrite or fool's gold). So, increased rates of pyrite burial would also oxygenate the world. We don't know exactly what caused burial rates to increase, but the oxygenation of the oceans is believed to have played a key role.

Whatever the cause, there are many who think that rising oxygen levels paved the way for life's explosion. The two go hand in hand. Large-bodied aerobic animals need enough oxygen to ventilate their tissues. Their virtual absence in the previous 3 billion plus years of life might just be because their size was impossible until oxygen levels rose high enough to sustain such bulk.

The Cambrian Period was a time of amazing evolutionary creativity. Within 20 million years we can read in the rocks the appearance of virtually every major animal group alive today. If periods are characterized by their most successful creatures, this was the age of the arthropods, distinguished by their external skeletons and jointed legs. Insects, crabs, lobsters, millipedes, spiders and the like are all their living descendants. But the now extinct trilobites really define the Cambrian Period better than any other group. Cambrian seas were full of a bewildering mix of these low-bodied, scuttling creatures that wore articulated armour of body plates and helmets that were sometimes flamboyantly ornamented, with spikes and spines and compound eyes with crystalline lenses. Trilobites had good reason to defend themselves for there were predators about. One explanation for the Cambrian explosion sees it as an arms race in which some animals scrambled to escape or repel predators, while others developed ever better ways to catch prey. Grazing and predation pre-date the Cambrian Period by hundreds of millions of years, but it was then that the ability to swallow large prey was firmly established. The foundations of modern food webs were laid in Cambrian seas, during a period that came to a close 488 million years ago.

As life diversified, a series of environmental crises lopped branches from the spreading crown of the tree of life. Just when the atmosphere had become more breathable, anoxia returned to the oceans and sulphidic seas reappeared.[31] What seems to have happened is that waters devoid of oxygen welled up on to shallow continental shelves, perhaps following rising sea levels, and snuffed out many trilobites and other creatures. Paradoxically, these crises may have driven atmospheric oxygen levels up. Ocean anoxia is linked to higher surface productivity and greater organic carbon burial in Cambrian oceans. Sinking faeces and dead plants and animals would descend faster out of the sunlit surface layers than the microbes of more ancient seas so more carbon would be buried, especially in anoxic waters where breakdown is sluggish. Free oxygen would get a boost.

The next leap forward in the construction of modern ocean ecosystems has again been linked to rising oxygen. It may come as a surprise to know that we owe the existence of sharks to the evolution of vascular land plants.[32] These plants were the first to develop differentiated organs like leaves, roots and stems, and a complex architecture of tubes to conduct water, nutrients and food throughout their systems. This group includes ferns, club mosses, conifers and flowering plants, although the latter would come much later. Vascular plants debuted about 420 million years ago, and within 30 million years they had sprouted roots to draw nutrients from the soil and stabilize bigger growth forms. By 370 million years ago, plants had clothed the continents in a lush tangle of green.

How does this relate to sharks, you might ask. Roots and other adaptations to extract nutrients speeded the rate at which phosphate, a key plant nutrient, weathered from the soil and washed into the sea; marine productivity boomed, enabling longer food chains and bigger predators. At the same time, land plants massively increased organic carbon burial in swamps and, through run-off, in marine sediments. This brought the amount of oxygen up to levels similar to the present (21 per cent of the air), where they have stayed, give or take 5 per cent, for most of the last 350 million years. Freed from the constraint of low oxygen, the stage was set for active predators with big oxygen demands.

Animals a metre long were already present in late pre-Cambrian seas 580 million years ago. They were probably sedentary grazers or detritus feeders. Eighty million years on, there was more diversity and hunters emerged that were the size of a small child. Fast forward another 100 million years and the seas had filled with primitive fish, including the heavily armoured placoderms. By 370 million years ago, some had become terrifying beasts, large as buses. The predatory megafauna had arrived. Over the next 120 million years they were joined by sharks and reptiles such as ichthyosaurs, plesiosaurs and turtles. The predatory expansion reached its zenith recently, almost yesterday in geological terms. During the Miocene Epoch, which spanned 23 to 5 million years ago, giant sharks the size of great whales patrolled the seas and whales ate whales. There were formidable sperm whales the same length as today's, but with teeth three times as big. They have been named *Livyatan melvillei* after Herman Melville's mythic white whale.[33]

The presence of these huge predators suggests seas more productive than our own. Indeed, the oceans and their life have had many ups and downs over the last 500 million years. The variety of life has gradually increased, but not in a steady rise, and not by the simple addition of new species to the existing inventory. The vast majority of species that have ever lived on Earth have disappeared. Five mass extinctions followed several crises that wiped out most of the trilobites during the Cambrian Period, possibly caused by the re-emergence of anoxic oceans.[34] The mother of all extinctions came 251 million years ago, at the end of the Permian Period. At this point, the fossils of more than 90 per cent of marine species and two-thirds of terrestrial species simply disappear from the rocks. Before the Permian extinction, seabed sediments were filled with lively burrowing animals that churned them over and over. Afterwards, the sediments were nearly stilled.

What caused this cataclysm remains a source of heated argument, but the most likely cause was a half-million-year episode of volcanism that discharged 2 million cubic kilometres of basalt across the surface of what today is Siberia.[35] The lava flooded an area of 1.6 million square kilometres to depths up to 3,000 metres. It passed through layers of carbonate rock and coal to release massive quantities of carbon dioxide into the air. Prior to this, carbon dioxide had declined more than ten-fold since the Cambrian Period, and the loss of this greenhouse gas had driven the world into a glaciation that had lasted over 60 million years.

Now it gets hot. As carbon dioxide levels climbed the world warmed, and as it warmed the poles melted. Global temperatures spiked 6°C higher than today's. The oceans rose and the Arctic reached a balmy 15–20°C.[36] For reasons I will return to, this temporarily slowed deep-ocean mixing and the stagnant water warmed all the way down to the deep seabed. The warm conditions melted huge polar and undersea deposits of methane to cause runaway global warming. Land plants withered and soils washed away. The stagnant sea became anoxic in all but surface waters. Meanwhile, dissolved carbon dioxide turned the oceans more acid, which, again for reasons I will come back to, was disastrous for chalky animals like corals, urchins, calcareous seaweeds and sponges.[37] The mass extinction at the close of the Permian Period almost wiped life's slate clean. But life rose again. Within 10 million years, the oceans had refilled with new species.

Sitting in my garden one day thinking about this book, my mind wandered back to the world's origin 4.5 billion years ago. It is hard to believe that this world is the same as our own – our planet, hurtling through the emptiness of space, so alien yet so familiar. I imagine hitting the fast-forward button to see continents ascend, meander, coalesce and crumble; oceans rise and fall; seas, deserts and ice caps come and go. Within the belly of these oceans creatures outlandish and familiar, bizarre and terrifying appear and disappear. Great reefs are built, destroyed and remade. Life flourishes, is choked off and resurrects itself again and again. And through the strangeness and inconstancy swim turtles and sharks, nautilus and jellyfish, scarcely altered, like a connecting thread that reaches back through time. A moment before the tape ends we appear. All of us carry this thread within ourselves. We are creatures of the sea with a lineage that stretches back to sponges and beyond, all the way through single-celled microbes to life's origin. The rest of this book tells the story of what happened to the oceans after we arrived.

2

Food from the Sea

Anthropology and archaeology have long been in thrall to an image of early humans as big game hunters of the open plains. Game-hunter thinking has us evolve from tree-dwellers into savannah-dwellers who started to walk on two legs. Sharp wits and ingenuity were necessary to thrive in the open, where we had to fend off dangerous carnivores, and our bipedalism freed us to hold tools and weapons. Our large brains later allowed us to develop language, which unlocked the possibility of technology and culture.

This view of human origins has a certain mythic ring to it, suggesting as it does that our plucky species succeeded in a heroic struggle against great odds.[1] But the story has holes. Baboons live on savannahs and have not become brilliant bipeds, and we have a mixture of adaptations that make little sense in the absence of water. We carry ten times more subcutaneous fat than other primates (we're about as fat as fin whales), which wouldn't particularly help endurance hunters run down their prey, but would insulate us from water. *Homo erectus* had dense bones more akin to those of diving mammals like manatees than fleet-footed plains hunters. We are prone to dehydration, an uncommon trait in savannah dwellers, and have an instinctual breathhold reaction when we plunge into water. The only other primate that is regularly bipedal is the proboscis monkey, which wades through swamp waters on its hind legs. Could our bipedalism have been an aquatic adaptation, developed by wading to gather shellfish?

Marc Verhaegen, a Belgian doctor, and his colleagues have recently revived the concept of man as an 'aquatic ape'.[2] Verhaegen believes that *Homo erectus* evolved at the waterside, and put the fruitpeeling and nut-cracking techniques developed by ancestral forest species to a new use by smashing open shellfish, turtles and crabs.

Shellfish are easy to find and gather, and, although not especially calorific (it would take 150,000 cockles to match the calorie content of a large deer), they are rich in protein and other nutrients needed for brain development. The nervous system, which evolved in the oceans half a billion years ago, was built in part from omega-3 fatty acids made by algae and plankton. These compounds are in short supply for land-dwellers, which helps explain why animals like the rhino, which weigh a tonne, have brains two-thirds smaller than our own, while marine mammals like dolphins have large brains. Savannah models of human evolution see us getting our brain food from scavenged or hunted brains of terrestrial mammals, but waterside sources such as shellfish, waterbirds, eggs and turtles would have been easier and more regular fare.

The idea of man as a big-game hunter has been hard to shake off. Museums freeze the thinking of the day into dioramas that endure for decades, impressing their stories into the minds of generations. I remember well from childhood the tableaux of stocky, hairy people wrapped in skins, triumphant in their slaughter of buffalo or antelope. Usually these scenes were set in open savannah, occasionally giving a nod to our dependence on water with a lake or river painted into the distant background. As Verhaegen points out, the savannah-hunter picture of human evolution pre-dates almost all fossil finds of early hominins.[3] In his accounting, all *Homo* fossils discovered before or since have been associated with lakes, rivers, deltas and coasts. The concept of the aquatic ape is widely contested in anthropology, but to me it seems more persuasive than the convoluted logic needed to explain our evolution on Africa's dusty plains.

The earliest remains of *Homo* species, dating from 2.5 to 2 million years ago, are from inland water bodies like the Gona floodplain in Ethiopia and freshwater springs near Lake Olduvai in Kenya. It was later on that we reached the coast. Around 2.8 million years ago, Africa, like the rest of the world, underwent major climatic upheavals, cycling between prolonged phases in which much of the continent was arid and inhospitable, and wetter phases with more varied climate. Wet phases greened the Sahara desert, perhaps opening up routes out of Africa, and arid phases pushed our ancestors into small pockets of land that remained hospitable. One such place was Southern Africa, where humans first developed their predilection for seafood. Our genes point that way, tracing our ancestry all the way back to Angola and Namibia.[4]

Pinnacle Point Cave 13B – the name sounds like a condo address – is on the coast of South Africa and was occupied on and off for 30,000 years. It has enormous significance for this cave was home to some of the earliest modern humans, *Homo sapiens sapiens*.[5] Today it lies beneath the ninth hole of the luxury Pinnacle Point Golf Resort and more modern condominiums line the cliffs. Deep inside the cave, thick deposits mark the passage of time and tell the story of our awakening humanity. At the very beginning, 164,000 years ago, fragments of bright red ochre signal the first use of pigment, probably for adornment. Scratch marks criss-crossed over one piece suggest the emergence of symbolic art and small sharp stone tools. At Blombos cave, further west along the coast, archaeologists found jewellery made 75,000 years ago from tiny seashells perforated by predatory snails.

The first occupants at Pinnacle Point gathered shellfish at low tide to eat and much of the cave floor deposits are drifts of shells. Shellfish were also consumed 140,000 years ago at Blombos and remained important for the entire time Blombos was used as a home.[6] (Having once spent a summer in the company of a student whose research involved rotting heaps of shells, I can assure you that the smell would have tested the sternest of modern constitutions.) Bones of black musselcracker and mullet, fish that follow the rising tides to catch their prey, appear 77,000 years ago in Blombos. Since no hooks have been found with these fish bones, people probably speared them or caught them by hand. Or they may have been lured into shallow water by the use of bait made of broken urchins or shells and then speared with bone-tipped spears. A slightly later intellectual leap made at Pinnacle Point at least 71,000 years ago was the discovery that treating rock with heat made it easier to shape into tools like spear points.[7]

Since the first appearance of our African ancestors long ago, people have hunted food in large packages. The earliest evidence of consumption of aquatic foods comes from northern Kenya, where the remains of butchered fish, crocodiles and turtles have been found alongside stone tools made by the predecessors of *Homo erectus*.[8] In South Africa's Blombos and Klasies River caves, alongside fish and shellfish, remains of penguins and seals complement catches of eland, antelope and buffalo.

Caves in Gibraltar occupied by Neanderthals over 30,000 years ago likewise contain the bones of monk seals and bottlenose dolphins.[9] The dolphins were probably scavenged carcasses of stranded animals, but the seals would have been hunted throughout the Mediterranean, falling relatively easy prey on their breeding beaches. Palaeolithic paintings on the walls of Cosquer Cave, near Marseilles, on the south coast of France, show monk seals being hunted with spears.[10] Today the entrance lies 37 metres below sea level, a testament to the rising seas after the last ice age, but the cave slopes upwards to a large chamber that remains dry. It was discovered by a French diver, Henri Cosquer, in 1985, but the paintings were not found until 1991 when he and two colleagues penetrated far into the cave and spotted images of human hands stencilled on the rock face. Subsequent expeditions revealed 177 paintings of bison, aurochs and giant elk, many of them exquisite artworks. Cosquer Cave is almost unique in its representation of marine species. As well as monk seals, there are images of the now extinct great auk and what seem to be jellyfish. The paintings date from around 19,000 years ago and were created by the ancestors of modern Europeans.

Jon Erlandson, an archaeologist from the University of Oregon, believes that seafood use and other adaptations to life on the coast were pivotal to human migration out of Africa, as our ancestors followed the coast to Asia and later crossed the Bering Strait to the Americas.[11] The richly productive coastal habitats found along the way, such as coral reefs, mangroves and kelp forests, would have provided abundant year-round food. From Asia, people dispersed through Indonesia, where land bridges joined many places that today are islands. At the time of this great migration the world was ice bound, and sea levels fluctuated between 30 and 60 metres below today's levels. People did face water barriers, the most daunting of which must have been the gap between Indonesia and Australia. Remarkably, they made the leap to Australia at least 50,000 years ago. A separate hop was made to the Ryukyu Islands south of Japan some 32,000 years ago. Colonization of the Bismarck Archipelago and Solomon Islands occurred between 35,000 and 28,000 years ago, requiring additional voyages of at least 90 kilometres, some of them out of sight of land.

The peopling of Australia provides the first concrete evidence of boat use by modern humans.[12] No physical remains of boats survive from this time, nor is there much direct evidence of fishing gear. Things made from wood or plant fibres rotted away long ago. Even more durable materials like shell, from which fish hooks were often made, can only survive under some conditions. In most cases they have simply dissolved or crumbled to dust.

For much of the last 125,000 years, the Earth was gripped by ice. Frozen ice sheets heaped up on continents drew sea levels down by 120 metres. When the world began to warm again, around 20,000 years ago, it triggered a rise in the seas that continued until present sea levels were reached about 6,000 years ago. Much of the physical evidence of our coastal lives between 15,000 and 120,000 years ago simply washed away or was submerged.

How did we catch fish at the very beginning? Early efforts to attract fish into the shallows with bait could have developed into the active construction of tidal traps from stones. Remnants of similar traps are well known from the Cape coast of South Africa, often built to block off natural gullies, though remnants of these traps date back no more than a century or so.[13] Tidal traps made of brushwood and stakes are widely known from the last few thousand years, but time has erased all traces of the earliest ones.

Some remarkable early evidence of our fishing prowess comes from East Timor, an island just north of Australia.[14] Uplifted coral terraces at the east end of the island are riddled with caves and fissures and were occupied for thousands of years. Fish bone remains in the Jerimalai Rock Shelter date back 42,000 years and include inshore fish familiar from coral reefs, like parrotfish, groupers and surgeonfish. But they also include fish that might have been caught offshore from boats, like tuna and shark. Although these animals could also have been caught from the shore, this interpretation fits with the idea that the ancestors of these people crossed the sea to Australia at least 8,000 years previously.

Some of the most detailed evidence for the development of fishing has come from caves and shell middens in California's Channel Islands. Over the last two decades their secrets have gradually been uncovered in excavations by Jon Erlandson and his colleagues.[15] Drifts of abalone, oysters and clams, half buried in blown sand, afford mute testimony to the Native American predilection for seafood. These mounds date back 12,000 years. So important was seafood in the early human diet that Erlandson and his colleagues suggested a ring of kelp forests that stretched from Japan to Mexico provided a 'kelp highway' that took seafarers from the Old World to colonize the New thousands of years ago. California's middens and cave shelters have yielded stone tools, fish gorges, hooks, harpoon points and woven seagrass artefacts. By the time of the first European contact, fishers also had nets and traps.

Fish gorges provide the earliest trace of line fishing. A gorge is a stick, bone or piece of shell that is sharpened at both ends and tied to a line in the middle. When the fish takes the bait the line is pulled and the gorge lodges at a right angle in its mouth or gut. Gorges are known from 30,000 years ago in Europe,[16] and may have been invented even earlier. Remarkably, they were still used by fishers on the Pacific Island of Palau, and probably many others, well into the twentieth century. Fishers there said that, even though they were less effective than a hook, gorges were used because they were so easy to make.[17] Early fishing lines were probably woven from animal hair or plant fibres. Until the twentieth century, Palauan fishers made a strong twine from braided coconut fibres in a method that had most probably been used for thousands of years.

We have no idea when nets were first invented. Fragments have been found on the Black Sea coast and in caves in South Africa dating back thousands of years. They are depicted in 5,000-year-old Sumerian wall reliefs and 4,500-year-old Egyptian wall paintings, but they had doubtless been used for much longer. Archaeologists believe the 42,000-year-old tuna catches in East Timor were made with nets.[18] Nets were probably made of flax, hemp or spun grass. The first evidence of domesticated flax comes from caves in the Czech Republic and dates to 30,000 years ago.[19] The first evidence of weaving, rather prosaically, comes from human bottom prints and bag impressions made on wet clay between 25,000 and 23,000 years ago, also in the Czech Republic.[20]

Most fishing methods were probably invented many times over. Single piece fish hooks pop up in archaeological finds all over the world several thousand years ago. The earliest found was a shell hook from Jerimalai cave in East Timor that could have been made 23,000 years ago. They appeared 3,000 years ago in California[21] and in Australia about 1,200 years ago.[22]

Against the vast panorama of human history, commercial fishing is a relatively recent development. Its first stirrings can be found in the Mediterranean and the Black Sea, but, for over 100,000 years, people fished to meet their own needs or those of their close kin. Early man developed and honed his fishing skills in the rivers and lakes of Mesopotamia and Egypt. Wall paintings and reliefs show nets, traps, hooks and lines, and even the first representation of a fishing rod.[23] Fishing appears to have developed as a specialist occupation here several thousand years ago.

By around 1000 BCE, evidence for commercial fishing gets stronger. The city of Gades or Gadir, today known as Cadiz, was probably the foremost fishing port of the ancient world. It lies on the Andalusian coast of Spain, just west of the Strait of Gibraltar. The town owed its early fishing expertise to the Phoenicians, highly skilled seafarers from the Levant who reputedly founded the city around the tenth or eleventh century BCE. Gadir was one of the first Phoenician colonies and became an important source of grain, silver, tin, textiles, dyes and salt fish. Over the next several centuries, Phoenician influence spread widely on trade routes that criss-crossed the Mediterranean. By 800 BCE, dozens of Phoenician colonies punctuated these routes at places like Carthage in Africa and Genoa, Marseilles and Palermo along the modern-day coasts of Italy and France.

Gadir was ideally placed for fishing. Local waters thronged with a great variety of fish, and it was close to the seasonal migration route of the bluefin tuna, a six-foot giant well known to the ancients that pops up repeatedly in literature. One author much taken by fish was Oppian, a poet from Corycus, a town in today's south-east Turkey.

His father apparently displeased a visiting Roman dignitary and was banished to Malta in the second century CE. There as a young man Oppian wrote a 3,500-line poem in Greek hexameters, *Halieutica – on the Nature of Fishes and Fishing of the Ancients*. The poem was so admired by the Roman emperor Marcus Aurelius that he paid him a gold piece for every line and pardoned his father. Here is Oppian's description of the migration of bluefin tuna into the Mediterranean:

> The breed of Tunnies comes from the spacious Ocean, and they travel into the regions of our sea when they lust after the frenzy of mating in the spring. First the Iberians who plume themselves upon their might capture them within the Iberian brine; next by the mouth of the Rhone the Celts and the ancient inhabitants of Phocaea hunt them; and thirdly those who are dwellers in the Trinacrian isle and by the waves of the Tyrrhenian sea. Thence in the unmeasured deeps they scatter this way or that and travel over all the sea. Abundant and wondrous is the spoil for fishermen when the host of Tunnies set forth in spring. First of all the fishers mark a place in the sea which is neither too straitened under beetling banks nor too open to the winds, but has due measure of open sky and shady coverts. There first a skilful Tunny-watcher ascends a steep high hill, who remarks the various shoals, their kind and size, and informs his comrades. Then straightway all the nets are set forth in the waves like a city, and the net has its gate-warders and gates withal and inner courts. And swiftly the Tunnies speed on in line, like ranks of men marching tribe by tribe – these younger, those older, those in the mid season of their age. Without end they pour within the nets, so long as they desire and as the net can receive the throng of them; and rich and secret is the spoil.[24]

Malta was then and remains to this day a major hub for tuna fishing. The accuracy of Oppian's description of the bluefin tuna's migration, breeding and capture reveals just how well the ancients understood this fish. It also shows the great antiquity of the almadraba method of catching tuna, still in use today – an elaborate chambered net trap to intercept their coastal migration. Oppian died of plague aged thirty, but his fame lives on for the remarkable insights he gave us into Mediterranean fisheries 2,000 years ago.[25]

It wasn't long before Eastern Mediterranean cultures developed methods to preserve fish with salt. By at least the fifth century BCE, salt fish was traded across the Mediterranean and the Black Sea. The importance of bluefin tuna in this trade is attested by the number of coastal cities that depicted the fish on their money. Coins from Gadir show two tuna, while those from Abdera in Spain show tuna as the pillars of a temple.[26] Tuna was cut into pieces, cured with salt, packed into amphorae and shipped to consumers hundreds, perhaps thousands of miles away. 'Canned' tuna has been around a very long time! Several other types of fish were preserved and shipped this way, including sea bream and mullet, and huge catfish and sturgeon from rivers and estuaries.

Oppian writes that the different kinds of nets are 'innumerable', and lists eight main varieties: the hand casting net, draw net, drag net, round bag net, seine (he used the word 'sagene'), cover net, ground net, ball net plus (my favourite) the hollow all-catching net. The art of fishing was clearly well advanced. He tells of a clever device for catching fish when they lie close to the sea bed:

> They have a stout log, not long but as thick as may be, about a cubit in length. On the end of it are put abundant lead and many three-pronged spears set close together; and about it runs a well-twisted cable exceeding long. Sailing up in a boat to where the gulf is deepest, mightily they launch into the murky deep the pine-log's stubborn strength. Straightway with swift rush, weighed down by lead and iron, it speeds to the nether foundations of the sea, where it strikes upon the weak Pelamyds [bonito tuna] huddling in the mud and kills and transfixes as many as it reaches of the hapless crowd. And the fishermen swiftly draw them up, impaled upon the bronze and struggling pitifully under the iron torture. Beholding them even a stone-hearted man would pity them for their unhappy capture and death.

It is hard to imagine today that a plank loaded with spikes thrown off the bow of a boat would catch anything at all. It could only work in an ocean crowded with fish or in water shallow and clear enough to see the target.[27]

Polychrome mosaics were developed around the first and second centuries CE, and marine scenes soon became popular. There is a magnificent floor mosaic of sea fishing from the Catacomb of Hermes at Hadrumetum in Tunisia (another former Phoenician colony). Fishermen in boats ply their trade amid a sea crowded with a wondrous diversity of fish and lobster. Some use basket traps, others hook and line, another a harpoon; one casts a net and two men work a drift seine net buoyed by bobbing corks. The boats are small and powered by oars.

The Mediterranean had long been plied by large, seaworthy craft but there was little point in fishing from a big boat. Only the very wealthy could afford ice (the Roman writer Galen mentions preserving fish in snow in the second century CE[28]) so fishermen had to work close to the coast to prevent their catches from going off on long journeys back to port.

In the fifth century BCE, large-scale salting works were built around the shores of the Mediterranean, especially in the productive western waters. They made salt fish, but increasingly turned to the manufacture of the fish sauces beloved by Greeks and Romans. Production varied over time but seems to have peaked in the fifth to fourth centuries BCE, serving mainly Greek markets, and again in the first and second centuries CE, when most of the region was under Roman rule.[29] Some salting works contain enough vats to hold more than a thousand cubic metres of fish and sauce.

Like Marmite or blue cheese today, fish sauces weren't to everyone's taste. In the first century CE, Pliny the Elder called fish sauce an 'exquisite liquid', and the best grades of *garum* sold for a price equal to perfume.[30] 'It's the overpriced guts of rotten fish!' Seneca railed at about the same time, and he wasn't far from the truth. Most fish salting works were smelly and located outside city walls, well away from residential areas.

Recipes for fish sauce can be found in a tenth-century collection of the works of ancient writers.[31] One directs you to take the intestines, blood and gills of a bluefin tuna, add salt and ferment the ensuing goo in a vase for two months. Another gives the method for a lower grade sauce made from a mixture of many kinds of fish, usually those good for little else. (Anchovies were a favourite.) To one measure of fish add two of salt and let stand overnight, instructs another recipe. Then place the contents in a clay vessel and leave uncovered and exposed to the sun for two or three months, occasionally stirring with a stick. It sounds unappetizing (all the more so when you consider that *allex*, another kind of fish sauce, was made from the residue scraped from the bottom of the empty fermentation vessel).

Spanish food technologists have recently created a version of *garum* using modern methods to mimic ancient processes.[32] The key is the protein-digesting action of enzymes within the fish guts, which break down the flesh to a slurry of amino acids, fats and nutrients. The resulting paste, rich in omega-3 fatty acids as well as many vitamins and minerals,[33] was not dissimilar to Asian fish sauces of today. It was an early health food.

As the Empire spread north through Europe, the Romans took their tastes with them. Fish sauce was transported in large quantities to their colonies – around a fifth of all amphorae found at one Northern European town were for fish sauce[34]– and production sites were established along the North Sea coast of Gaul. With the collapse of the Roman Empire, these factories fell into disrepair and fish sauces dropped off the menu. Sea fishing dwindled, except in the chilly extremities of Scandinavia where the climate was too harsh for most crops and people hunted and fished to survive. Remains of fish bones from eighth- and ninth-century domestic waste dumps in England, Flanders and other parts of Northern Europe were dominated by species that lived permanently or seasonally in fresh waters, such as sturgeon, salmon and whitefish. That all changed in the eleventh century, when a sea-fishing revival marked the onset of a process of expansion and industrialization that continues to this day.

Archaeologists discovered the Northern European sea-fishing revolution by sifting through over a hundred kitchen garbage heaps.[35] Within the space of a few decades around the middle of the eleventh century there was a dramatic shift in the source of fish from around 80 per cent from freshwater to 80 per cent saltwater fish like cod, haddock and herring. At the time, demand was increasing rapidly thanks to a combination of population growth, urbanization and the spread of Christianity (Christian practice required periodic or complete abstinence from the meat of quadrupeds). Fresh fish supplies were in freefall as a result of human-caused habitat change in rivers, lakes and estuaries. With the spread of agriculture, forests were felled and land deep ploughed for crops. Soil erosion soon turned fast running, cool, clear waters into sluggish, warm and turbid water which species like salmon did not enjoy. Nor did salmon, and other fish that migrated from sea to rivers to spawn, benefit from the construction of thousands of dams across Europe's rivers to supply power for corn mills and other industry. With their migration routes blocked, freshwater fisheries' production collapsed.

Most fish in the medieval period were caught and eaten locally, but around the thirteenth century, long-distance trade resumed, this time of air-dried and salted fish.[36] Fish caught from the prolific Lofoten spawning grounds of Arctic cod were dried in the frigid air into metre-long pieces as hard as wood. They could be carried hundreds or thousands of miles to supply towns and cities all the way to the Mediterranean. Stockfish, as it was called, lasted for two or three years – a perfect convenience food for an age that lacked refrigerators.

Sometime around the fifteenth century, an improved method of salting was developed in Amsterdam to preserve herring in brine[37] (I should say redeveloped, since the ancients were expert fish salters). Fisheries expanded swiftly after that, first with Dutch, then British fleets. By the seventeenth century, more than 2,000 Dutch boats pursued herring from the Shetland Isles to the Baltic. By this time, interest in cod had shifted west to the shores of Canada and New England, where abundant fish of extraordinary size immediately attracted European interest.

Fishing methods in Europe developed gradually as new ways to capture fish were invented. Beam trawling appeared in the fourteenth century, if not before,[38] and involved dragging a net held open by a beam of wood across the seabed. The idea probably came from smaller dredges that were towed over the seabed to catch oysters.

A French nobleman, Henri-Louis Duhamel du Monceau, compiled a meticulous catalogue of sea-fishing methods in the 1750s, lavishly illustrated with engravings.[39] They included intertidal and subtidal long-lines bristling with hooks, baited lift nets manned from platforms over shallow water, drift nets like those used by herring fishers to snare fish by the gills at night, and many others. With the exception of large-scale cod and herring fishing vessels, most people still fished from small boats close to coasts. Fish spoilage remained a problem, and most of the catch was consumed in towns and villages close to the sea. Well-boats, another reinvention of a classical Greek technology, could operate further afield. They carried a water-filled tank and brought live halibut, cod, turbot and other fish from the central North Sea into towns like Hamburg and London. Markets for fish underwent dramatic enlargement between the end of the eighteenth and the middle of the nineteenth centuries, in both the Old and New Worlds. Toll roads and railways speeded up transport between coast and inland cities so that fresh fish could reach more people. In turn this spurred a rapid increase in fishing effort. But these fisheries still used medieval technologies of sail and oar, hook, net and trap.

In other parts of the world, developments in fishing technology matched those of industrializing countries in their ingenuity. Perhaps the most remarkable fishing method I know of comes from Melanesia and islands of the West Pacific.[40] A kite made from a dried breadfruit leaf stiffened with the midribs of coconut leaf-blades is flown on string made of coconut husk fibres. The kite trails a mat of spiders' webs in the water and a skilled fisherman will make it skip from wave to wave, so exciting the interest of garfish. When they grab the lure, their teeth become hopelessly entangled. I have found myself wondering whether the first kite was invented by a fisherman after a floating leaf tangled in a line and lifted off.[41] But I digress.

The addition of steam power to boats in the 1880s heralded the beginning of the modern era in commercial fishing. Unleashed from the bonds of wind and tide, fishing fleets multiplied and spread forth across the world's continental shelves. Engines freed people to pursue fish further offshore and deeper, and to work around the clock in worse weather. Engine power and rising demand expanded vessel size and made possible larger trawls and nets. In the early twentieth century, steam yielded to diesel and fishing intensity climbed as vessels multiplied and their captains embraced new technology. Then, after the Second World War, there was another burst of industrialization that lasted into the 1970s. During this period, boats increased in size and power and monofilament nets replaced hemp and cotton, enabling people to fish far more effectively with larger gear. The destructive potential of fishing was hugely increased by these developments. Longlines were extended to tens of kilometres with thousands of hooks; drift nets became walls of death of similar enormous length; huge engines made it possible to tow mid-water trawl nets large enough to engulf cathedrals; and bottom trawl nets spread to a hundred metres wide and were fitted with heavy steel balls on the footrope to ease their passage into areas of rough seabed.

Up to the 1950s, fishing technologists had mainly improved upon ancient traditional methods. They substituted materials, refined designs, added engine power, and deployed more or bigger catching gear. The introduction of echo sounders in the 1950s marked the onset of a new fishing revolution. Echo sounders revealed the presence of fish beyond the dreams of the most skilled captains from prior generations. Electronics, augmented from the 1980s by computers and satellites, ramped up the stakes yet again.

What strikes me, looking at the grand vista of human history, is how technological development has speeded up with time. There are gaps of tens of thousands of years between major innovations made by our earliest ancestors. After the end of the last ice age 10,000 years ago, things speed up, and new methods of fishing were introduced at intervals of a thousand years or less. In the last thousand years, the tempo accelerated as people invented clever ways to catch and preserve an ever-expanding variety of fish and shellfish. In the last hundred years, the addition of engine power and modern materials vastly increased the reach of fishing, and in the last thirty years the lethal edge of fisheries has been sharpened by the addition of computer and satellite technologies. In the next two chapters, I will explore how this upward-racing curve of progress has affected life in the sea.

3

Less Fish in the Sea

'There is a large-sized fish called Hallibut, or Turbut: some are taken so big that two men have much a doe to hall them into the boate; but there is such plenty, that the fisher men onely eate the heads & fines, and throw way the bodies.'
Captain John Smith, first governor of Jamestown, *History of Virginia*, 1624[1]

It is early morning and the sea is smooth and dark. The peace is soon broken by an engine echoing off the hills that rise directly from the sea. A fishing boat heads straight for the shore, where it turns and backs up into water scarcely deeper than its draught. Figures on deck manoeuvre two sets of four dredges over the side and drop them into the water with loud splashes. The dredges consist of heavy steel frames bristling with vertical teeth that dig into the bottom to knock scallops into attached chain-mail bags. The boat roars into gear and the tow begins. Half an hour later, far out to sea, the dredges are swung back on deck where they spill their loads of stones, seaweed, starfish and scallops. While one man sorts the catch, the boat turns back and heads shoreward to begin another tow. Only desperation would drive a captain to risk his boat among the shallow rocks of Scotland's Firth of Clyde, but fishing has been bad in recent years, so he must feel there is no alternative but to scratch his living along the shore.

For all their technological brilliance, modern fishing fleets operate at the margins of profitability. Fishing has always been a hard living: wet, backbreaking, messy and often dangerous work. But in the past, fishermen could at least guarantee a good catch. Nineteenth-century photographs of dockside scenes show landings of fish that appear almost miraculous today. Quaysides are stacked with boxes overflowing with outsize fish, while giant halibut, cod, wolffish, ling, turbot and others too large to fit in crates cover stone floors. But even in the late nineteenth century, fishermen had begun to grumble of declining catches. A British commission of enquiry was convened in 1883 to investigate complaints. Unable to resolve the issue of falling stocks in the absence of fishery statistics, the commissioners recommended that the government start to collect data, and records duly began in 1889.

Strangely, until one of my graduate students, Ruth Thurstan, dusted off the old volumes a couple of years ago, most fishery statistics collected before the European Union's Common Fisheries Policy began in 1983 were neglected. Perhaps they were felt to be too antiquated to be of value. Methods of data collection change with time, so comparing old figures with new was difficult. But it was worth the effort, for these old charts paint a stark picture of the present state of fisheries, and one much worse than is suggested by modern statistics.

The graph of landings from bottom trawlers – boats that drag large bag nets across the seabed to scoop up bottom fish – looks like a steep-sided mountain incised with two deep valleys.[2] From 1889, catches rose steeply and peaked in the mid-twentieth century, when they levelled off briefly before a dramatic collapse to the present day. In 1889, more than twice as many bottom fish (cod, haddock, plaice and the like) were caught in British waters as we catch today. That is an astonishing fact given the technological gulf between then and now. The peak came in 1938, when the fleet landed over five times more fish than we do now. The two valleys cut into the graph correspond to the world wars, when catches plummeted because it was too dangerous to fish and boats were put to other uses such as laying mines.

Landings tell only part of the story, because they depend both on the number of fish in the sea and the time spent fishing. To get a better picture of the availability of fish, you have to divide the catch by the amount of power expended. This index of landings per unit of fishing power brings you closer to a real measure of how many fish there are in the sea.

Fortunately, fishing records include the number and size of boats that landed each year's catch. But how can you compare a sail boat to a trawler? Here a long dead nineteenth-century fisheries scientist called

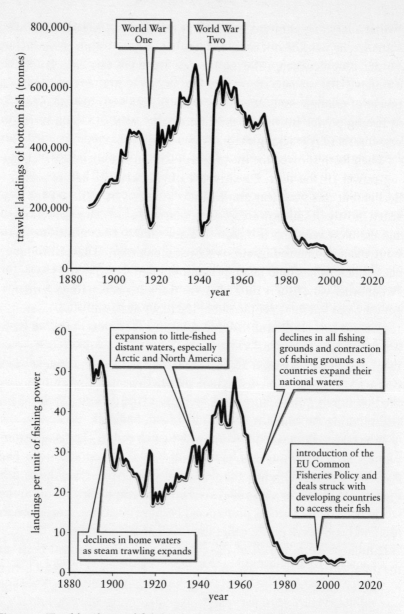

Figure 1: Total landings of fish by the English and Welsh bottom trawl fleet from 1889 to 2007 (*top panel*), and landings of fish per unit of fishing power (*bottom panel*). Redrawn from R. H. Thurstan et al. (2010) The effects of 118 years of industrial fishing on UK bottom trawl fisheries. *Nature Communications*, 1: 15.

Walter Garstang came to our aid. Using the first decade of official statistics, he worked out that landings per unit of fishing power fell by 50 per cent between 1889 and 1899. To make this calculation, he estimated that the new steam trawlers had more than twice the fishing power of sailing vessels. We used similar methods to estimate increases in fishing power by vessels throughout the twentieth century to the present, thereby tracking each advance in technology.[3]

When Ruth showed me the graph of landings divided by the changing power of the fleet, I nearly fell off my chair. I had expected a decline but this was near annihilation. A fleet that in the 1880s consisted mostly of sail-powered boats open to the elements was far more successful at wresting fish from the sea than we are now. For every hour spent fishing today, in boats bristling with the latest fish-finding electronics, fishers land just 6 per cent of what they did 120 years ago. Put another way, fishers today have to work seventeen times harder to get the same catch as people did in the nineteenth century. The simple reason for this stark contrast between past and present is that there are fewer fish in the sea. When we broke the figures down by type of fish, for some the contrast between the nineteenth and the twenty-first centuries was even more extreme. Landings per unit of fishing power are down by 36 times for plaice, over 100 times for haddock, and a breathtaking 500 times for halibut.

What were the seas like before the bite of industrial fishing? Eyewitness accounts from past centuries tell of waters off the coasts of America teeming with fish. Here is a report from the *Gloucester Telegraph*, a Massachusetts newspaper, from 4 June 1870:

> Accounts from New Jersey say that bluefish came in at Barnegat Inlet last week, sweeping through the bay, over flats as well as through the Channel, driving millions of bushels of bunkers before them and filling the coves, creeks, ditches, and ponds in the meadows full. At Little Egg Harbor Inlet they drove shad on shore so that people gathered them up by wagon-loads. Fish lie in creeks, ponds, etc., along the meadows two feet deep, so that one can take a common fork and pitch them into a boat or throw them on the bank.

The 'bunkers' referred to were menhaden, an oil-rich fish that seemed to have been created to feed all of the oceans' predators, ourselves included. A 1913 report on the US menhaden fishery found that more than a billion fish were caught that year, producing 6.5 million gallons of oil and 90,000 tonnes of fertilizer.[4] If all these fish were placed nose to tail, they would have stretched six times around the Earth at the equator.

Meanwhile, on the west coast, nineteenth-century photographs show men thigh deep in salmon pulled from Puget Sound. Some catches yielded 30,000 fish. In 1915, 400 million pounds of salmon (180,000 tonnes) were taken from Alaskan waters alone. If they had been packed in barrels, each containing 200 lbs of fish, and placed one on top of another, they would have formed a column 1,200 miles high.[5] In the Gulf of Mexico, whole fleets would set forth in sailing boats and test the water with baited lines until they hit upon a concentration of red snappers. At some sites as many as 2,000 fish could be taken in a day.[6]

In 1819, the Reverend Lewis Anspach, a resident of Newfoundland, described the appearance of Conception Bay during the capelin season.[7] The capelin is a small shoaling fish that spawns close inshore. It is prey to a bewildering variety of animals.

> It is impossible to conceive, much more to describe, the splendid appearance of Conception Bay and its harbours on such a night, at the time of what is called the Capelin Skull. Then its vast surface is completely covered with myriads of fishes of various kinds and sizes, all actively engaged either in pursuing or avoiding each other; the whales alternately rising and plunging, throwing into the air spouts of water; the codfish bounding above the waves and reflecting the light of the moon from their silvery surface; the Capelins hurrying away in immense shoals to seek a refuge on the shore, where each retiring wave leaves countless multitudes skipping upon the sand, an easy prey to the women and children who stand there with barrows and baskets ready to seize upon the precious and plentiful booty.

While some fish were deemed to be useful, others were considered pests and great efforts were sometimes taken to annihilate them. Incredible as it may seem now, an extermination campaign was mounted in Western Canada as recently as the 1950s and 1960s against basking sharks, the second largest fish in the sea. These docile plankton feeders often got tangled in gill nets set for salmon, so they were hated by fishermen. The fishery protection vessel was mounted with cutting equipment on its bow to slice through sharks feeding at the surface. Several thousand basking sharks died in nets or were slaughtered this way.[8] In Europe, dogfish and porpoises were similarly regarded as pests because they stole fish from nets and hooks, and eyewitnesses recorded devastating accounts of porpoise massacres going back to the eighteenth century.[9]

With the sole exception of Alaskan salmon, which have been well managed, the species described in these historical vignettes have plummeted since their historic highs. Puget Sound's salmon runs have dwindled to a trickle. Red snapper, bluefish and menhaden are all overfished in US waters today, while capelin numbers are far below the abundances witnessed in the nineteenth century. In 2010, a quarter of commercial fish stocks assessed in the US were considered overfished, meaning that they lie below target levels that are themselves set far below historic abundances.[10] But this misses the real scale of overfishing. The sad fact is that the status of over half of 528 US fish stocks examined that year could not be determined: they were too uncommon today for it to be worthwhile collecting data, or so rare that data were unreliable. Many are rare because of past overexploitation; they would have been familiar fare on the tables of nineteenth-century Americans. The picture is much the same throughout the rest of the world. Things are now looking up in the US due to a welcome shift in direction of fisheries management made in 2007 that I will come back to later, but elsewhere the trends still seem to be in the wrong direction.

We find it hard to believe these descriptions of extraordinary past abundance because it has been so long since such scenes were commonplace. It is a human trait to give greater weight to personal experience than to others' descriptions. The result, as I described in the opening of this book, is an intergenerational shift in the way we perceive the world. Science is particularly susceptible to these shifting baselines, as scientists work at the forefront of knowledge and are always in hot pursuit of the latest ideas.

When we published Ruth's study on the decline of UK trawl

fisheries, many journalists asked me why these old fisheries statistics had been ignored for so long. It perplexed them. The answer is that because they were old, the figures were no longer thought to be relevant to the question that fisheries scientists are asked: how many fish is it safe to catch next year? The answer to that depends on how many fish there are at the moment, which is usually settled by the latest survey, not by reaching back decades or centuries. But it is only by going back many years that a full perspective can be gained on the fortunes of fish stocks and so a judgment reached about their sustainability.

The balance of power between fishers and their quarry has been weighted in favour of the hunters for the best part of 150 years. While in the last few decades we have established regulations to restrain fishing, they have failed to give most species the time and space they need to reproduce. Life is extraordinarily dangerous for a fish or a 'bycatch species', as those that are caught or killed alongside the target species are called. Fishing intensities are now so high that, once some species reach a size where they can be caught, their chance of death from fishing in any given year ranges from 30 to 60 per cent, or even more.

Estimates suggest that trawlers sweep an area of seabed equivalent to half of the world's continental shelves every year.[11] Together with dredgers like the Clyde's scallop boats, they have transformed life on the seabed, converting three-dimensional, complex habitats rich in coral, sponge, seafan and seaweed to endless monotonous expanses of shifting gravel, sand and mud. While trawls and dredges decimate the ocean floor, the hand lines of old have evolved into longlines, whereby enormous lines, some of them tens of miles long and studded with thousands of hooks, ravage the upper layers of the sea. The length of hook-studded longlines set every night in our seas is enough to wrap around the world 500 times over.

Although the scale of the oceans is hard for the human imagination to grasp, life is concentrated into the surface layers, continental margins and other hotspots where nutrient-rich currents well up to the surface. We focus our fishing on these places, to devastating effect. Even migratory species like tuna and swordfish, which undertake oceanic odysseys of tens of thousands of kilometres each year, can be intercepted in a few areas of particular importance to them, such as feeding and breeding grounds.

Throughout history, fishing has followed a familiar pattern. At first, we meet all our needs from local sources. But fishing eventually reduces populations and so, to sustain catches, fishers have to develop better ways to outwit their quarry, or they move to places that have been less exploited, or switch to less favoured species. James Bertram recognized this tendency as early as 1873, when he lamented: 'We are continually, day-by-day despoiling the waters of their food treasures. When we exhaust the inshore fisheries we proceed straightaway to the deep waters.'[12]

This pattern can be read in the changing fortunes of the UK trawl fishing fleet. From 1889 to the outbreak of the First World War, British trawlers fished mainly on home grounds. Landings per unit of fishing power fell steeply as those stocks were depleted. After the war, people undertook long voyages to Iceland, the Arctic and West Africa in search of virgin stocks and landings per unit of power climbed. From the 1960s, these grounds in turn became depleted and the rewards from fishing fell steeply. In the 1970s these losses were compounded by the enlargement of sovereign waters and the establishment of exclusive economic zones reaching out 200 nautical miles, forcing boats back to home grounds again, where the rewards from fishing were greatly reduced.

Geographic expansion and substitution of species has served fishers well for most of history. In the nineteenth and early twentieth centuries, the US oyster fishery worked its way around American estuaries in sequence, moving south from New York and north from San Francisco, exhausting stocks along the way. But, in the last century, fishing has spread ever wider and deeper until it has today reached the limits of the oceans. After the Second World War, Japanese longline boats took to the high seas and fanned out across the world's oceans, and by the 1970s fished almost everywhere. Soviet fishing boats targeted deep seamounts one after another from the 1960s and 1970s. They have been joined today by fleets from Spain, Taiwan, Sri Lanka, China and other countries and they have diversified from tuna and swordfish to sharks, mahi-mahi, opah and high-value deep-sea beasts like the 2-metre-long Chilean seabass found in the freezing depths of the Southern Ocean. Fishing boats now pursue fish to the edges of polar ice shelves and down thousands of metres into the abyss. The impact of the fishing industry goes even deeper than the maximum limits of our gear.[13] Animals that migrate up and down from underlying layers are also caught, and the palls of mud thrown up by trawls settle further down. Estimates suggest that most of the world's major fishery species have been reduced in numbers by 75–95 per cent or more. Figures compiled by the UN Food and Agriculture Organization, the FAO, show that two-thirds of the species we have fished since the 1950s have experienced collapse, and the rate is accelerating.[14] It doesn't take a genius to see that with a track record like this we are squandering life.

Life begins at forty, so the saying goes, but for most of us our reproductive years are behind us by this time (although wealthy old men seem to buck the trend!). For fish and shellfish, however, life really does get better with age. Big, old, fat fish produce far more offspring than their young, lithe and smaller brethren. Their size and experience give them an edge. Their eggs are better provisioned than those of small fish so more of them survive the dangers of early life. But, in the span of a century or two, fishing has dismantled the dominion of the old and evolution has begun work in a different direction.

A universal rule of fishing is that when you exploit a population, the average size of animals gets smaller. Most fishing methods are size-selective, which is to say they catch animals that have bodies or mouths larger than the size of a mesh or a hook. Even hand-gathered fish and shellfish are susceptible to this, as people tend to pick out the largest and juiciest ones first. Over time, fishing therefore alters the balance between young and old in a population. Modern-day fishers are not the first to have caused their quarry to become smaller. You can read the signals of increasing fishing intensity in the ancient shell middens of California, where mussels fell in size by over 40 per cent over a period of 9,800 years, while the average red abalone eaten decreased from nearly 20cm long to under 7cm.[15]

Evolution works to maximize the number of descendants that an animal leaves behind. Where the risk of death from fishing increases as an animal grows, evolution favours those that grow slowly, mature younger and smaller, and reproduce earlier. This is exactly what we now see in the wild. Cod in Canada's Gulf of St Lawrence begin to reproduce at around the age of four today, when previously they had to wait until six or seven to reach maturity forty years ago. Sole in the North Sea mature at half the body weight they did in 1950.[16] Surely these adaptations are good news for species hard-pressed by the onslaught of fishing? Not exactly. Young fish produce many times fewer eggs than large-bodied mature animals, and many industrial fisheries are now so intensive that few animals survive more than a couple of years beyond the age of maturity. In combination, this means there are fewer eggs and larvae to perpetuate future generations. In some cases, the amount of young produced today is a hundred or even a thousand times less than in the past, putting the survival of species, and the fisheries dependent on them, at grave risk.

Over the course of decades, climate cycles like the North Atlantic Oscillation shift the probability that young will survive up and down. Many species have adapted to these cycles by developing long lives over which they reproduce many times. By hedging their bets this way, they maximize the chances that some of their young will make it. Excessive fishing ramps up the risk of reproductive failure and population collapse by contracting the number of years over which species can reproduce. Fishing drives up the risk of population crashes today by pushing evolution in a direction counter to that of the environment, and in doing so unravels adaptations developed over thousands of years. Long-term surveys of marine life off the coast of California show that the species we catch now have more variable reproduction than the species we don't.[17]

Over long timescales, intensive exploitation leads to the progressive loss of larger animals. This phenomenon is so pervasive it has been given a name: 'fishing down the food web'. It is brother to the process of sequential overexploitation – the removal of species in sequence from large to small, predator to prey, high to low value – as they tend to go hand in hand. We begin with the pursuit of large and valuable animals. Big species are often predatory by nature and as a consequence have firm, succulent flesh that we prize. As predators, they tend to be bold and voracious and they readily succumb to the lure of the baited trap, hook or net. When they have been depleted, we switch to smaller, less-favoured, lower-value species.[18]

Where once vast sawfish churned the waters of estuaries in the southern US now much smaller fish raise puffs of mud as they feed in the shallows. When rich new fishing grounds were discovered for cod on the Rockall Bank off Scotland's west coast in the 1860s, one report said, 'The crews caught the fish as fast as they could bait and haul and it is stated that when any of the cod broke from the hook, great monstrous sharks as blue as if painted with a brush darted round the ship's side and swallowed them in an instant.'[19] Today the blue sharks are almost gone and the cod has been humbled.

In all but the most extreme environments, the web of life is highly complex. Animals and plants interact in a multitude of ways. They hunt or hide from each other, or use the living space others create. When species are reduced in numbers or removed, the effects on other species cascade through the web of life, with unpredictable consequences. We know from farms that simplified ecosystems suffer problems. Reducing diverse plant communities to a handful of species raises the likelihood that there will be outbreaks of pests and diseases. On land, we control these unwanted effects with chemical sprays or elaborate husbandry. We have no such recourse when things go wrong in the sea.

Think of life as a game with many players. Some players fulfil similar roles and circumstances dictate their fortune. One set of conditions will favour one species or group of species. When circumstances change, fortunes shift and others take their place. But the role itself endures. A higher diversity of species and greater complexity in ecosystems gives them the resilience to adapt when the environment shifts. Ecosystems that have lost species due to overfishing or other forces such as pollution will be less able to cope.

The Firth of Clyde, where this chapter began, offers a bracing illustration of what can go wrong. This narrow arm of the sea penetrates 100 kilometres into southwest Scotland, ending where mountains shade the head of Loch Fyne. In the eighteenth century its banks and bays abounded with fish, shellfish, whales and porpoises, none more abundant than the herring which migrated each year into the loch to breed. One eighteenth-century traveller marvelled at how, when the herring shoals came in, the sea seemed to consist of two parts fish, one part water. Huge basking sharks and whales feasted upon plankton that bloomed in the tide-churned waters, while packs of hunters savaged the herring. Such abundance and vitality attracted a lively fishery. Henry Beaufoy, a British Member of Parliament, described the fishing in 1785:

> On conversing with [one of these fishermen], he gave me accounts of the quantities of fish he catched, that appeared altogether incredible; for one article, he assured me, that when he baited his small line for the smaller flat fish, which line contains 400 hooks, it was not at all an uncommon thing for him to take in, at one haul, 350 fish from the 400 hooks. These consist of turbot, soal, and large fine flounders, about two or three pounds weight each. As to skait, he said he seldom fished for them, as they are not a saleable article there; but he could easily fill his boat with them, when he chose it, at one haul of his lines – and from the accounts of other people, I had no reason to doubt the man's veracity.[20]

In the late nineteenth century, nets and trawls were introduced that could touch bottom on the shallow spawning banks at Ballantrae upon which herring laid their eggs in the southern Firth of Clyde. Within a couple of decades, the herring fishery there collapsed. This catastrophe spurred fishers to establish a ban on bottom trawling throughout the Firth of Clyde that stood until the early 1980s. While it was in place, landings of many bottom fish held up well, although herring and saithe, a relative of cod, were wiped out in the 1960s and 1970s through the invention of electronic fish-finders and mid-water trawls.

With the herring gone, pressure then mounted from prawn trawlers to open up the closed areas. Following repeal of the ban in 1984, cod, plaice, haddock, whiting – in fact all of the Firth's productive fisheries – declined to virtually nothing within twenty years. Today, the seabed is barren and the only fisheries left are for prawns and scallops, and even these are overfished. It is a marine wasteland.[21]

In the Firth of Clyde we can glimpse the endpoint of overfishing, a time when nothing worth catching is left. Unfortunately, this story is far from unique. The combination of increased fishing power and dwindling areas of refuge is repeated in all of the world's seas and oceans. The Firth of Clyde gives us a stark vision of a future without fish.

When Ruth Thurstan and I published our research on the Clyde in 2010, many fishermen and others familiar with the area agreed with our conclusion. Indeed, we thought the sorry state of Clyde fisheries was, to paraphrase Jane Austen, 'a truth universally acknowledged'. But industry leaders were not pleased. In one newspaper report, the heads of two Scottish fishing organizations between them deployed the three favourite tactics of those in denial: attack the science, slander the scientists and blame something else (pollution, seals and climate change were the straws at which they clutched).[22] If the consequences of their denial were not so devastating, I would have found more humour in their last comment: 'We don't need to do much about this report . . . we are letting proper marine scientists rubbish it for us.'

Two sea anglers who were interviewed for the same piece were not afraid of saying what everyone who lived there recognized: 'There's only mackerel left,' one said, 'and then only in the summer. You used to see people angling from the beach at the West Bay there, even at night. There's no point now because there's nothing to catch.' The other said: 'It's not a one-man-band or a guy with a rod that's causing the problem. And it's not down to the Clyde being dirty – it's cleaner now than it was in the 70s – it's the trawlers.'

Over the years I have come across spectacular levels of denial among fishing industry representatives. In parliaments and senate buildings and committee rooms across the world I have seen them dig their heels in to resist regulations that could help fish stocks recover. Politicians too willingly believe their claims that greater regulation would cause unnecessary hardship. In reality, failure to acknowledge and deal with the problem represents a far more serious risk to their livelihoods. In the European Union, the relationship between politicians and the fishing industry has become like that of a doctor assisting the suicide of a patient. For the last twenty-five years, politicians have given the industry a third larger catch quotas on average than those that scientists have recommended to them as being safe.[23] The only outcome possible from such a policy is the collapse of fish stocks and the fishing industry. If you don't believe me, imagine a farmer who takes ten more sheep to market each year than his flock produces. You can't cheat nature, however good you may be at spinning a story.

To close this tale of demolition, allow me to return to the Atlantic bluefin tuna, for this majestic fish has recently become emblematic of the worst of humanity's destructive tendency. Since 1970, their abundance has declined by at least two-thirds and the total decline over the course of the last century is probably closer to 95 per cent. By the 1970s, bluefin tuna had already disappeared from the North Sea, perhaps the result of the decimation of a highly migratory sub-population of the species.[24] By the 1980s, the Black Sea population had all but gone, and is now considered extinct. By any metric, this is a species that we have taken from abundance to the verge of extinction. The scientific evidence is unequivocal. Yet the badly misnamed International Commission for the Conservation of Atlantic Tunas continues to award its members catch quotas far above those that would enable this fish to recover. Corporate greed has triumphed over human decency.

Daniel Pauly, a charismatic fisheries scientist from Canada's University of British Columbia, describes world fisheries as a giant Ponzi scheme. In this type of scam, fraudsters pay investors from the capital in a fund rather than from the returns made on their investments. Ponzi schemes collapse when the flow of new capital dries up. From the nineteenth century, when fisheries were first industrialized, landings have been sustained by fishing ever further afield and deeper. The fishing industry has been dependent on a constant input of new capital. Whenever fish began to run out, fishers moved on or switched to other species. Over time, fisheries have eaten up their capital stocks rather than lived within the limits of annual production, and they are now failing because, like a Ponzi scheme, they are running out of new capital. We now hunt fish to the furthest limits of the oceans and to depths where productivity slows to a trickle. There is nowhere else to go and few species worth eating remain untouched by fishing. The solution is not for a handful of people to stop eating fish. We need to set up new regulations and police them well. But I will return to this later on in the book when it is time to consider solutions.

Hunting and fishing are the oldest of human influences on the sea. Arguably, fishing also remains the most serious. But, in the last century, climate change from greenhouse gas emissions has been building in the background. It has burst to prominence in the last twenty years as it has begun to affect our daily lives. In the next four chapters, I explore the most important of the many ways in which greenhouse gases are changing the oceans.

4

Winds and Currents

Benjamin Franklin is celebrated as a founding father of the United States, an astute diplomat, the co-founder of the first public lending library and a prolific inventor whose contraptions ranged from the lightning rod to bifocal spectacles. He is known to have had an exceptional range of interests and abilities, but his pioneering contributions to oceanography are not widely recognized outside scientific circles.

Before the American Revolution, Franklin was postmaster general for the British colonial mail, and as such he applied his formidable intelligence to the problem of how to speed up deliveries between the Old and New Worlds. He wondered, more specifically, why it took ships two weeks longer to sail from London to New York than it took them to travel in the opposite direction. His cousin Timothy Folger, a whaling captain, told him that British ships battled a 3-knot current, while those coming from America sailed with it. Together the pair tapped the knowledge of experienced American whalers to produce, in 1770, the first map of what is now known as the Gulf Stream.

The Gulf Stream is a fast-flowing surface current that jets into the Atlantic through the Florida Strait and then runs north along the east coast of America. It turns out to sea just south of Cape Hatteras, off the coast of North Carolina, at which point it crosses the Atlantic to Europe where it dissipates into a general northerly flow called the North Atlantic Drift. It is part of the 'global ocean conveyor', a system of currents that circles the planet and loops from the surface to deep sea.

The idea that the seas might overturn, with water pouring from the surface to the deep and then back again, was first proposed by

Benjamin Thompson, a contemporary of Franklin who, unlike his more famous counterpart, was a loyalist and fled revolutionary Massachusetts for London. In Europe he made a name for himself as a scientist and statesman. He is best remembered today for his discovery of how to increase the draw of fireplace chimneys, thanks to which he helped cure the centuries-old curse of smoke-filled rooms. Thompson, who in 1791 became Count Rumford (after the New Hampshire town where he grew up, today's Concord), based his proposal that the shallow waters of the oceans overturned and mixed with the deep on his knowledge of the properties of heat. What set him thinking was a single measurement of the temperature of the deep ocean made 1,100 metres below sea level by a British slave ship in the tropical Atlantic in 1751.[1] It was less than 12°C, a big contrast to the 29°C at the surface.[2] This measurement led him to suppose that in polar latitudes water,

> deprived of a great part of its heat by cold winds, descends to the bottom of the sea, [where it] cannot be warmed where it descends, [and] as its specific gravity [i.e. density] is greater than that of water at the same depth in warmer latitudes, it will immediately begin to spread on the bottom of the sea, and to flow towards the equator; and this must necessarily produce a current at the surface in an opposite direction. There are most indubitable proofs of the existence of both these currents.[3]

This down-up current is sometimes called the 'thermohaline circulation', because it is driven by differences in temperature and salt content between water masses. The reasons for downwelling differ between Arctic and Antarctic. In the Antarctic, frigid conditions cause sea ice to form. Freezing separates freshwater from salt and leaves behind a more briny sea. This bitterly cold and saltier water is denser than normal seawater and therefore sinks in a process known, rather unimaginatively, as 'deep bottom water formation'.

In the far north of the Atlantic, the Gulf Stream supplies water to polar seas that is more salty than local waters, and is made even saltier by evaporation from intensely cold, dry winds blowing off Greenland and Europe. At both poles, as cold, salty and dense water sinks it pulls in surface water to replace it. This pull is one of the

Figure 2: Simplified view of the global ocean conveyor current. Grey circles show places where surface water sinks to create deep bottom water.

energy sources for the Gulf Stream and North Atlantic Drift (the other being wind). When this water sinks it begins a deep sea journey that may last 1,500 years before it upwells again into the sunlight far away.

You can create a loop current a little like the global ocean conveyor by blowing on a cup of tea to cool it. Your breath pushes the tea across the cup until it hits the edge, at which point it flows down the inside of the cup, crosses the bottom and upwells again at the side you are blowing from (if you are ever short of a teaspoon, this is a handy way to mix milk into your tea).

Like the ideas of many gifted thinkers, Rumford's theory of ocean circulation was not fully accepted until long after he had died. He never returned to America, although his British sympathies were later forgiven and he endowed a professorship at Harvard University. Oddly enough, despite the fact that they were born just twelve miles apart and had such similar interests – Franklin also invented a stove that Rumford later improved – there is no evidence that they ever met or corresponded. Theodore Roosevelt would later declare that Franklin,

Rumford and Thomas Jefferson were the three greatest minds America had ever produced. So he was clearly forgiven for being on the wrong side.

In the 1960s, atmospheric nuclear tests produced radioactive materials that were washed into the Arctic and gave us a way to trace the speed of water flow to the deep sea. Scientists discovered that the global ocean conveyor moved at just over a metre per hour.[4] This sinking rate means that 15.5 million cubic metres of water is displaced per second, which is equivalent to the flow of eighty Amazon rivers, or twelve times the flow of all the world's rivers combined.[5] As powerful as this river in the sea is, it would take nearly 2,800 years to circulate all of the oceans' waters through what oceanographers call the 'North Atlantic pump'.

As I intimated, there are two other key areas of deep bottom water formation, in the Ross and Weddell Seas of Antarctica. They transfer another 21 million cubic metres of water per second from the surface to deep sea, reducing turnover time for water in the deep oceans to under 1,200 years. These 'pumps' are critical to the vertical mixing of water in the world oceans. They carry freshly oxygenated water into the deep sea, helping sustain life there. They also transfer carbon dioxide from the atmosphere to the deep ocean, a point I will return to later on.

One thing that fascinates me is that water bodies in the sea, especially the deep, tend to retain characteristic signatures of temperature and salt content for very long periods. Oceanographers can build up a picture of the three-dimensional structure of the sea from simple vertical profiles of salinity and temperature measured using instruments dangled from boats. It turns out that the oceans are made up of many parcels of water in constant motion, driven by wind and density differences, and their origins and movement around ocean basins can be tracked. At the mouth of the Mediterranean Sea, for instance, water flows in through the Strait of Gibraltar at the surface. Beneath there is a deeper flow back into the Atlantic of dense, salty water concentrated by evaporation as it travels around the Mediterranean Basin. This pours out across the shallow Gibraltar sill and forms giant turbulent loops that travel west to the Azores and north to Ireland. Similar measurements reveal that the southward flow of deep water from the North Atlantic is not compact, like the surface Gulf Stream, but broad and sluggish as it creeps along the eastern seaboard of North America.

The North Atlantic and Southern Ocean engines of the global ocean conveyor have recently been identified as possible climate 'tipping points' by a multinational team of climatologists,[6] meaning that a certain threshold of water temperature or density must not be passed if the currents are to remain in a stable state. If that critical point is passed, we could see a radical and rapid shift to a different state. Since major ocean currents move at the behest of wind and water density gradients, there is a good chance they would shift in a world of changing climate. The possibility that this might happen is now a cause of great concern to many scientists.

Unlike past changes, whose causes were geologic or celestial in nature, today's are down to us. People have multiplied and spread to every corner of the globe. We have shifted from being a species governed by nature to one that can harness nature to its own ends. Unwittingly, our ingenuity has unleashed forces over which we have little control and that threaten the way we live.

The concept that some form of 'global warming' might be triggered by the burning of fossil fuels was first proposed in 1896 by a Swedish chemist, Svante Arrhenius.[7] At that time, people were fascinated by the ice ages, whose world-shaping influence had only recently been understood. Arrhenius argued that variation in the amount of atmospheric carbon dioxide could have played a key role in glaciations.[8] But he took the idea further. Perhaps his thoughts coalesced while tramping the icy streets of Uppsala one evening as the coal smoke from thousands of hearths wrapped around him. He made the connection that the enormous quantities of fossil fuel being burned could eventually lead to planetary warming. From his rather chilly perspective, he thought warming would be a good thing:

> We often hear lamentations that the coal stored up in the earth is wasted by the present generation without any thought for the future . . . We may find a consolation in the consideration that here, as in every other case, there is good mixed with the evil. By the influence of the

increasing percentage of carbonic acid [carbon dioxide] in the atmosphere, we may hope to enjoy ages with more equable and better climates, especially as regards the colder regions of the earth, ages when the earth will bring forth much more abundant crops than at present, for the benefit of rapidly propagating mankind.[9]

By Arrhenius's calculation, doubling atmospheric carbon dioxide would increase average temperature by 4°C, a figure since revised by the Intergovernmental Panel on Climate Change (IPCC) to 2–4.5°C. Not bad for a nineteenth-century chemist!

Today we don't share Arrhenius's optimism about the benign influence of global warming. A little warming will not simply help shrug off the winter cold or lengthen growing seasons. Now we understand that warming alters patterns of wind, cloud cover and rainfall, changing conditions in ways that are hard to predict and will be difficult and expensive for us to adapt to. Thousands of studies attest to the reality of global warming, now usually called 'climate change' because of the far-reaching influences warming has on climate. A record-breaking snowfall or unexpectedly fierce tornado or monsoon flood is as much a product of global warming as scorching summers and prolonged droughts. Measurements from mountains and lakes, glaciers and ice sheets, deserts and rainforests, coasts and oceans all confirm that temperatures are on the rise.

As I explained in the opening chapter, Earth's atmosphere is like a blanket that traps warmth from the sun but also shields us from the ferocity of the sun's heat and harmful ultraviolet rays. If Earth didn't have an atmosphere, temperatures would be like those on the moon, where there are wild swings between extremes.[10] In sunshine, the moon's surface can top 100°C, while on the dark side it can plunge close to –150°C. (The moon averages –23°C compared to Earth's clement 15°C.) The warmth of the atmospheric blanket depends on its content of heat-trapping gases. Collectively, these 'greenhouse' gases slow the radiation of heat from Earth back into space. The most important are water vapour, carbon dioxide, methane and ozone, but there are several others in the mix.

Since pre-industrial times (before 1750), carbon dioxide concentration in the atmosphere has risen by 38 per cent, from 280 parts per million (ppm) to 388 ppm.[11] Over the same period, the concentration of methane has gone up by 150 per cent from 700 parts per billion (ppb) to 1,745 ppb. Although there is much less methane than carbon dioxide, it is a far more potent greenhouse gas. You will recall from my description of the early Earth in Chapter 1 that each molecule of methane has twenty-five times the warming potential of a molecule of carbon dioxide.[12] The main source of carbon dioxide is fossil-fuel burning, while methane comes from livestock (cows and other ruminants fart endlessly), landfill sites and rice paddies. It is also a by-product of warming, as trapped methane is released by melting tundra near the poles.[13] Just as Arrhenius predicted, as we have burned more fossil fuels, temperatures have risen. Averaged across the globe, they have increased by 0.7°C since pre-industrial times. What is alarming is that the rate of rise has now reached 0.2°C per decade. We are on a slippery slope.

Much of the heat trapped by greenhouse gases so far has been taken up by the sea. If it hadn't we would all be sweltering by now. The oceans have sucked heat from the atmosphere because the heat capacity of water is several thousand times greater than air.[14] It is for this reason also that temperature increases in the oceans have been less than on land. Averaged from the surface to seabed, temperatures have risen just 1/25th of one degree centigrade since 1955.[15] It doesn't sound like much, but most warming has been near the surface, where average temperatures have increased by 0.6°C in the last century.[16] This heating has not been even: some places have warmed quickly, others are little changed. The tropics have warmed least while temperatures at the poles have shot up, which is why polar bears and penguins are icons of the changing climate.

On the Antarctic Peninsula, where air temperatures have warmed by 6°C in the last fifty years,[17] Adelie penguins that once nested on frozen ground now huddle in pathetic groups, ankle deep in mud. The chicks' downy feathers are well adapted to snow, but they lose their insulation in sleet and drizzle. Soaked through and frigid, the chicks die. While it seems obvious that they should move to colder rookeries, it is hard to abandon traditional sites used for generation after generation. As a result, Adelie penguins on the Peninsula have declined by 90 per cent in the last thirty years (though those further south are still doing fine). At the other end of the world, polar bears and seals depend on sea ice, the seals to breed on and the bears to hunt from. Polar bears are frequently seen these days swimming in open water up to 100km from the nearest coast or ice. Many now drown far out at sea as exhaustion overtakes them when ice cannot be found.[18]

A two-degree rise in global average temperature is expected to produce a four- to six-degree warming at the poles. Summer sea ice cover has declined fast in recent years. In the summer of 2000, Jim McCarthy, a Harvard oceanographer who was then a leading member of the IPCC, was left gasping in amazement when the icebreaker he was on found open water at the North Pole. Since then, summer sea ice has shrunk further and the region is expected to be regularly ice free by the summer of 2030.[19]

Polar oceans are highly affected by climate change. Less sea ice formation means less saltiness of surface water at the poles, and warmer water is also less dense than colder water. (The word 'warm' is relative – even in a greenhouse world you couldn't enjoy a swim at these latitudes!) Together, these changes will slow the rate of sinking and the replenishment of deep bottom water. Another factor related to climate change conspires to slow the engine of the global ocean conveyor further. More rain and the thawing of frozen tundra have swelled rivers that pour into the Arctic Ocean. This freshwater reduces surface water density and so also inhibits deep bottom water formation.

Ever since Benjamin Franklin's time, people have generally understood that the Gulf Stream carries the heat of the Caribbean far into northern latitudes. It is held to be responsible for mild, wet winters in Britain and France and for the warmth of the seas from Long Island to Boston. Without the Gulf Stream, people have come to believe, a deep freeze will descend on London and New York.[20]

The central premise of *The Day After Tomorrow*, a blockbuster thriller released in 2004, was that the Gulf Stream might suddenly shudder to a halt. In the movie, eastern North America and Europe were plunged into ice age conditions by the sudden failure of the Gulf Stream. While the idea may seem far-fetched, and the near-instant freeze in the film was definitely science fiction, there is legitimate cause for concern. Oceanographers have already seen deep currents slow in the North Atlantic. Between 1957 and 2004, there was a 30 per cent reduction in deep water flow (although this does not yet seem to have slowed the shallow Gulf Stream and the more diffuse North Atlantic Drift). It is too early to link the slowing conclusively to climate change.

Over 120,000 years of climate records held in annual layers of snow cored from within the Greenland ice sheet suggest that this part of the global ocean conveyor has stopped many times in the past. It seems that breakdown can occur rapidly, in as little as a few decades. The trigger for these stoppages in the past was a sudden drop in the salinity of Arctic seas.[21] The period covered by the Greenland ice cores lies within the last glaciation, from 110,000 to 10,000 years ago, when much of North America and Europe were ice-bound. The ice sheets grew in thickness as snowfall accumulated over thousands of years. Eventually it grew top-heavy and became unstable. Ice surged through the Hudson Strait into the North Atlantic, where it melted, freshened the sea and switched off deep bottom water formation at the northern extremity of the global ocean conveyor current. More ice poured into the Eastern Atlantic from the Baltic Ice Lake.[22] Each event is marked in bottom sediments far out at sea by the appearance of ice-rafted rubble swept offshore as the ice sheets collapsed.

At the end of each shutdown, of which there were at least seven in the last 60,000 years, ocean circulation switched on again abruptly. We know this because sea surface temperatures suddenly rose several degrees as water flowed in from warmer latitudes. This effect extended south at least to Bermuda, where cores through deep sea sediments show sudden rises in surface temperature from about 16 to 21°C (today they are just over 22°C).[23] Nobody is exactly sure what caused these currents to shift. There are no abrupt changes in any of the factors that might drive the shift, like wobbles in the tilt of the Earth's axis (which change the heat balance of the planet), or fluctuations in atmospheric carbon dioxide that correlate with the timing of North Atlantic temperature flip-flops. However, scientists have come up with several good explanations.

One theory is that the global ocean conveyor current has three different stable states in the North Atlantic: 'on', 'off' and 'partly on', with deep water formation taking place further south during times when ice sheets were more extensive.[24] Most of the time, one of these states predominates and the current varies over time as conditions fluctuate. A change between states occurs when a critical threshold has been exceeded and the system flips into a different regime that itself might remain stable for hundreds or thousands of years before it flips into a different state again. According to this way of thinking, a small change in some driver, like carbon dioxide or methane concentration, could cause a large change to the entire climate system because of its sensitivity to jumps between different states.

If this sounds confusing, a simple analogy might help. Imagine a drunkard walks home late at night along a riverside path. He lurches from side to side but remains on the path. Halfway home someone brushes past him from the opposite direction and shifts him slightly nearer the river. The bank now lies within the ambit of his more extreme lurches. A hundred metres along the path the drunk falls into the river and so passes from one stable state (on ground) to another (in the water).

Ice cores from Greenland, Antarctica and glaciers the world over spell out the reality of abrupt climate change. These cores contain an extraordinary wealth of information on snowfall, atmospheric gas concentrations and dust levels. The chemical isotope concentrations in ice tell us what the temperature was, while trapped dust layers reveal the extent of deserts and the frequency and size of forest fires and volcanic eruptions. They allow us to stroll back through time, at first season by season, then year by year and, deep down where the ice is compressed, decade by decade. How far back these records go depends on the depth of the core, from a few thousand years in some glaciers, to over 700,000 years in the deepest Antarctic cores. If there is one lesson it is that the climate of the last few thousand years of human history has been remarkably stable when set against a long-term background of steep fluctuations and sharp flip-flops. Abrupt changes revealed by the ice record were not so sudden that you wouldn't have had time to buy a polar jacket and upgrade your home boiler. But they were sharp enough that, were one to happen now, we would really struggle to adjust. If the mighty river that is the global ocean conveyor falters, it will have profound impacts on life above and beneath the sea. Radical regional shifts in marine climate alone would precipitate the wholesale reorganization of ecosystems and human societies. Those of us who live in Europe can take some consolation from the fact that the sage advice of the IPCC is that the North Atlantic arm of the global ocean conveyor is unlikely to fail in the next hundred years. But other shifts are already upon us, like the rapid warming of the Arctic Ocean.

Water flows from the surface to the deep ocean in polar seas at a rate equivalent to 20,000 Niagara Falls. Since the oceans do not expand by the same volume, this flow must be matched by an equal and opposite upwelling towards the surface somewhere else. Part of that counterbalance is generated by gentle upward mixing in the tropics. In other places, upwelling is assisted by wind-driven surface pull. These upwellings are concentrated on the eastern boundaries of the oceans, where winds blow parallel to the coast, like off the western United States.

This upwelling explains why the seas of northern California are at their most bracing in spring and summer and why San Francisco then wears a foggy shroud. Cool deep water upwells under the influence of winds from the north. In a physical quirk of life on a spinning sphere, the average direction of water movement is at a right angle to the direction of the wind.[25] In the northern hemisphere the twist is to the right, while in the southern it is to the left. This effect is called the Coriolis force after Gaspard-Gustave de Coriolis, the nineteenth-century French mathematician who worked out the interplay of forces acting on rotating objects like water wheels. As far as we know, Coriolis never thought to apply his ideas beyond the energy production of rotating machinery, but the Coriolis force has a profound influence on how water and air circulate in both oceans and atmosphere. Indeed, it applies to everything that moves across the surface of the Earth. If they did not correct for the Coriolis force, airline pilots would miss their destinations and artillery gunners would shoot wide of their targets.

Winds that blow from the south along the coast of South America drive the world's largest upwelling in the sea off Peru. Upwellings have enormous significance. Although they cover just 1 per cent of the area of the oceans, they generate nearly half of the global fish catch. Peruvian anchovy alone account for a tenth of world fish landings in good years. This sleek silver fish lives in schools of almost incomprehensible size and feeds mainly on phytoplankton, micro-scopic plants that live in the open sea. Upwelling fuels spectacular blooms of these phytoplankton (which, as I will come to later, are also caused by nutrient enrichment from human sources).

Over most of its surface area, the ocean has two layers, surface water and the deep sea. Surface waters are warmed by sunshine. Since warm water has a lower density than cold water this layer tends to float on top of the much cooler deep waters. There is little mixing across the boundary between the two, which is called the thermocline because of the abrupt temperature gradient. This creates a problem for marine life. Sufficient light for plant growth only reaches to depths of 10 to 100 metres, depending on water clarity. In the crystal-clear waters of the mid-Pacific, light will sometimes penetrate as far down as 200 metres, but even that is just a small percentage of the average depth of the sea (3,680 metres). Plants grow in this sunlit zone by tak-ing up dissolved nutrients and carbon dioxide. These plants are either eaten by animals or microbes, in which case the nutrients are recycled, or they sink. When they pass through the thermocline, which is typic-ally around 30–100 metres deep, their nutrients are lost to the deep sea.[26] The bright surface layer thus leaks nutrients to the deep, with the result that plant growth is usually limited by a lack of fertilizer. The deep, meanwhile, has plenty of nutrients but no light, so plants cannot grow. Animals living there exist on table scraps that sink down from the surface layer. Life is sparse. Where deep waters are pushed or pulled up to the surface, as they are off Peru, they flush nutrients into the sunlit surface layers to fuel explosive plant growth.

The world's most intense upwellings occur where winds that blow parallel to the coast are reliable and strong, like off West Africa, both north and south of the equator, and the Pacific coast of South Amer-ica. These winds are a consequence of differential heating between land and sea. The land heats up more than the sea during spring and summer in sub-tropical latitudes and year round in the tropics.[27] Hot air rises above the land, carrying water vapour with it. As it ascends it cools and the water condenses into those spectacular thunderstorms you can sometimes watch boil upward while sipping piña coladas on a tropical beach. Condensed water falls in showers so intense you feel as if you were underwater.

This condensation of water leaves the air at cloud level warmer and lighter so it rises even higher, eventually stopping when it is thoroughly chilled. That high altitude cold air is now pushed sideways by more air welling up from below. North of the equator it blows north, while south of the equator it heads south, carrying heat from the tropics to higher latitudes where it cools and sinks back down around 30° north or south. If you glance at an atlas, you will see that many of the world's great deserts are concentrated around these latitudes because the downdrafts are very dry, having had all the moisture squeezed out as the air rose: the Mojave, Sonoran and Chihuahuan in North America, the Sahara and Arabian desert in Africa, the Takla Makan in China, the Atacama in Chile and so on. Since air cannot rise at the equator without being replaced by other air, surface winds blow air back from these drier latitudes to the tropics to complete the loop and drive upwellings as they do. At temperate latitudes, surface winds blow poleward, where tropical circulation loops intersect like cogs with others that carry heat poleward and draw cool air back to warmer latitudes.

As greenhouse gas concentrations rise, the tropics will trap more heat, which will drive a faster redistribution of warmth and strengthen the winds that power upwelling.[28] The productivity boost this would give the oceans could benefit fisheries and help the sea absorb more carbon dioxide (although this is a double-edged sword, as I will explain later). Thumbing his nose at Aristotle, Oscar Wilde once quipped, 'Moderation is a fatal thing. Nothing succeeds like excess.' He found out to his dismay that it was the other way round, and so it is with upwelling. Excessive upwelling can make unpleasant things happen in the sea. In Southwest Africa, we may have a taste of the future of upwelling in a greenhouse world.

The coast of Benguela, offshore from the vast red dunes of the Namib Desert, as scorched and thirsty a place as you are likely to find on Earth, sustains one of the planet's most intense upwellings. The lack of plant life on shore contrasts vividly with its abundance in the sea. Copiously supplied with nutrients, phytoplankton flourish, turning the sea green. Normally, those plant cells would be consumed by zooplankton, which in turn would be eaten by fish and other predators. The trouble is the wind is so strong it pushes water offshore before

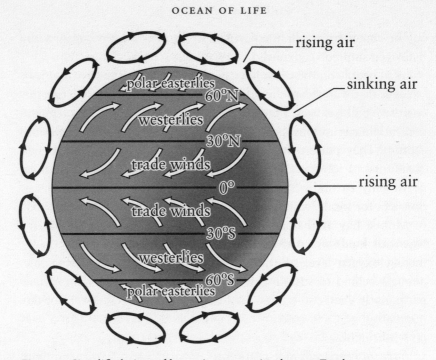

rising air

sinking air

rising air

polar easterlies
60°N
westerlies
30°N
trade winds
0°
trade winds
30°S
westerlies
60°S
polar easterlies

Figure 3: Simplified view of how air masses circulate on Earth.

zooplankton, which take longer to develop than the plant cells, have time to complete their life cycles. While phytoplankton turnover at high speed and continue to bloom right in the midst of upwelling, zooplankton are swept offshore so they cannot curtail the explosive growth of the phytoplankton. The result is devastating, as decaying phytoplankton sink to rob the sea of its oxygen deep down.

Beneath Benguela's upwelling plumes, the seabed is thick with swirling drifts of dead and decomposing plankton. It can sustain only the ancient microbes that first evolved in the similar toxic broth of the early oceans. This metres-thick sludge is anoxic or nearly so and gurgles forth methane and hydrogen sulphide, the waste products of microbial metabolism. Hydrogen sulphide compounds the problem of anoxia because it reacts with any oxygen present and strips it from the water as it rises towards the surface. From time to time, the sludge effervesces so fast that the rising bubbles rush upwards in an eruption of sulphide and methane, turning the sea bright turquoise.[29] The colour comes from sulphur particles in suspension, produced when hydrogen sulphide is oxidized.

Local people on the coast have become inured to the stench of rotten eggs that wafts from the sea. In the Namibian port of Lüderitz, the corrosive pall has been a fact of life for as long as people can remember. 'Sulphur eruptions' have been reported here since the nineteenth century. They poison the sea, cause mass death of fish and trigger 'walkouts' of lobsters onto beaches as they try to flee one form of death only to find another. These walkouts provide a short-lived bonanza for locals, who pile them into every container they can find. In Mobile Bay, Alabama, similar walkouts are known as 'jubilees', when all kinds of creatures are forced to the shoreline by plummeting oxygen levels. But they decimate populations, leaving few animals behind to sustain catches later on. In the early 1990s, one particularly energetic eruption off Namibia killed 80 per cent of the population of Cape hake, a bottom-dwelling fish, and led to a collapse of catches.

In recent years, sulphide eruptions have increased in frequency and intensity, a phenomenon that is most likely linked to overfishing. A few decades ago, sardines (also known as pilchards) were to Namibia what anchovies are to Peru. These tiny fish are filter feeders and sweep phytoplankton and some zooplankton from the water as they dart around, mouths agape. In good years, sardines shoaled by the billions. Unlike zooplankton, which are poor swimmers and easily swept offshore and away from the heart of phytoplankton blooms, sardines are powerful swimmers and remained in the thick of it. Voracious sardines ate the phytoplankton as fast as it was produced, so dead plankton did not usually accumulate on the seabed and sulphide eruptions were rare. Sardines in turn were eaten by other predators, like tuna, swordfish and seabirds, and the Benguela upwelling sustained one of the oceans' most spectacular concentrations of wildlife.

Such productive seas soon drew the attention of the fishing industry and sardines were targeted heavily from the 1960s. In the 1970s, the fishery for sardines collapsed and has never recovered, and coastal residents can now blame overfishing for the corrosive sulphide fogs that overwhelm their towns and irritate their eyes and throats. Namibian seas reveal what can happen when we knock holes in the fabric of an ecosystem, through fishing or by other means.

Andrew Bakun from the University of Miami is the scientist who first made the connection between upwelling intensity and global warming.[30] He predicts that warming will intensify upwelling in other regions of the world, so that they will match or exceed the rates seen in Benguela today. If there are insufficient fish to control phytoplankton blooms, they too could experience the poisonous combination of anoxia and sulphide eruptions. One such place is northern California. Fifty years from now, there is a real possibility that the homely charm of Mendocino will be tarnished by the stench of rotten eggs and its beaches heaped with putrid fish.

There is a further marine twist to the climate change story. The ocean has two layers, as I mentioned: a warm surface layer floats on top of cooler, denser water below. The additional heat absorbed by the surface layer as the world warms increases the temperature contrast, and hence the density difference between these layers. The steeper the density gradient at the thermocline, the band separating surface water from deeper sea, the less water mixes across it. The result is that there are fewer nutrients transferred from deep sea to shallow water, and less oxygen mixed from shallow waters to deep, as warm water holds less oxygen than cool water. As the oceans heat up they will thus release some of their oxygen into the atmosphere.

Dissolved oxygen is plentiful at the surface, but drops sharply below the thermocline and sags further as you descend, eventually rising again towards the bottom. This pattern has a simple explanation. The surface layer is well ventilated and the deep sea gets its oxygen from the water that sinks down at the poles. In between, there are regions where the life-giving gas is scarce.

Low oxygen zones form in places where the water column is strongly separated into layers by density differences (caused by differences in temperature and salinity). Dead plankton and animals sink from the surface – marine snow, as oceanographers like to call it – and are then eaten or broken down by microbes, which use up oxygen. Most of the oxygen in the middle layers of the sea has been there since the water was last in contact with air at the surface (in some cases hundreds to a thousand years or more ago). The longer ago that was, the more time there has been for animals and microbes to use up its oxygen. There is no sunlight in these inky depths, so no oxygen can be produced to replace that which is used up. Deep waters in the Pacific and Indian oceans are 'older' than those of the Atlantic, meaning that they have been submerged for longer, and so have less oxygen. They also have more carbon dioxide, which has been breathed out by the animals and microbes that live there.

The eastern Pacific and Northern Indian Ocean have enormous slabs of water in which oxygen concentrations are less than 30 per cent of those at surface levels. Fully a quarter of the Pacific is affected by low oxygen at some depth or another.[31] These oxygen-starved zones impose physiological stresses that few organisms can tolerate for long. In the eastern Pacific, the warm ventilated surface layer is thin and the low oxygen zone lies just 200–600 metres below the surface. Fast-swimming tunas are confined to well-aerated surface waters, which is why they are easy to catch there. The upper region of low oxygen water is home to many planktonic animals that migrate towards the surface to feed under cover of darkness and then retreat by day to a place where few predators can follow. Sharks and tunas can penetrate low oxygen layers only for minutes or hours before they must return to the surface to catch their breath and warm up.

Every animal is different of course, and some are better able to tolerate low oxygen levels than others. But, to put the conditions into a human context, to enter water with less than 30 per cent oxygen saturation would be like a climber venturing into the death zone at the extremes of altitude. At heights of greater than 8,000m (26,245ft) oxygen levels are a bit over a third of those at sea level – too low to sustain human life. Without bottled oxygen, we can only make brief sorties into this zone, because we use oxygen faster than our bodies can take it in.

Aquatic animals face different challenges to those living on land. Many creatures are neutrally buoyant in water, which means they neither rise nor sink. The major constraint on their movements is not gravity but the frictional drag of water. Other things being equal, big animals can swim faster than smaller ones and use relatively less energy to do so. But they face a problem extracting oxygen from water. Oxygen demand scales with body volume, which is the cube of fish length (think of the volume of a box: length x breadth x height). On the other hand, the surface area of the gills over which oxygen can be exchanged increases less fast as body size increases, as the square of fish length (the area of a box: length x breadth).[32] While being large-bodied is advantageous hydrodynamically, it is a disadvantage when it comes to breathing. This simple fact explains why predatory fish like tuna don't simply follow schools of their prey, snacking on them when the mood takes them until all the prey are gone.[33] In a long-distance race, it is the bigger predators that run out of puff first. And, when a fish is left gasping for breath, it could easily fall prey to something else, like a shark. So, marine animals must be very careful not to fall into oxygen deficit.

The huge Humboldt squid is one of the only large predators able to withstand low oxygen for extended periods.[34] These beasts can reach more than 3 metres long with their tentacles extended, and weigh more than a man. Within the crown of muscular tentacles is a parrot-like beak made of two interlocking blades, sharp as butchers' knives. They swim by jet propulsion, squirting water through a tube, but most of the time they probably just wait for prey to come to them.

In a warmer world, low oxygen zones will expand as surface warming reduces mixing across the thermocline. The effect will compound as heat absorbed by the oceans penetrates deeper, which means they will hold less oxygen. In combination, the volume of the sea capable of supporting life will shrink. Already, the upper limit of low oxygen waters has moved up to 100 metres closer to the surface off North America's west coast.[35]

Since 2002, the west coast of the US has seen repeated mass deaths of ocean life on the shallow continental shelf, events unprecedented in the previous fifty years.[36] It turns out they are due to the strong winds that blow parallel to the coast and drive upwelling, a harbinger of the dark side of strengthening upwelling I described above. As these winds push surface waters offshore, a tongue of deep, oxygen-starved water creeps up over the shelf to replace them and snuff out life. Jane Lubchenco, Head of the US National Oceanic and Atmospheric Administration, described the scenes she viewed through video cameras towed through a dead zone off the coast of Oregon in 2002:

We saw a crab graveyard and no fish the entire day. Thousands and thousands of dead crabs were littering the ocean floor, many sea stars were dead, and the fish have either left the area or have died and been washed away . . . seeing so much carnage was shocking and depressing.[37]

One of her colleagues, Francis Chan, said:

One of the areas sampled is a rocky reef not far from Yachats, Oregon. Ordinarily it's . . . swarming with black rockfish, ling cod, kelp green-ling, and canary rockfish . . . [but] the fish are gone, and worms that ordinarily burrow into the soft sediments have died and are floating on the bottom.

Such mass die-offs of marine life will become a regular feature of future oceans unless climate change can be halted.

The Humboldt squid seems to have benefited from expansion of its low oxygen habitat (and loss of big predatory sharks from overfishing).[38] It used to be common only in the seas off Baja California, but in recent years it has expanded north to the Gulf of Alaska. They increasingly come into contact with divers as the oxygen minimum layer reaches closer to the surface. To meet a squid in the wild is one of life's great pleasures. They follow your movements with intelligent, saucer eyes. Their emotions are written on their skin in quick-fire colour changes that pulse and ripple in incandescent waves across their bodies. The Humboldts' uncertain temper adds a little frisson to any encounter. Divers have been roughed up by squid, or had their masks and gear tugged. After the adrenalin rush has passed, most divers feel the animals were more curious than aggressive. If they had really wanted to hurt them, with their huge strength they could have done far worse.

The trade-off between body size and oxygen demand explains a pattern that has long been known but never well understood: fish tend to grow larger in cooler water. This is because they have lower metabolic rates, so they can afford to grow bigger. This suggests that we will have more problems ahead as the seas warm under climate change. Fish under temperature and oxygen stress will reach smaller sizes, live less long and will have to devote a bigger fraction of their energy to survival at the cost of growth and reproduction. From our own rather self-centred perspective, their response to the new conditions will be bad because it will reduce production of body mass, which will limit catches. Of course, some exploited species have already seen their body size and growth compromised by overfishing. Thus we see increasing temperature, decreasing oxygen and high fishing intensities combine to reduce the productivity of fish stocks and undermine the long-term resilience they so desperately need to cope with the changing world around them.

The deep past affords us a bleak warning of what might happen if we fail to control greenhouse gas emissions. The end-Permian mass extinction 251 million years ago, which I described in Chapter 1, was caused by a planet in the grip of runaway global warming. Life in the sea suffered the one–two punch of anoxia and high carbon dioxide and took 5 million years to recover. However, one group of animals that lived within open waters of the sea, the ceratid ammonoids, creatures similar to the beautiful nautilus of our day, rebounded more quickly, returning within a million years of the catastrophe. It is possible that they were adapted to low oxygen conditions. Although the entire group is now extinct, their nearest living relatives are squid, cuttlefish, octopus and nautilus. Indeed, one of the oldest living forms of squid is *Vampyroteuthis infernalis*, the Latin name of which means 'vampire squid from hell'. It is a blood-red beast that hunts deep within the low oxygen zones of today's oceans. Its success as a predator in these extreme conditions may be a modern legacy of past catastrophes. Several other mass extinction events in geological history hacked off large branches of the evolutionary tree. The toxic alliance of high carbon dioxide and low oxygen appears to have wielded the axe on more than one occasion.

The medieval Scandinavian and English King Cnut the Great is reputed to have had his throne set down by the sea, where he commanded the tide to halt. The tide, of course, paid no heed, proving that the sea was moved by powers that were beyond the influence of people, even kings. A thousand years later, humanity has inadvertently gained the power to move ocean currents. But our powers remain circumscribed and we must now live with the consequences, or find a way to keep the forces we have unleashed in check.

5

Life on the Move

Jean-Baptiste de Lamarck is famous today, or perhaps infamous, for his wrong-headed theory of evolution. He believed that organisms could pass on to their offspring characteristics that they had acquired during their lifetimes. Giraffes grew long necks, by his account, because generations of ancestors had stretched further and further to reach leaves high on trees. While his evolutionary ideas were discredited by Darwin, Lamarck was in his day a keen observer of life and a distinguished and influential French scientist. One of his other equally mistaken ideas has taken considerably longer to dismantle.

By the end of the eighteenth century it had become clear that people could deplete some species of wildlife to the point of extinction. The aurochs, a huge European wild cow, had been killed off by 1627 and the dodo by 1662. Many animals had been erased from large parts of their ranges, like beavers, wolves and bears in Europe, and panthers in North America. Aquatic animals seemed more resilient, and Lamarck was convinced that they were in effect impervious to extinction because no natural barriers could contain them. As he wrote in 1809 in his landmark work *Philosophie Zoologique*:

> Animals living in the waters, especially the sea waters are protected from the destruction of their species by Man. Their multiplication is so rapid and their means of evading pursuit or traps are so great that there is no likelihood of his being able to destroy the entire species of any of these animals.

The limit to how far a species will spread from its place of origin depends on its physiology and habitat, and on predators and competitors. A few species do well over wide geographic ranges – tigers, for instance, can survive from torrid jungles to frozen mountain passes – but most species do well in just one or a few habitats. As they move away from their comfort zone to places where environmental conditions are sub-optimal, interactions with better-adapted competitors and predators keep them in check.

On land, a species' range will often be restricted by physical barriers such as mountains, deserts or large bodies of water. For some time now we have generally accepted Lamarck's contention that few such barriers exist in the oceans. Most marine species release eggs or larvae into the open sea, where they drift for days or even months before finding a new home, and many swim hundreds of thousands of miles in their lifetimes. This received wisdom had scarcely been challenged until my wife, Julie Hawkins, and I started to plot the ranges of species living on coral reefs. Julie spent years amassing lists of species from all over the world, among them a set of exquisitely illustrated fish lists collected by the naturalist Sir Peter Scott on his travels around the world in the 1970s and 1980s.

It was painstaking work and, after ten years and several thousand species, we decided to call it a day. We found that many coral reef species were not in fact widespread, and some were limited to remarkably small fragments of habitat. The dark indigo Limbaugh's angelfish, to take one example, is known only from Clipperton Reef, a pinnacle of coral in the eastern Pacific that is isolated from neighbouring reefs by nearly a thousand kilometres of blue water. Or there is the pale dottyback, a shy and ghostly fish found only in the Gulf of Aqaba, a narrow finger of water in the northern Red Sea. Or the splendid toadfish, the only known haunts of which are reefs of the resort island of Cozumel in Mexico. In recent years, a remarkable catalogue of coral reef species with small ranges has built up.[1]

A similar picture is forming for marine species in other habitats and regions. Submerged seamounts, rocky oases amid the vast mud plains of the deep sea, sustain extraordinary numbers of species found nowhere else. The great trenches carved into the ocean floor where tectonic plates grind back into the Earth's mantle provide other unique habitats, with their own distinctive plants and animals isolated by thousands of kilometres of inhospitable terrain. Brachiopods – they look like bivalve molluscs but are unrelated – have knocked about on this planet since the Cambrian Period 500 million years ago[2] and are mostly scarce and solitary today, but in the fjords of Patagonia they build mounds that are an echo of the reefs they made in their heyday hundreds of millions of years ago, before the end-Permian mass extinction.

What will happen to all these creatures as their environment alters under the influence of global climate change? Unless they can adapt to the new conditions or move, extinction beckons. Adaptation proceeds via natural selection, as Darwin taught us, since those best suited to prevailing conditions leave more offspring who carry their genes forward. Genetic mutations take place over generations and the species imperceptibly shape-shifts into something different. The problem is these evolutionary changes take time; they require many generations to take hold. If conditions change quickly, the only option may be to move. If you can't move, you die. The history of our planet is punctuated with periods of mass extinction, when geologic or climatic upheavals pushed species beyond their endurance. The comings and goings of life can be read in seaside cliffs where different fossils are found only in certain layers of rock. While a handful of species seem immortal, most last between just 1 million and 5 million years.[3]

Polar bears and penguins leap off the pages of every circular from an environmental organization that comes through the letterbox, and television correspondents reporting on climate change are packed off to icy latitudes to be filmed with them. They are the most telegenic sufferers of global warming, but there are countless other less familiar species whose existence is also in flux. In the last twenty-five years, the North Sea has warmed by 1.25°C and life there is on the move. Fishermen on the coast near where I live in Yorkshire have seen growing numbers of new species pop into their catches over the last decade, like grey gurnard and cuttlefish. Surveys made since 1977 using bottom trawl nets towed across the seabed show how geographic ranges are shifting under the influence of warming.[4] The John Dory, for instance, an odd-looking narrow-bodied fish with a long thin jaw, used to be common only in south-west England, but they have moved north around Scotland and colonized the North Sea.

Over the last thirty-five years, fifteen of the thirty-six species surveyed in the North Sea shifted latitudes. The average shift was 300 kilometres north. (The temperature difference between the equator and 80°N, roughly the limit of sea ice, is about 30°C; if you divide this difference by the distance, temperature falls by about 1°C for every 300 kilometres travelled north.[5]) When shifters were compared to those that had stayed put, an interesting difference emerged. Species that responded quickly to climate change tended to be small-bodied animals that reproduce early in life, while those that did not move were larger-bodied and late to mature. This difference echoes one common throughout nature: disturbed, fluctuating or rapidly changing environments are dominated by weedy, opportunistic species. The rats and cockroaches of the sea may cope well with climate shifts, while the tigers and elephants are left floundering as their habitats shrink around them.

One reason why animals must move as the climate changes is that their physiology is tuned to temperature. Enzymes that regulate bodily processes work best within a relatively narrow range of temperatures and species suffer when those limits are breached. But there are other reasons. Species need food and habitat and these are undergoing their own moves. In the North Sea, for example, many kinds of microscopic plankton that drift in open water have shifted north over the last forty years.[6] Northern zooplankton communities are dominated by a copepod called *Calanus finmarchicus*. This diminutive relative of crab and shrimp is fat and juicy by copepod standards and reaches 3–4mm long. Southern zooplankton are dominated by a more puny species, *Calanus helgolandicus*, that measures just 2–3mm. This difference may not seem like much but to larval cod it is crucial. They thrive on fat *finmarchicus* but starve on scrawny *helgolandicus*. The northward march of zooplankton has been linked to poor replenishment of North Sea cod, a fact that has compounded decline from overfishing.

It might seem a simple matter to shift with one's food sources but most species are driven by another urge – the need to find mates and reproduce. Over thousands of years, many have developed intricate life cycles in which they migrate between predictable feeding and spawning areas, tracking patterns in food availability to maximize survival. Scientists playing with equations have discovered that group phenomena like gathering to spawn appear to respond very slowly to shifts in prey availability.[7] Instead of following their food, many fish continue to gather in the same places in order to fulfil their primordial urges. A gap thus opens between what they feel compelled to do by their instincts and what they should do to meet their needs. This underscores a point that applies to many of the changes underway in the oceans today: while species can track creeping shifts in conditions, rapid change could overwhelm them.

There is another way for species to stay within their comfort zone: they can go deeper. Water cools with depth and, at the present rate of warming, they will soon have to go down as much as 3.5 metres per year.[8] Some North Sea fish in the survey were found to go deeper as well as heading north. But light falters quickly as you go deeper. Herbivores will find little to eat beneath the sunlit surface layer, and those species that prefer hard bottoms will struggle to find a home in the muddy expanses of the deep.

Warming seas will change marine habitats as the species that build them die off or are compelled to move. Creatures like oysters and mussels, mangrove trees and saltmarsh grasses, seaweeds and corals are ecological engineers. They develop slowly and take a long time to establish themselves in new terrain, and they are likely to lag far behind the needs of their inhabitants to find more agreeable places to live. The warming seas have already had a profound impact on one such habitat: coral reefs. Shallow water coral reefs, confined to tropical seas, are the richest of all marine habitats. To drift with a gentle current over a coral reef is one of life's great thrills. The world that unfolds below is crowded with bright dappling colours in constant motion. It feels like gazing at a Monet painting where every brush stroke has come to life and is swimming around the canvas.

Warm water corals live on a knife edge; wherever they occur they adapt to local temperatures. If temperatures rise more than a degree or two above the typical maximum, they bleach. Bleaching happens when there is a breakdown in the relationship between corals and the microscopic plants that live within their tissues and provide them with much of their food. At higher temperatures these microscopic plants, or zooxanthellae as they are called, begin to harm their hosts, at which point the corals kick them out. Since it is the zooxanthellae that give corals their colour, the tissue becomes transparent and colonies turn deathly white as the chalky skeleton beneath is revealed. Bleached corals are starving; they die within weeks unless temperatures revert to normal.

In 1998, sea temperatures soared across the tropics. In some places, like the Seychelles and Maldives, entire reefs turned white and almost all corals were killed. If this had been an isolated event it would have been bad enough, as corals can take decades to recover. But such bleaching events have become common. Scientists are now predicting that corals will be exposed to temperature spikes of the same magnitude every year by the middle of this century.[9] If corals cannot adapt, whole reefs will die, taking with them a host of species that call them home. Adaptation is clearly possible. Some of the world's most northerly reefs are found in the seas off Kuwait, where temperatures swing wildly. When winter winds blow hard from the desert, water temperatures plunge as low as 12°C. In the dead calm of stifling summer they top 35°C. Despite these tough conditions corals have built reefs large enough to support islands. Fast adaptation will be crucial to reef survival, but the prospects are not good. The rate at which species evolve depends heavily on generation time. Species with high turnover rates evolve rapidly (think of flu viruses), while long-lived species evolve more slowly and will struggle to adapt. Like the ancient trees of the High Sierras in the US, some corals live hundreds of years, while a few reach a thousand or more.

The Galapagos Islands offer a harsh warning of what could be ahead. In late 1982 a powerful weather phenomenon called El Niño triggered a sharp rise in water temperatures that lasted five months. By the end, Galapagos reefs were all but dead. Thirty years later, they still haven't recovered and structures built over thousands of years have now crumbled to rubble and dust. A few years ago I swam through the remains of what had once been a glorious reef there. It was a scene of devastation. The seabed was strewn with chalky fragments like a field of bones in the aftermath of a massacre. Here and there a mound rose above the wreckage where once some mighty coral had stood. It had survived centuries of lesser temperature spikes only to die in this one. A few young corals struggled to gain a foothold but were quickly overcome by black-spined sea-urchins that swarmed over them like ants on a carcass. What that experience and others like it tell us is that today's bleaching events are unprecedented on millennial timescales.

From our warm-bodied mammalian perspective, it is hard to get too animated about a couple of degrees (although there are ample reasons for anxiety from our perspective as well). But if you live at your thermal limits, as many species do, even a few degrees could be intolerable. Warming seas will trigger a global diaspora that will lead to a massive reorganization of life as long-separated species invade each others' ranges. There will be a march towards the poles, as we have already seen for fish and plankton in the North Sea. These invaders will alter polar ecosystems in profound and unpredictable ways. But where do you go if you already live at the poles and prefer your habitat with ice? The answer, some scientists predict, will be a surge of extinctions among polar residents as their waters heat and ice melts away. Perhaps they will find refuge in deeper layers of the sea, but for some that is not an option. This is why the polar bear and walrus, icons of the realm of floating ice, are now on the World Wildlife Fund's top-ten list of species in peril.

This global diaspora is not all about loss. Some species will do well and some countries may prosper from the changes. Others will suffer. In an elaborate thought experiment, William Cheung from the University of East Anglia and his colleagues created a global ocean model to try to predict the consequences of life's reorganization between now and 2055.[10] As they cranked up the heat, species shifted poleward; those that could not move, for want of suitable habitat or because the way was blocked, declined, in some cases to the point of extinction. The movement of all these species left some places with fewer inhabitants than before, mostly in the tropics, while the poles suddenly became more populous. There will be a knock-on effect on fish catches. Low latitude nations like Indonesia, Senegal and Nigeria could see catches dwindle by 10 or 20 per cent, compounding the loss of agricultural productivity that terrestrial climate models predict. In the US, fish production will have to shift as most of the lower continental States could lose 10 per cent of their catch, while Alaska's catches could potentially expand by a quarter.

Although there are many reasons to worry about the impact of rising temperatures on the sea, the good news is that life can and will move, often quickly. The eggs and larvae of many marine species can travel tens to hundreds of kilometres, and we have erected few barriers to prevent their dispersal. The seas off Greenland experienced a warming spell that lasted from the 1920s to the 1960s. Cod, haddock and herring expanded their ranges north very rapidly. Cod spread more than 1,000 kilometres poleward in less than twenty years and kick-started a fishery that at its peak landed 400,000 tonnes of fish a year.[11] A host of other species shifted alongside these commercial fish, including bottom dwellers and marine mammals.

Cheung's predictions hold many uncertainties. They assume that the production of food by ocean plants will either not change or might even increase a little. For reasons I will return to in later chapters, how these phytoplankton react as the seas change is largely guesswork right now. There are equally good arguments that productivity will fall. If so, a few countries may break even, but most will see significant reductions in their fisheries.

By now it will be clear where my views lie on climate change. Like thousands of other scientists, I am convinced it is underway. There is abundant evidence of such variety and strength that those who reject it are increasingly isolated. It is important to remain sceptical about every new study and to weigh the solidity of its methods and rigour of its sampling. Scepticism is the stuff of science. Old theories are overthrown when new data and new ideas provide better explanations for what we see around us. But then there is denial. It is easy to spot those in denial. They use evidence selectively, acknowledging only those findings that support their prejudices and rejecting, or simply ignoring, findings that don't. As evidence mounts against their ideas, their explanations for the discrepancies become ever more tortuous and far-fetched.

Those who claim there is some elaborate conspiracy among scientists to delude politicians and the public should ask themselves a simple question: why? What could anyone possibly gain from such a hoax? If the answer to 'why' is not sufficient, there is the question of how. Science is a highly competitive pursuit. There are no prizes for coming second. Scientists constantly put each other's ideas to the test.

A hypothesis is cast aside as soon as it is found wanting or is reworked on the forge of new evidence. No conspiracy could resist such relentless scrutiny for long.

Much of the controversy around climate change stems from the variability of the Earth's climate over long timescales. We live on a dynamic planet where yearly climatic variations are overlain by multi-year or decadal cycles, known as oscillations. Cores taken through deep sea sediments in California's Santa Barbara Channel show how marine life tracks these changes. In the aftermath of frenzied predatory attacks on shoals of sardines and anchovies, a rain of glittering scales makes its way down to the seafloor. Scales trapped in sediments allow us to trace the ups and downs of shoaling fish through thousands of years as they respond to climate events like El Niño and the swings of a cycle known to oceanographers as the Pacific Decadal Oscillation.

Modern measurements have been made on timescales of a few decades to a century, so it can be hard to see the signal for the noise. Much of the change to date may fall within the range of natural variation if you extend the timeline out far enough, but the drivers of our current change – greenhouse gas concentrations – have moved far above historic levels. If you are pushing a broken-down car, you have to apply a lot of force before it starts to move. The force of humanity's push on the climate system has been building for over a century and the consequences are beginning to become obvious.

Life is on the move. This is nothing new. Records locked in ice, seabed sediments and rocks show that change is normal for life on Earth. As conditions shift new habitats and ecosystems constructed by better-adapted species will reform in the place of old ones. It is not change that we must worry about but its speed. If conditions alter too quickly they could exceed the pace at which ecosystems can respond. It could take centuries for life to adjust, and we can't afford to wait that long. In a world of growing human need, that is a big concern.

6

Rising Tides

As I strolled down Ocean Drive in Miami Beach one evening a few years ago, I could almost imagine myself transported back to the 1930s. The party atmosphere from the nightclubs and bars mingled with the smell of seaweed and distant boom of breaking waves. Today, the Art Deco splendour of South Beach is overshadowed by towering hotels and apartments that line the seafront to the north. It is hard to imagine the place any other way, but in the nineteenth century Miami Beach was little more than a windblown barrier island backed by dense mangrove swamps. Its potential was first appreciated by Henry Lum, a veteran of the California Gold Rush of 1849 who had a keen nose for a bargain. He bought the land for 25 cents an acre and planted coconut palms. But Miami Beach was not really born until 1915, when a bridge was laid across Biscayne Bay to connect it with Miami.

Lum fell in love, as many other Americans have after him, with the wild beauty of the barrier islands. These low-lying islands are made of wave-washed sand, and run, on and off along the East Coast, from Long Island to Miami. The twentieth century saw a coast-wide building boom that covered many of them with houses and hotels, in places like Kitty Hawk, Surf City and Wrightsville Beach. But, after the glory days of speculation and development, many have begun to fear that the ocean is about to reclaim these lands.

We are reaching the end of an era of around 5,000 years of unbroken stability in the volume of water in our oceans. This followed a period of rapid increase that came after the last ice age, when melting ice sheets pushed the height of global seas upward by 130 metres. Since then, some coasts have seen levels rise or fall due to coastal uplift or subsidence, but the overall amount of water in the oceans has been very stable.

Stable sea levels promote the formation of wetlands where rivers meet the sea. Rivers spread sediment with every flood and regularly discharge mud into coastal waters. Deltas, mudflats, marshes and barrier islands form where the rate of new sediment deposition exceeds the rate of erosion and subsidence. When plants bind and stabilize the sediment, the land inches out to sea. Over the course of history, people have given wetland formation periodic spurts when they have cleared land to grow crops and build cities.[1] Denuded lands shed their topsoils into valleys and rivers far more rapidly than those bound by forests or other vegetation. In parts of Europe, soil erosion during the early Middle Ages choked estuaries and built wetlands that stranded once busy ports far inland, such as Luni in northern Italy and Chester in England. In North America, a pulse of wetland creation followed forest clearing for agricultural purposes between the seventeenth and nineteenth centuries.

Today rivers are again running thick with soil, but less of it reaches the coast than in the past. Dams, which have been constructed along almost all of the world's large rivers and countless smaller ones, hold back mud that would once have reached the sea. In arid regions, crop irrigation often diverts water and sediment that would have replenished wetlands. Australia's legendary Murray and Darling rivers, for instance, lose so much water to irrigation that they now reach the coast as mere trickles.

When sediment supplies are cut off, the balance shifts in favour of the sea and wetlands retreat. Deltas naturally subside over time as sediments settle and the earth sinks below them in a process called isostatic adjustment. In many places today, from Miami Beach and Venice to Tokyo and New Orleans, sea level rises have combined with coastal subsidence to threaten life and property.

The Earth is ruled by certain immutable physical principles. One such law is that warm substances take up more space than cold ones.[2] To us, it hardly matters if soil and rock expand as our planet heats up. But it matters a great deal when that heat expands the oceans and the sea begins to reclaim our most valuable coastal land. Ancient Roman fishponds built at sea level 2,000 years ago have helped define

when sea level rise began in earnest.[3] At the time they were constructed, the Mediterranean was thirteen centimetres lower than
today. It didn't begin to rise until about a hundred years ago, when
global temperatures started to respond to greenhouse gas emissions
from the Industrial Revolution. Since then, the process has sped up
considerably. Sea levels rose by approximately twenty centimetres
between 1870 and 2000.[4] We know this from thousands of tide gauges
across the world. Since 1993 those recordings have been supple-
mented by satellite observations that measure sea-level fluctuations
with millimetre accuracy. Averaged over 130 years, the rise comes to
1.7mm per year, but in the last twenty years the rate has accelerated
and now tops 3.3mm per year.[5] Three-quarters of the rise in sea levels
since 1900 has been caused by global warming from carbon dioxide
emissions. Water expansion from sea surface warming contributed a
quarter of the rise since 1960, and 30 per cent from 1993 to 2009.
The rest came from melting glaciers and ice caps, and groundwater
pumping for irrigation.[6]

Sea levels would have risen half a millimetre per year faster if we
had not embarked on wholesale dam building from the mid-twentieth
century, which locked up freshwater on land. However, this benefi-
cial effect of dams has been rather undermined since they have also
blocked the passage of vast quantities of mud, an effect that has hastened erosion around river deltas from the Yangtze to the Mississippi.
Parts of the Nile Delta have retreated by as much as 150 metres per
year since the Aswan Dam was built in 1964.[7]

So far, only the upper layers of the sea have begun to warm up. As
heat penetrates deeper, the sea will expand and sea levels will continue
to rise. In 2007 the IPCC predicted that sea levels could rise by
another 18–59cm by 2100, depending on how fast we gain control
over the drivers of climate change. Their forecasts combine thermal
expansion of seawater with ice loss from mountain glaciers and the
Greenland and Antarctic ice sheets. But their figures seem increasingly
at odds with the recent acceleration: sea levels have risen faster than
the most rapid rate of change they predicted a decade ago.

New research from the icecaps of Greenland and Antarctica suggest
that we are close to, and perhaps past, a tipping point for the rapid
melt of land-based ice, which will become the major source of sea

level rise in future. If the Greenland ice sheets were to thaw in their entirety, they would add six metres to the height of global seas and trigger a mass exodus from low-lying coasts and cities. The thaw of the West Antarctic ice sheet would add another three metres.[8] A six-metre rise would wipe out most of Florida from just north of Miami.[9] It would obliterate the Mississippi Delta and drown a third of New York. It would flood much of London and Hamburg, turn Lagos into a lagoon and Bangladesh into a swamp.

Sea level records preserved in fossil coral reefs in Central America suggest that we have crossed this tipping point at least once before, 121,000 years ago, when sea levels rose by three metres within the space of a hundred years (ten times the present rate).[10] Some argue that this period is an analogue for what a warmed world might be like. Temperatures then averaged 2°C above today's and the seas levelled off six metres higher than they are now,[11] drowning deltas and plains in many of the low-lying coastal areas that are now densely populated. Of course, such extreme levels would not be reached instantaneously. It will take time for sea levels to respond to a warming planet. A range of estimates made since the 2007 IPCC assessment suggest an upper limit to the rate of rise of about two metres in a hundred years.[12]

The melting of Arctic sea ice is perhaps the most striking manifestation of global warming, but, even if it were to disappear in its entirety, Arctic ice would not add to the sea level because it floats in the sea. (If you aren't convinced, add an ice cube to a glass of water and you will see how the level remains the same as it melts.) It is the ice sheets sitting on land, mainly over Greenland and Antarctica, that we need to worry about. They have proven less durable than the IPCC had hoped. Ice heated from the surface melts and forms pools and streams. Runnels turn into rivulets that coalesce into blue rivers which course across the frozen surface until they find a crevasse. They then plunge in muted waterfalls that would be magnificent if only we could see them. Beneath the ice sheet, this running water lubricates the stuttering creep of ice across rock and accelerates its slide towards the sea.

At their margins, ice caps are often held back by ridges of undersea rock on which the ice sheet has grounded. Global warming is gnawing away these barriers too. Tongues of warm water (at least by polar standards) are creeping under the ice and freeing it from its anchors. The West Antarctic ice sheet has already been loosened from one key holdfast.[13] Measurements from planes and satellites show the seaward edges of both Greenland and Antarctic ice sheets slipping ever faster into the sea. Ice sheet thaw seems to be accelerating through positive feedback loops. Blue pools and rivers of melted water at the surface mean that less of the sun's heat is reflected and so more ice melts. It is hard to predict the future contribution of ice sheets to sea level rise because climate change has also increased snowfall over their interior regions, helping counteract losses from the margins. How the balance between ice sheet growth and melt develops over the next century will be critical to countless people who live on low-lying coasts.

Another positive feedback to global warming is newly emerging in the Arctic Ocean. The seabed has begun to spew forth bubbles of the greenhouse gas methane on a colossal scale as permafrost melts at the bottom of the ocean where ice sheets are in retreat. Russian scientists announced the findings at a San Francisco conference in December 2011, where the leader of the team, Professor Igor Semiletov, described fountains of methane more than half a mile wide.[14] He said they saw hundreds of such fountains but estimated there could be thousands more. So methane emissions from the seabed are now adding to those from melting permafrost on land to accelerate global warming.

According to recent estimates, even the more modest sea level rise predicted by the IPCC – a rise of up to 59cm by 2100 – could displace hundreds of millions of people and inundate over a million square kilometres of the world's agricultural land and coastal cities. Ten per cent of the world's population lives near coasts on land less than ten metres above present sea level. Eleven of the world's sixteen megacities, each home to more than 15 million people, are built on coasts or estuaries: Tokyo, Guangzhou, Shanghai, Mumbai, New York, Manila, Jakarta, Los Angeles, Karachi, Osaka and Kolkata. As the world's population increases that number will grow, with a further four coastal megacities predicted by 2025: Buenos Aires, Dhaka, Istanbul and Rio de Janeiro.[15] When rising tides threaten there are three possible ways we can respond: retreat, adapt or defend. Defence is our usual reaction to threats from the sea.

*

1. *above* Catch of goliath groupers landed from a recreational fishing charter boat into Key West in Florida, USA, in the 1950s. The swollen bellies of the fish show that they are full of eggs and were caught from a spawning aggregation.

2. *left* Recreational fish catch from Key West in the 1970s.

3. *left* Recreational fish catch from Key West in 2007. The fish and catches have become smaller over time. Most of today's anglers are unaware of the extraordinary decline of Florida's reef fish revealed by these images.

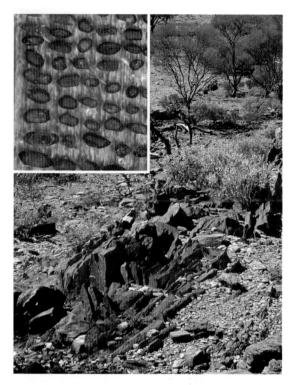

4. Rocky outcrop in the Jack Hills of Western Australia. These rocks were laid down approximately 3 billion years ago and include zircon crystals eroded from other rocks that were up to 4.4 billion years old, almost as ancient as the Earth itself.

5. *left inset* Zircon crystals separated from the Jack Hills rocks. Individual crystals measure ⅕th to ⅒th of a millimetre across. They were rounded by wind and water abrasion and date to 3 to 4.4 billion years ago and include the oldest known rocks from Earth. Microscopic inclusions in these crystals are made of materials that formed at temperatures characteristic of coalescing granite in contact with liquid water, suggesting a world with continents and oceans far earlier than once believed.

6. A slice of 2.5-billion-year-old seabed from Australia. Wavy layers of rust brown, yellow and orange silica alternate through the slab with dark black and grey stripes of iron-rich magnetite. Thicker layers of shimmering tiger eye fill gaps where the rock was later twisted and deformed under pressure. This rock is thought to reveal the emergence of free oxygen in Earth's atmosphere, which reacted with sulphides in terrestrial rocks that washed into the sea as sulphates. The sulphates reacted with dissolved iron in the water which was deposited at the seabed as iron oxides. Rocks like this now yield some of the richest iron ore in the world.

7. View from the entrance of Pinnacle Point Cave 13B in South Africa. This cave was occupied on and off by the ancestors of modern humans from around 140,000 years ago. Deposits laid down in the cave at different times reveal the earliest remains of seafood dinners as well as evidence of our emerging humanity, including shell beads, heat-treated stone tools, ochre for body adornment and abstract art.

8. Fishing methods were well developed long ago in the Mediterranean, as this second century CE mosaic panel from the Catacombs of Hermes in Hadrumetum, Tunisia, shows. It illustrates the use of throw nets, octopus traps made from perforated amphorae, drift nets and hook and line, as well as the wonderful diversity of marine life present.

9. *The Fishmarket* by the Dutch artist Frans Snyders (1579–1657). Although the composition is fanciful, the animals and their sizes were representative of those available to customers of the day. A huge wolffish gapes from the centre of the painting and lies across enormous sturgeon and halibut; to the left are large cod, ling and salmon steaks, while turbot hang from hooks at the back. Other creatures that might surprise present-day shoppers include lamprey, porpoise, seals, common skate (now extremely rare in spite of the name) and pike.

10. *The Gulf Stream* by Winslow Homer (1836–1910). Eyewitness testimonies from hundreds of different sources show the seas of this time to have contained far greater numbers of large sharks and other predatory fish than they do today.

11. *below* Auction scene at Grimsby fish market, UK, from the early days of steam trawling around 1900. Not only is the size of the catch extraordinary by today's standards, but so is the size of the fish. The long, slim fish in the background are mainly ling, while those in the foreground are predominantly cod.

12. Fiorello H. LaGuardia, mayor of New York City, posing with a 300-pound halibut at the Fulton Fish Market in 1939. By this time, giant halibut had become scarce in the North Atlantic, but they were common at the time the Jamestown colony was founded in Virginia in the early seventeenth century.

13. Late nineteenth-century catch of 30,000 salmon from Puget Sound, Washington State, USA.

14. *right* Catch from a prawn trawler in the Torres Strait, Northern Australia, showing a large weight of unwanted bycatch, mainly small fish and invertebrates, that will be thrown away (mostly dead) when the prawns have been removed. Bycatch often outnumbers prawns by 10 or 15:1 in trawl catches like this.

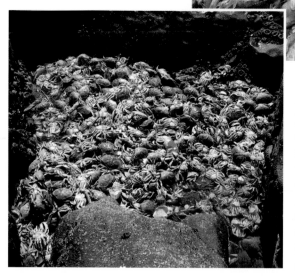

15. *left* Dungeness crabs killed by the encroachment of a deep layer of low oxygen water onto the continental shelf off Oregon, USA. Such events are linked to stronger winds blowing parallel to the shore, a probable result of climate change. As these waters also contain more dissolved carbon dioxide, they are more acidic than the shallow water layer that usually bathes coastal seas off the US west coast.

16. Caribbean coral reefs were separated from those of the Pacific Ocean between 3 and 4 million years ago, since which time their life has evolved in isolation. Caribbean reefs are immediately recognizable today by their distinctive communities of seafans.

17. Corals gain their colours from microscopic algae – zooxanthellae – that live within their tissues. Normally these algae provide food for the coral but when the sea warms to a degree or two above the maximum temperatures normally reached at a locality, the corals expel their zooxanthellae and the coral bleaches white. Corals can regain these algae later on but can't survive without them for more than a few weeks if temperatures remain high. This photograph from Palawan in the Philippines captures a fleeting moment in the process of bleaching when the colours of the corals brighten before they disappear. Three months later, all these corals were dead.

18. *above left* Hotels at Ocean City in Maryland, USA, have been constructed on low-lying coastal barrier islands made of shifting sands. Developments on islands like this are under extreme threat from rapidly rising sea levels.

19. *above right* Part of the Maldive archipelago seen from space. Nowhere in the Maldives is more than a few metres above sea level and the nation is one of the most vulnerable countries on Earth to sea level rise.

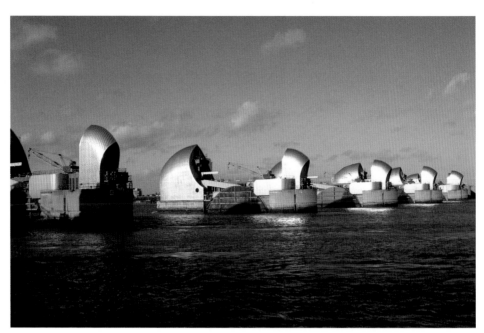

20. The Thames Barrier is a movable floodgate across the River Thames that protects London from flooding during periods of extreme high tides and storm surge. Rising sea levels will demand similar protective measures for many other cities across the world in the coming century.

21. Severe floods inundate much of Bangkok, Thailand, in 2011. The city lies on a river delta that has sunk several metres in the last century due to pumping of groundwater for irrigation and aquaculture, leaving the city in growing peril from rising seas and extreme rainfall.

For centuries Holland has been one of the most densely populated countries in the world. Its flat lands and low elevation mean the Dutch have always been intimate with the North Sea. Storm surges would periodically push the sea far inland to drown farmland and ruin crops. It is hardly surprising then that the Dutch long ago became masters of the science of coastal defence, constructing elaborate dykes as early as the fifteenth century to hold back the sea and increase fertile land. Contemporary illustrations show complex protection schemes of sea-walls and groynes (walls or fences built at right angles to the shore to intercept the movement of sand) that reveal a detailed understanding of the forces that build and destroy coasts. By the seventeenth century, Dutch engineers were in demand across Europe and they exported the technologies of coastal protection, or 'hardening', that remain in use across the world.

In 1953, the Dutch learned through a harsh lesson that sea level was only half of their worries. An intense storm surge from the north combined with an extreme high tide to devastate low-lying areas, killing 1,800 people and destroying 10,000 houses. To prevent a repeat they embarked on the world's most ambitious programme of coastal defence, raising dykes and levees to close off the North Sea from the Rhine–Meuse delta. Today, a quarter of Holland lies below sea level and owes its existence to these defences. But recent sea level rise threatens once again, and their coastal defences have come under renewed scrutiny.[16] The new plans foresee a sea level rise of up to 1.3 metres by the century's end, and work is underway to upgrade coastal defences to meet the threat.

The Dutch are better equipped than just about anyone to protect their coast. The problem is sea level rise poses an urgent threat to many developing countries that don't have half the resources the Dutch can muster. Nowhere is the battle to protect land more pressing than on low-lying oceanic islands such as the Maldives, the Phoenix Islands and Tuvalu.

Flying into the Maldives' International Airport is like landing on the ocean itself. The runway seems such an insubstantial sliver of tarmac amid the endless sea. The capital city of Malé is made up of a clutter of concrete buildings that completely conceal the coral island upon which it perches. A rampart of bleak concrete tetrapods defends the city from the waves of the Indian Ocean. Like Malé, the airport is edged in concrete, a stark reminder that the sea which gave birth to these islands could soon claim them back.

The Maldives is a waterworld. None of its 1,200 islands rises more than a metre or two above the sea and each sits atop one of the countless reefs that make up this nation of fish and coral. Seen from the window of a seaplane, the islands look like a scatter of emeralds set within white bands of coral sand. Collectively, they cover less than 1/30th of 1 per cent of the country. The rest is ocean.

Not surprisingly, Maldivians seem as much at home on their boats as on land. The captain who ferried me from resort to dive site on my last visit looked every bit the sea gypsy as he coaxed his boat through the hazard-filled coral labyrinth, foot on the tiller, a mobile phone pressed hard to one ear in an effort to hear above the roar of the engine.

The reefs in the Maldives have grown over the course of thousands of years on top of platforms planed flat by the sea during the last ice age. These platforms, now 40–50 metres below sea level, stretch north to south for 800 kilometres in a chain of atolls that follows an ancient wrinkle in the Indian Ocean seabed. Corals grow by absorbing carbonate, the main ingredient of chalk, from seawater. They crystallize carbonate into hard skeletons that form the backbone of reef growth. Swimming across a Maldivian reef crest last year I came upon a coral head freshly smashed open by waves from a passing squall. By outside appearances, stony corals like this one give an impression of immovable solidity, but inside the skeleton was riddled with tubes and cavities, bored and dissolved by a host of creatures seeking refuge from the predatory world outside. This hidden community represents the other side of the reef balance: the erosional yin to the depositional yang. Reefs only grow where deposition outpaces erosion.

The Maldives and thousands of other oceanic islands scattered throughout the tropics exist only because corals have flourished sufficiently to win the race against erosion. Climate change is tipping the balance the other way. As I mentioned earlier, Maldivian reefs bleached badly when temperatures soared in 1998. By the time of my visit twelve years later, corals were on the ascendancy again, but like the regrowth of a forest after wildfire, there were fewer species than before and most were shrubby branching forms that grew quickly. Most of the giants of old, corals that might have begun life when Europeans were just setting forth on voyages of global discovery, were dead.

Healthy coral reefs are naturally repairing breakwaters, better at protecting coasts than any concrete wall we can throw up to defend cities, resorts or agricultural land, and far cheaper. The combination of faltering coral growth and rising seas means that nations whose fortunes are founded on coral have begun to contemplate the unthinkable. Tuvalu in the South Pacific has made an arrangement to relocate its people to New Zealand when sea levels rise beyond their capacity to adapt. Already, the highest tides wash over islands in Tuvalu that were dry a generation ago.

At least Maldivians can take comfort from the fact that their nation rests on a solid foundation.[17] In places where coasts are subsiding, rapid relative sea level rise has been a reality for decades, even centuries. For them, it is not the twenty centimetres by which sea levels have risen in the last century that matter, it is the amount by which land has sunk. There are several reasons why coasts sink. During the last ice age, vast ice sheets up to three kilometres thick lay across much of North America, Europe and Asia. They pressed the Earth's crust deeper into the mantle, creating depressions. A bit like the hollow on a cushion that is left when you stand up, these indents have gradually disappeared as the Earth returns to its former shape. As it is a geological process, it is monumentally slow, and is still underway today, 18,000 years after the peak of the last ice age. Northern Scotland, for example, is still rising, while southern England, which was not covered by ice, is sinking. This helps explain the difference in the amount of hard coastal defences between these countries. Eight per cent of Scotland's coast is protected by seawalls, rock armour and the like, while half of England's is defended.[18]

Another major contributor to subsidence or sinking is sediment supply. Where supplies are cut off behind dams or levees, coastal marshes and mudflats are starved of the material they need to build upward as sediments settle beneath. In the southern US, for instance, the Mississippi Delta now loses fifty square kilometres of land each year as a result of subsidence and sea level rise. In years of heavy snowfall, the Colorado River used to disgorge upwards of 200 million tonnes of mud into its delta at the northern end of the Gulf of California. After the Hoover Dam was built in the 1930s, supplies were choked off and the river today carries a few million tonnes at most to replenish delta wetlands. There are now more than 45,000 large dams around the world, which together trap a quarter to a third of all the suspended sediment carried by rivers.[19]

Sediment supplies can also be cut off by structures that interfere with coastal currents. Some coasts consist of soft rocks or soil and naturally shed sediments to the sea. That sand and mud is carried by waves and currents to other places, where it settles into barrier islands, lagoons, sandy beaches and dunes. Such areas are often stabilized by plants like mangroves, seagrass and saltmarsh grasses that bind mud together. So coastal development can have far-reaching and unintended consequences as it disrupts the natural flow of tides and currents, leading to erosion and loss of sandbanks out at sea.

Coasts also subside where we pump water or oil from beneath the ground. Much of Houston in Texas lies barely a metre above sea level today, having sunk from the removal of both groundwater and oil. Some districts have sunk three metres since the 1930s and in parts of the city recent sinking rates exceed five centimetres a year.[20] It is on the frontline of cities threatened by sea level rise, although it doesn't make the top ten global list of cities with uncertain futures. That list is topped by Miami, while New York/Newark and New Orleans are second and third. Miami's Art Deco district survives today only because of regular applications of dredged offshore sand to hold back the sea. The lower floors of high rise buildings to the north are regularly splashed by salt spray from Atlantic storms.

It is inconceivable that we would abandon to the ocean some of the world's most populous cities – New York or Nagoya, Japan, for example. Doubtless we will girdle them with walls to hold back the ocean. London has already been defended since 1984 by a barrier across the River Thames whose sail-like gates swing shut when high tides combine with strong onshore winds to lift the sea to dangerous levels. The record of how many times the gates have been closed tells its own story. They were shut once a year for the first three years of operation, but now typically close more than twenty times a year.

New Orleans gave us a striking example of the risks faced when we choose defence over retreat. If defences are breached, as New Orleans' levees were during Hurricane Katrina in 2005, the ensuing floods are catastrophic. Bangkok has sunk several metres in the last century due to groundwater pumping in the Chao Phraya River delta. Much of the city had to be evacuated for the first time in late 2011 as flood waters combined with high tides to overwhelm the ring of dykes thrown up around the city.

Defence becomes increasingly expensive and risky as sea levels rise. If the thaw of ice sheets adds more than a metre to sea levels in the coming century, we will have to abandon parts of many cities and coasts to the sea. A metre doesn't sound like a big deal, but any sea-wall would have to fend off the worst the seas could throw at it, and the extremes are expected to rise faster than the average.[21] New York, for example, may soon find itself within the Atlantic hurricane belt as warmer seas sustain hurricanes into more northerly latitudes. New Yorkers got a shock in 2011 when lower Manhattan had to be evacuated due to flooding from Tropical Storm Irene. This storm may be a harbinger of things to come as there is now broad agreement that hurricanes will intensify over the coming decades.[22]

The great river deltas of the world are humanity's breadbasket. They support production of staple crops like wheat, rice and corn, and are highly vulnerable even to modest rises in sea level. A survey of forty of the planet's largest deltas – areas that collectively receive over 40 per cent of runoff from the world's landmass – showed rates of effective sea level rise of up to 12.5mm per year, nearly four times the rate at which the sea is rising.[23] The higher rate is due to land subsidence in places where the supply of sustaining sediments has been blocked by dams, or where there has been groundwater or oil abstraction, or a combination of the two. This has led to a rapid loss of land around the world's great deltas such as those of the Nile, Ganges, Mekong and Mississippi, and so ending a long period of delta growth that followed the stabilization of sea level 5,000 years ago.

There is little doubt that sea level rise will provoke mass migration in the coming century as tens of thousands of square kilometres of delta lands are at risk of flooding and submergence.[24] Relative sea level rise, accelerated by subsidence, could displace tens of millions of people from these lands and would cut off agricultural production from some of the world's most fertile soils just at a time when we will have the greatest need for it because of population growth.[25]

The human cost of climate change in deltas is becoming more obvious with each passing year. Take Bangladesh for example, a land of floodplains and swamps better known to the world as a land of natural disasters. Formed at the confluence of two immense rivers, the Ganges and Brahmaputra, more than a third of the country barely rises above water. The people who inhabit this aquatic landscape live and often die by water. Monsoon rains and Himalayan snowmelt nourish their fields and rice paddies, but water the giver also takes away as typhoons or deluges periodically bring catastrophe. It is hard to argue with the logic of dykes, levees and seawalls when faced with images of desperate people stranded on the roofs of their houses in the midst of lakes that stretch from horizon to horizon. Yet, ironically, if Bangladeshis build embankments to defend themselves from riverborne floods they would become more vulnerable to inundation from the sea as their now sediment-starved lands would sink faster. We cannot sustain life and society for long on coastal floodplains where those life-giving floods have been quenched.

When we harden coasts with engineered defences, rises in sea level will narrow and steepen shorelines and increase the erosive force of waves. Some of the most productive marine habitats will therefore suffer. The shifting matrix of wetlands that separates land and sea on low-lying coasts will dwindle as it is squeezed between elevated sea and seawalls. Estuarine mudflats, mangroves and saltmarshes will find themselves in retreat with nowhere to go, their natural adaptability blocked by our handiwork. What happens to them will seal the fate of thousands of species that depend on them, such as the wading birds which use coastal wetlands as migration refuelling sites and winter habitat. The Humber Estuary, not far from my home, supports over 150,000 such waders in winter. They enliven the mudflats and lift the spirits as they twist and turn in shape-shifting flocks across the evening sky. Already a hundred or so species of wetland birds are at risk of extinction worldwide as a result of wetland loss. Sea level rise does not bode well for them.

Like coral reefs, wetlands and dunes are living barriers between land and sea. They can be battered and breached by tempests but the living matrix renews itself with time. I once sat out a terrible hurricane in the US Virgin Islands (in the toilet, whose block-built walls had a comforting solidity!). The wind raged all night with a dreadful roar that left my ears ringing. The ground shook, the house trembled, and through the darkness came muffled crashes as our neighbours' homes disintegrated. When dawn finally came I emerged to an island whose verdant green had been stripped back to greys and browns of shattered branches and twisted trunks. Along the shore, mangrove trees stood like skeletons, their crowns smashed by wind and wave. But still their gnarled and knotted trunks, deep-rooted in mud, held the line against the sea.

Mangroves in Indonesia and Sri Lanka likewise seem to have helped dissipate the destructive force of the terrible tsunami of 2004.[26] (Because mangroves colonize sheltered shores that were already likely to be less affected by the tsunami, the correlation between reduced wave impacts and the presence of mangroves has been challenged. But where would you prefer to be if a tsunami struck: on a shore fronted by shrimp ponds, or behind a dense natural stand of mangrove trees?) Having seen the incredible debris flows pushed by the Japanese tsunami of 2011, even mangroves will do nothing to stop the worst of the waves. But they could save many lives in places at a greater distance from the epicentre of destruction.

The protective function of wetlands has not been properly appreciated, nor have their many other values, such as the filtration and purification of runoff, or their role as nurseries for commercially important fishery animals. Instead, they have been, and still often are, perceived as mosquito- and disease-ridden swamps or wastelands. As a result, wetlands have been drained, filled and cleared for agricultural land, marinas, homes, shrimp ponds, airports, malls – you name it – pretty much any use that can be made of flat coastal land.[27] Developers in search of prime seafront see wetlands as easy wins in planning applications. As a result, the phase of wetland expansion that has prevailed since sea levels stabilized has in most places been reversed. Around 30 per cent of the world's mangrove forests have been cleared since the mid-twentieth century, and some countries, like the

Philippines, have destroyed half to three-quarters of their mangroves.[28] Nearly a third of the global tally of seagrass beds has gone since the late nineteenth century and the rate of decline has increased to 7 per cent per year since 1990.[29] A similar tale of loss can be told for saltmarshes.

There is a physical relationship between sea level and erosion on open sandy coasts that says the shore will retreat two orders of magnitude more than any rise in sea level. A barrier island might lose a couple of hundred metres to the twenty-centimetre rise in sea level we have experienced in the last century, but it could lose a thousand metres to the one-metre rise that is predicted in the coming century. This will not bring comfort to those who live on coastal barrier islands that fringe the eastern and southern coasts of the US. The American geologists Orrin Pilkey and Rob Young predict that it will become impossible to sustain communities on these barrier islands into the twenty-second century if sea levels rise as predicted.[30] Seafront streets and marinas beloved of vacationers in Nantucket, Long Island or Martha's Vineyard could be washed away. Seafront towns like Ocean City and Atlantic Beach will have to defend themselves or disappear, while to the south, the sea will probably erase all homes and infrastructure from Cape Hatteras and Sea Island, Georgia. On less dynamic coasts, there is an alternative to expensive construction or abandonment and it goes under the name of 'soft engineering': nature, in other words.

Like coral reefs, wetlands are self-repairing breakwaters. Coastal engineers are just beginning to realize their value even as their area declines. The complex matrix of roots and stems traps and binds sediment and stabilizes the coast. If the health of these habitats is maintained and the input of sediment is sufficient, they can keep pace with sea level rise by trapping and consolidating a layer of mud and peat whose thickness will increase with the sea level. Vigorous wetlands are like a wall that builds itself higher as the need arises. Replanting mangroves in Vietnam after they were destroyed by herbicides during the war reduced the cost of dyke repair from $7 million per year to just over $1 million.[31] Likewise, where storm surges are a problem, broad and dense saltmarshes require little additional coastal protection, while narrow, scrappy ones need to be backed by dykes.[32]

Future wetland growth is expected to get a boost from higher carbon dioxide levels in the atmosphere, since plants use this gas for photosynthesis. Seagrasses and mangroves are among only a handful of marine species that could benefit from higher carbon dioxide concentrations. The new Dutch coastal protection plan calls for much more soft engineering of this kind, even to the point of removing existing dykes to promote the restoration of estuarine habitats and tidal regimes.

Scientists forecast that the sea level rises of up to 59cm predicted by the IPCC by the century's end will lead to the loss of another third of our coastal wetlands. This loss is comparable to or less than the direct destruction that has already been wrought by the conversion of wetlands to other uses. Given their growing value in ameliorating the impacts of sea level rise, our attitudes towards wetlands urgently need a rethink.

7

Corrosive Seas

Approached by boat in spring, wrapped in the vapours of early morning, the island of Ischia appears to float atop a sea of perfect aquamarine at the entrance to Italy's Bay of Naples. Its intoxicating beauty and therapeutic thermal springs attract over 6 million tourists a year. The Victorian war correspondent Sir William Russell captured well the contradiction between the island's tranquil appearance and its volcanic birthright:

> If one could have been aware of the terrible forces which were at work beneath that smiling surface, how delusive would the whole of that bright pageant – the charming little villas nestling in their gardens, the country houses white as snow, with their green jalousies, and the small spires of chapels piercing the mass of foliage – have appeared.[1]

At the east end of the island, the Castello Aragonese grips the top of a craggy rock, its ancient ramparts blending into grey cliffs that plunge to the sea. Under these battlements, volcanic gases – mainly carbon dioxide – bubble up from underwater springs. Molluscs living close to these springs have paper-thin shells, so weak they can be crushed between thumb and finger. These champagne seas, with their high concentration of carbon dioxide, are a worrying portent of the difficulties that could soon face corals, lobsters, clams, oysters and other shelled forms of life all across our planet.[2]

Ocean chemistry and that of our atmosphere are inextricably linked. Gases dissolved in seawater are in equilibrium with those in the air, which means that as concentrations of carbon dioxide go up in the atmosphere, so does the amount of carbon dioxide dissolved in the sea. When carbon dioxide dissolves in seawater, it produces carbonic acid (which gives fizzy drinks their tang). This liberates bicarbonate and hydrogen ions and decreases carbonate ions (a union of carbon and oxygen and a key ingredient of chalk).[3] The hydrogen ions make the sea more acidic, and the decrease in carbonate ions is bad news for anyone or anything that depend for a living on making chalky shells and skeletons.

The oceans have absorbed around 30 per cent of the carbon dioxide released by human activity since pre-industrial times, mainly from fossil fuel burning, conversion of forests and swamps to cities and agriculture, and cement production.[4] Over that period, the pH of seawater, a measure of its acidity, has fallen by 0.1 units. Most of this drop has taken place in the last few decades. Since pH is measured on a logarithmic scale, where one unit equals a tenfold change in acidity–alkalinity, this means acidity has risen by 30 per cent. If carbon dioxide emissions are not curtailed, acidity is expected to rise 150 per cent by 2050,[5] the fastest rate of increase at any time in at least the last 20 million years and probably as long as 65 million years, which takes us back to the age of the dinosaurs.[6] As Carol Turley, an expert on ocean acidification from Plymouth Marine Laboratory, put it, 'the present increase in ocean acidity is not just unprecedented in our lifetimes, it is a rare event in the history of the planet'.

With a pH that currently stands at 8.1, today's oceans are actually slightly alkaline. Dissolving carbon dioxide is bringing their pH closer to neutrality (defined as the pH of pure water, which is 7.0). While our great-grandchildren won't exactly get a chemical peel from a paddle in the sea, animals with shells might. By way of comparison, black coffee is several hundred times more acidic than the sea, and a soda drink over a thousand times more acidic than the pH of 7.6 that could be reached in some places by 2100.[7] But even this small change could completely transform life in the sea.

Ocean acidification is en route to becoming one of humankind's most serious impacts on the sea, yet it has inexplicably been overlooked until very recently. One of the first marine biologists to realize just how bad more acidic oceans could be for marine life was Joanie Kleypas, an American expert on coral reefs. At a meeting to discuss climate change in 1998, she experienced one of those rare 'eureka' moments. The realization that coral reefs would, by the end of the twenty-first century, be bathed in water corrosive enough to destroy them was suddenly so overwhelming that she excused herself and ran to the bathroom to be sick.

Despite decades of attention to rising levels of carbon dioxide in the atmosphere, the problem of ocean acidification was first highlighted only in 2003, in a study published in the journal *Nature* by Ken Caldeira and Michael Wickett, American scientists based at Stanford and The Lawrence Livermore National Laboratory. They concluded that, if we were reckless enough to burn all of the world's fossil fuel store, it would cause acidity in the oceans to rise in the next 300 years to levels not seen in the last 300 million, save from during rare catastrophic events like the end-Permian mass extinction.[8] For the scientific world this study was an epiphany. Since then, scientists across the world have mobilized to study ocean acidification and hundreds of papers have been published. They make sobering reading. Increased acidity is the last thing marine life needs given all of the other ways in which we are making oceans a tougher place to live.

Many marine plants and animals take up dissolved carbonate minerals from seawater and secrete it in the form of a calcium carbonate skeleton, often with small amounts of magnesium mixed in. Pteropods (tiny swimming snails), microscopic coccolithophores and foraminifera do so among the plankton, while corals, snails, crustaceans (crabs, lobsters, shrimp), sea urchins and coralline algae do so on the seabed, alongside filter-feeding clams, mussels and oysters. As the pH of seawater falls, it takes extra energy for all of these forms of marine life to secrete carbonate structures. This is because acidification alters their internal chemical make-up, making it harder to crystallize carbonate out of solution. So when atmospheric carbon dioxide rises, corals and other calcifying organisms will produce less robust skeletons.

The industry of countless generations of corals over periods of hundreds of millennia has produced some of the world's most spectacular and diverse marine habitats. Coral reefs form vast geological structures, some of which, like Australia's Great Barrier Reef or the Maldives, can be seen from space. I gained a keen sense of the enormous solidity of coral reefs when I once stood in the calm waters of the lagoon of a Pacific coral atoll just inside the reef crest. Great ocean swells expended themselves impotently in angry foam against the coral buttress while calm water gently lapped my ankles. No castle could enjoy greater security from harm than is provided by these self-renewing breakwaters.

Just as Joanie Kleypas and her colleagues predicted in 1999, recent experimental findings suggest that coral reefs and other habitats built from calcium carbonate could cease to grow within our lifetimes. Scientists are suddenly waking up to the possibility that acidification, in combination with coral bleaching due to global warming, will cause the destruction of reefs worldwide. Indeed, the damage has already begun. The skeletons of corals on Australia's Great Barrier Reef have weakened measurably in the last twenty-five years and now contain 14 per cent less carbonate by volume than they did before.[9] Since corals contain annual growth rings, much like trees, changes in skeletal strength can be measured over time. The results suggest that recent weakening has been unprecedented in the last 400 years. Experiments with deep sea corals from the Mediterranean indicate that they lay down half as much carbonate today as they did before the onset of the Industrial Revolution.[10] Because of this skeletal weakening, ocean acidification has been dubbed 'osteoporosis for reefs'.

But corals alone do not make a reef. Without the cement of coralline algae, a group of calcareous seaweeds, to bind corals together into solid structures, reefs would not develop nearly so well. We see this in the deep sea where, far below the reach of light, coralline algae are absent (so are the algae that live harmoniously within coral tissues and boost their growth rates in shallow water). Deep sea corals grow at glacial rates: at best, they build mounds of loose rubble that over thousands of years usually only reach a few metres thick. Healthy shallow-water corals, by contrast, can grow by as much as twenty centimetres per year. Coralline algae are especially susceptible to a fall in pH because they secrete a form of calcium carbonate that is rich in magnesium and is more soluble than the form of calcium carbonate produced by corals. In addition to the role played by coralline algae, reefs are held together by carbonate cement that precipitates chemically from the water that trickles through the interstices of the reef. This form of cementation also weakens as carbonate saturation falls.

Under normal conditions, large regions of the oceans are 'super-saturated' with carbonate, which means that it is relatively easy to crystallize it to form skeletal structures. Mineral carbonate structures that are not protected by a covering of animal tissue cannot form or survive in seawater that is not saturated or supersaturated with carbonate ions. We know this because carbonate saturation changes with temperature and pressure. At low temperatures and high pressure, water dissolves more carbonate. As you go deeper in the sea, carbonate concentrations fall below saturation once you pass depths of around three to four kilometres in the tropics, shallowing to depths of a couple of hundred metres in polar regions. Below this, solid unprotected carbonate dissolves. These effects are enhanced by dissolved carbon dioxide. Cold water thus holds more carbon dioxide than warm, and the deep sea has more than the shallows because of respiration by the creatures that live there.[11]

As carbon dioxide levels in the sea rise, carbonate saturation will fall and the depths at which carbonate dissolves will become shallower. Recent estimates suggest that this horizon is rising by one to two metres per year in some places. So far, most carbon dioxide added by human activity remains near the surface. It has mixed more deeply – to depths of more than a thousand metres – in areas of intense downwelling in the polar North and South Atlantic, where deep bottom waters of the global ocean conveyor current are formed. Elsewhere, the sea has been stirred to just a few hundred metres deep.

All tropical coral reefs inhabit waters that are less than 100 metres deep, so they will quickly come under the influence of ocean acidification. If carbon dioxide in the atmosphere doubles from its current level, all of the world's coral reefs will shift from a state of construction to erosion. They will literally begin to crumble and dissolve as erosion and dissolution of carbonates outpaces deposition. What is most worrying is that this level of carbon dioxide will be reached by 2100 under a *low* emission scenario of the IPCC. The 2009 Copenhagen negotiations sought to limit carbon dioxide emissions so that levels would never exceed 450 ppm in the atmosphere. That target caused deadlock in negotiations but even that, according to some prominent scientists, would be too high for coral reefs.[12]

Just as Ischia's carbonated volcanic springs provide a warning of things to come, bubbling carbon dioxide released beneath reefs in Papua New Guinea give us tangible proof of the fate that awaits coral reefs.[13] Reef growth has failed completely in places where the gas bubbles forth vigorously, reducing pH there to levels expected everywhere by early in the twenty-second century under a business-as-usual scenario. The few corals that survive today have been heavily eroded by the corrosive water. The collapse of coral reefs in the Galapagos Islands following El Niño in the early 1980s was hastened by the fact that eastern Pacific waters are naturally more acidic due to their deep water upwelling than those in other parts of the oceans.[14] Corals there were only loosely cemented into reef structures and collapsed quickly.

Acidification is an even greater problem at the poles, where carbonate dissolves at only a couple of hundred metres deep. Polar seas are renowned for these enormous productivity, which is why many whales trouble themselves to swim thousands of kilometres every year to gorge themselves in these waters. Pteropods, tiny snails that swim with a foot that expands into a pair of delicate wings (the name means 'wing foot'), are keystone animals in polar food webs. They have been called the potato-chips of the sea because of their critical role as a food source, but the name does them no justice. They live within shell castles sculpted from transparent crystal whose cold beauty seems perfectly fitted to the icy seas. They range from the size of a lentil to a fingernail and reach densities of up to 10,000 per cubic metre of water. To visualize such a density, imagine each snail in a space equivalent to about the volume of a kiwifruit. For a predator, hitting a patch like this would be like being caught in the whiteout of a snail snowstorm. If you are a fish lover and have ever eaten salmon, cod, pollock or any number of other species from frigid waters, you have tasted pteropods second hand.

In less than fifty years, pteropod 'chips' could be off the menu as polar seas become undersaturated with carbonate. Experiments with captive pteropods show their shells dissolve at acidity levels that may soon be reached. In fact, parts of the sea off northern Canada have already become corrosive to pteropod shells as a consequence of ocean acidification enhanced by sea ice melt.[15] The Southern Ocean around Antarctica is predicted to follow by 2030, and the Bering Sea by 2100.

*

One of the great unknowns in the future life of acid seas is how fast species will be able to adapt as their environment shifts. The geological record holds clues but does not reassure. Fossils record long stretches of vigorous reef development interrupted by periods with few reefs. Charlie Veron, a veteran Australian coral scientist, believes that periods of elevated carbon dioxide correlate with the periods in the geological record when reef formation ground to a halt. Others are less sure and think that episodes of reef growth are less closely linked with ocean chemistry. They believe that reefs have developed in the past in seas that were more acidic than the levels we are on target to reach by the end of the century. But even they will concede that periods of high carbon dioxide in the atmosphere and oceans are associated with mass extinctions, followed by bursts of evolutionary diversification. This alone rings alarm bells for me.

In the opening chapter, I charted the emergence of life in the sea. A billion years ago, atmospheric carbon dioxide levels were much higher and the oceans more acidic than they are today. The evolution of simple photosynthetic life forms in the pre-Cambrian reduced atmospheric carbon dioxide levels as these early life forms withdrew carbon dioxide from the air and released oxygen. Around 570 million years ago, 30 million years before the Cambrian explosion of life, carbon dioxide concentrations seem to have come down enough for a signal event in the evolution of marine life. As the oceans became less acidic, dissolved carbonate reached saturation levels in shallow seas with sufficient consistency to make it possible to construct shells. Fossil reefs in Namibia, hundreds of metres across, date from this time and are made from giant heaps of tiny calcified tubes produced by *Cloudina*, possibly an early antecedent of polychaete worms, some of which make shelly reefs today.

The reduction in acidity of the world's oceans heralded an evolutionary explosion of animals that produced carbonate skeletons and shells. To begin with, most secreted calcite or calcium phosphate shells. Aragonite, the form of calcium carbonate produced by pteropods and modern reef corals, is more soluble than these compounds. Corals which secreted aragonite began to build reefs about 230 million years ago. They yielded to calcite producers in a period of global warming and high carbon dioxide that prevailed between 145 and

65 million years ago. Modern corals evolved about 40 million years ago, as carbon dioxide levels fell once more. There is obviously a link between atmospheric carbon dioxide, ocean acidification and the fortunes of animals with chalky carbonate shells. Some managed to survive and even thrive in more acidic seas in the deep past. But that does not mean they can adjust fast enough to survive the more rapid acidification of modern times.

The last time marine life was subjected to seas as corrosive as those anticipated by 2100 was in a period geologists call the Palaeocene-Eocene Thermal Maximum, 55 million years ago. Sea surface temperatures spiked around 5–9°C higher than today's due to a massive release of methane and possibly carbon dioxide.[16] We aren't sure exactly where these gases came from or why they were released. Theories include volcanic eruptions, offgassing of peat and a massive belch of methane hydrates from the seafloor. A methane belch seems the most likely explanation, given the chemical signature of sediments laid down at the time. Methane would have been rapidly oxidized to form carbon dioxide, which in turn acidified the seas. The amount of carbon released at that time is about the same as the amount that would come from burning all of the world's fossil fuel reserves, but it took place more slowly – over a few thousand years rather than the few hundred years in which we are burning fossil fuels today.[17] You can see this event written in sediment cores drilled from the deep sea. Chalky sediments give way suddenly to a brown layer of clay, which almost as quickly returns to chalk. The whole event is over in just ten or twenty centimetres of core, corresponding to the hundred thousand years or so it took the Earth to swallow back the excess carbon it had released.[18]

The Palaeocene-Eocene ocean acidification spike does not seem to have led to any great cataclysm for marine life. Looked at against the sweep of geological time, it might be reported as 'Minor disaster strikes: not many dead'. Yes, the few coral reefs around seem to have suffered badly,[19] but the only outright extinctions were of deep sea foraminifera. These tiny single-celled organisms – neither animal nor plant – succumbed when corrosive water swept over them as the oceans acidified.[20] All the bottom-living foraminifera became extinct, while those from surface waters survived.[21] Nonetheless, conditions for their existence and that of other calcareous plankton became harsh and sediments from the time are full of stunted and malformed shells. If any creatures are capable of adapting to higher acidity, we would expect to find them among single-celled plankton, for which a generation can be measured in hours or days. They do seem to have toughed it out pretty well. Although there were major changes in their composition, there was no apparent shut down in the construction of chalky body structures.[22] Some species fared well under conditions of higher acidity. We can take some comfort from this, but have to remember that the rate at which we are adding carbon dioxide to the sea is much faster today, so the impacts will likely be far more severe. Carbon dioxide will accumulate in the surface layers of the ocean that are home to most marine life long before it is mixed into the deep seas. It takes a thousand years or more to flush carbon dioxide into the deep sea, where it is eventually neutralized by dissolving chalky sediments.

Coral reefs, like polar bears, have come to symbolize the perils of carbon dioxide emissions and climate change. They are sentinel habitats that forewarn of dangers ahead. If they perish, millions of people will be left homeless. But I have another concern that gets to the very heart of the planetary life-support system: what will acidification do to phytoplankton? Phytoplankton, or 'plant' plankton, produce roughly half of the oxygen that we breathe and are the foundation of nearly all ocean food webs. Life itself depends on them. We depend on them.[23]

Many phytoplankton produce chalky skeletons, like coccolithophores. These ball-shaped microscopic plants armour-plate their bodies with exquisitely ornamented discs of chalk that overlap like porcelain shields. When the days lengthen, spring blooms form, fed by nutrients stirred from deeper water by winter storms. Countless trillions of coccolithophores then turn the sea a milky white, producing swirls that can be seen from space. A single bloom can produce a million tonnes of calcium carbonate in just a few weeks.[24] These blooms do their bit for our climate as they reflect heat back into space from the oceans. The effect of acidification on coccolithophores and other photosynthetic plankton is not straightforward. Carbon dioxide boosts photosynthesis, just as it does with land plants. In experiments at higher levels of ocean acidity, many coccolithophores seem to produce just as much carbonate as under normal conditions, sometimes more.[25] Another plus is that not all phytoplankton produce calcium carbonate skeletons. Perhaps diatoms, with their silica shells (the basic ingredient of glass), will take up any slack caused if carbonate-producing phytoplankton struggle – or cyanobacteria, or dinoflagellates. We had better hope so.

Coccolithophore blooms are often stopped in their tracks by follow-on explosions of viruses that eat them. Viruses are critical to nutrient cycling and phytoplankton production in the sea, liberating nutrients to be used again by other plants and animals. Probably the most astonishing fact I have ever heard about the sea concerned viruses. A litre bottle of crystal-clear seawater contains some 4 billion viruses.[26] Added together, there are something like 4,000,000,000,000,000,000,000, 000,000,000,000 viruses in the sea (4 nonillion if you want an easier way to say it). They outnumber all other marine life forms combined fifteen-fold. To get an idea of what that enormous number means, imagine placing all the viruses end to end. They would form a thread less than 1/200th the thickness of the finest spider gossamer that would stretch for 200 million light years. You read that right: the thread would reach so far across the universe it would pass by sixty galaxies and countless millions of stars. The microbial world of our ancient seas is still alive and well. In all the research I have read on ocean acidification, I haven't come across a single study of the effects it might have on viruses. As it happens, viruses have slowed the rate at which the oceans are acidifying. Because they are so good at recycling nutrients in sunlit surface waters, less carbon is exported to the deep sea via the sinking bodies of animals and plants. This means less carbon dioxide is sucked from the atmosphere by the ocean: good for acidification, bad for climate change.

Most scientists' attention to date has concentrated on what will happen to the production of chalky carbonate body parts in more acidic seas. But pH change in itself can alter the availability of micronutrients that phytoplankton need to thrive. Iron is a key nutrient that limited phytoplankton productivity in ancient oceans, and still does throughout much of the high seas. Under more acidic conditions, iron will be less willing to combine with organic matter and will therefore be less available to organisms that need it.[27] Ocean acidification could expand areas of the sea where iron limits productivity, which would ultimately reduce the amount of food the sea could produce for us.

The more we look at the effects of acidification, the more ways we find in which it will disrupt life at sea. Dozens of experiments on captive invertebrates show increased acidity affects almost every aspect of life, from growth and behaviour to reproduction. One surprise is that predator avoidance in larval fish can turn to fatal attraction.[28]

The orange clownfish lives among the stinging tentacles of large sea anemones on coral reefs. Clownfish hide their eggs under the anemones' frilled edge to keep them safe. After a few days, newly hatched larvae head for the open sea. Weeks later they return under cover of darkness to find an anemone of their own. The most dangerous moment of their lives comes as they thread their way through a thicket of predatory mouths in search of a new anemone. Larval clownfish can smell their predators, which helps them avoid being eaten. But clownfish larvae raised under levels of ocean acidity that will be common in 2100 if we fail to curtail carbon dioxide emissions lose this ability, and instead they are attracted to predators. Other experiments have shown that they also lose the ability to smell their anemone hosts.[29] Perhaps they will be able to spot the anemones when dawn arrives, but by then they will be seen by a bevy of other predators that take over from the nightshift.

Unless orange clownfish can adapt quickly, acidification will likely bring their time on this planet to a close. We can only guess at how common such responses are among the many hundreds of thousands of other species with planktonic larvae. These larvae may be more sensitive to acidification than adults, which have been the subjects of most studies to date. What we do know is that the use of smell to detect habitats and predators is widespread among marine life. Once again Ischia offers a glimpse of what the future may hold. Hardly any larvae of species with chalky skeletons set up home or survive near the underwater carbon dioxide vents.[30] Experiments on sea urchins have shown reduced sperm vitality and fertilization at acidity levels that could be reached by the century's end.

Oyster farms on the Oregon coast give one indication of how hard life will get.[31] Oysters there periodically experience a complete failure of reproduction linked to more acidic water. I described previously how upwelling recently flooded the continental shelf with deep water so low in oxygen it caused mass mortality of animal life. Upwelled waters are also more acidic than surface water. When farms were flushed by these more acidic waters, larval oysters died at the point when they had to form their first shells.

Overfishing has greatly increased our reliance on shellfish from the wild as well as from farms. Prawns, lobsters, clams and scallops are relatively more resilient to overfishing than fish, and have come to dominate intensively fished seas. All lay down carbonate shells. In 2006, half the value of US fisheries came from shellfish. The fishing industry is badly exposed to risk from more acidic seas. Not only that, acidification threatens the important role that filter-feeding shellfish play in cleansing ocean waters.

Ocean acidification affects a fundamental aspect of almost all existence: the ability of marine life to process oxygen. When carbon dioxide levels increase, it becomes harder to extract oxygen from water. Dissolved carbon dioxide quickly diffuses across gills and into cells, reducing the amount of oxygen in circulation and suppressing metabolism. Animals stressed in this way expend more energy to maintain basic bodily functions, which means they can't afford to give as much to growth and reproduction. More acidic oceans may increase the severity of problems resulting from low oxygen or anoxia, something I will return to in the next chapter.

Coral bleaching used to be what kept me awake at night; now it is ocean acidification. Warming seas have devastated reefs worldwide by weakening the sunlit coalition of corals and algae. Acidification is a punch in the gut to reefs that are already on their knees. It is chilling to think that, within the space of 100 years, humanity could reverse a process of coral reef formation that has flourished since the end of the last ice age.

More acidic seas will not be devoid of life; there will be winners as well as losers. But life will change. The seabed around Ischia's champagne hot springs is covered by lush stands of seagrass, their vivid green leaves unsullied by the chalky crusts and tubes of coralline algae and worms that cover plants further away. Brown and red seaweeds elbow seagrasses aside wherever rocks punctuate the sand. These plants benefit from the fertilizing effect of carbon dioxide and the lack of sea urchins to eat them. Good news, at least, for the Mediterranean's hard-pressed green turtles. They love seagrass.

8

Dead Zones and the World's Great Rivers

British parliamentarians enjoy a long summer break that lasts from July to October. What could be so important to draw them away from the business of running the country for so long? The Houses of Parliament squat in Gothic splendour on the north bank of the River Thames, and it is to this river that we must look for the answer. Like many cities, London grew organically over the centuries, with little or no central planning. By 1815 it was the largest city in the world, inhabited by 1.4 million people, but it was a modern city with a decidedly medieval approach to sewage and refuse, much of which was dumped in the street or in cesspits. Ironically, it was an invention intended to improve hygiene – the flush toilet[1] – that helped create the problems in the Thames for which nineteenth-century London became notorious, and which culminated in the Great Stink of 1858.

In the late eighteenth century, flush toilets began to replace chamber pots. They greatly increased the volume of waste discharged into cesspits, which regularly overflowed into the streets. To remedy the problem a law was passed in 1815 that allowed the discharge of sewage into the Thames through a sewer network then under construction. By the mid-nineteenth century, the Thames had become choked with sewage and refuse that sloshed back and forth with every tide past the Houses of Parliament. As bacteria went to work on the sludge, they sucked up dissolved oxygen and turned the river into a stinking sewer that belched poisonous gases into the heart of London. In 1858, the stench became so bad that the windows of the Houses of Parliament had to be hung with sheets soaked in bleach to drive off the smell. Eventually summer sittings of Parliament were abandoned.

London's problems (mercifully since solved) have been repeated across the world and continue to plague most cities in developing countries today. Many of the world's most populous cities straddle mighty rivers on which people depend to bring water and carry away waste. By the time these rivers reach the sea, they are loaded with sewage, toxins and rubbish from cities, as well as industrial and agricultural runoff.[2] The problem increased sharply after the Second World War when farms became dependent on chemical fertilizers to boost yields. Fertilizers now fuel coastal plankton blooms that can cover thousands of square kilometres and are often visible from space. Satellite photographs of the Gulf of California in Mexico show the telltale green of phytoplankton growth around the mouth of the Yaqui River soon after coffee growers spray their crops with fertilizer.[3]

Plankton blooms are a natural seasonal feature in spring and early summer, when the days lengthen, and must have occurred ever since the seas were first populated with microbes more than 2 billion years ago. They are mostly seen in temperate and polar seas, where nutrients are stirred from the bottom by winter storms, and at the mouths of large rivers. When plankton die their decay takes oxygen from the water. As a result, beneath pollution-fuelled plankton blooms the water often becomes anoxic, spreading a shroud of death. Dead zones are now permanent or seasonal features at over 400 places worldwide, all of them coastal or in enclosed seas like the Baltic.[4] At the last count in 2008, collectively they covered a quarter of a million square kilometres, or about 1 per cent of the area of the world's continental shelves. This may not sound like much, but these are some of the most productive and biologically rich areas of the sea. We can ill afford their loss.

One of the world's largest dead zones forms every year off the mouth of the Mississippi Delta in the Gulf of Mexico.[5] The Mississippi River catches the runoff from 40 per cent of the land mass of the lower forty-eight states of the US and discharges an average of 580 cubic kilometres of water into the Gulf of Mexico every year, enough to fill Lake Mead, the largest reservoir in the US, sixteen times over. Its water is loaded with particles of organic matter and dissolved nutrients that plankton need to thrive. In the past, this annual nutrient bonanza sustained rich fisheries around the delta, but in recent years nutrient overload has tipped the balance from benefit to blight.

The dead zone begins to form in the spring, after winter snowmelt and rains swell the mighty Mississippi and flush its nutrient-rich water into the Gulf of Mexico. When stultifying summer heat anaesthetizes human life in Louisiana, Alabama and Tennessee, the plankton begins to decay and oxygen levels plummet. Those that can flee do so as a suffocating blanket engulfs the ocean. Many don't make it and the corpses of fish, crabs and shrimp litter the seabed. Others, like clams, snails, worms and starfish that cannot leave die where they are. As the shroud of death expands, fishers are forced to travel further offshore to make a catch. They pass through waters that are eerily empty – no leaping fish, nothing showing on their fish finders, nothing to interest the birds that pass overhead. Researchers like Melissa Baustain from Louisiana State University dive in the dead zone quite often. She reported, 'The deeper we go down in the water, it gets kind of scary, because there's nothing there. There's no fish, there's no organisms alive, so it's just us.'[6] At its peak, the dead zone off the Mississippi River chokes life from more than 20,000 square kilometres of sea.

Some people believe that the dead zone is a natural feature of the Gulf of Mexico, but much of the Mississippi watershed is farmland and the evidence points firmly to agricultural fertilizers. The history of the dead zone can be read in the bodies of microscopic animals trapped in sediments laid down on the seabed near the mouth of the Mississippi and far beyond.[7] The relative numbers of species that depend on oxygen-rich water compared to those that thrive when oxygen is scarce tell the story of when the dead zone first formed and how it has grown with time. Chemical signatures in the sediments add detail to this history.

The Mississippi watershed has seen drastic changes since the colonization of the Midwest and the conversion of prairie to farmland. That led to a steep change in the rate of soil loss around the beginning of the nineteenth century, which can be seen in core samples through a corresponding rise in the productivity of diatoms, a type of phytoplankton. Sediment loads fell again through the early twentieth century as dams were built and trapped mud. The application of artificial fertilizers to cropland soared after 1950, along with the number of sinking diatoms, and has not yet peaked. The dead zone only took hold after this, although it was too small to attract much notice until the 1970s. Since then, it has formed nearly every year like a spreading bruise. Inexorable and inescapable, it chokes life from a place that was, within living memory, the most vigorous fishing ground in the Gulf.

A Sumerian creation myth inscribed on clay tablets 5,000 years ago appears to describe an early effort at dam building:

> Ninurta, the son of Enlil, did great things
> He built a big mass of stones in the mountains . . .
> He constructed a barrier at the horizon . . .
> The stones competed with the powerful water
> Now the water of the mountains will not flow down to the valley forever

The first experiment in major dam building took place 4,600 years ago, when ancient Egyptians tried to control Nile floodwaters at Sadd el-Kafara.[8] Their construction was destroyed by a flood before it was completed, but this early failure didn't hold us up for long. Since then, we have transformed freshwaters, raising over 35,000 major dams. Nearly half of the world's largest rivers are dammed and over 40 per cent of river flows to the sea are intercepted by reservoirs.[9] Collectively, they exert a growing influence over life in the sea. The attractions of low-carbon hydro-electric power and greater water security (at least in the medium term, before reservoirs clog with silt) are unstoppable, even if we now have a far better understanding than we once did of the costs of blocking rivers.

Before the Aswan Dam, when sediment-laden Nile floods reached the Mediterranean they triggered plankton blooms that fed huge shoals of anchovies and a lively fishery. The dam now blocks most of these nutrients, but a new source has been found in the nourishing broth of effluents that flow into the sea from almost every river and stream in the country. The anchovy fishery, which floundered after the dam was built, is back in business today as the Nile now carries the waste of a population of 80 million as well as fertilizer runoff from the delta – not that this thought would tempt me to a feast of anchovies the next time I visit Alexandria!

Dams and crop irrigation cut off sediments and divert water, reducing nutrient flows into coastal seas. Their impact may in fact be more far-reaching than one would imagine. The Yangtze River carries vast quantities of soil from the Chinese highlands. But even this great river supplied less than a tenth of the phosphorus and a third of the nitrogen – two key plant nutrients – to the East China Sea.[10] The rest upwelled from deep water beneath the river plume. (Buoyant plumes of river water flowing offshore at the surface can drag deeper water in the opposite direction, sucking nutrients into sunlit waters where phytoplankton thrive.) After the Three Gorges Dam was built across the Yangtze, peak primary production by phytoplankton in the East China Sea fell by 86 per cent due to loss of nutrients.[11]

Dead zones form most readily where water movement is sluggish and where it is stratified into layers, cutting deeper water off from life-giving oxygen. If water slops around for long enough, pools of stagnant water near the seabed lose their oxygen as the sinking 'snow' of decomposing plankton builds into drifts. Despite their massive nutrient flows to the ocean, the Orinoco and Amazon rivers have no dead zones because water is pushed offshore too fast for oxygen to be lost. Where the flows of once great rivers have been staunched to a trickle after dams, cities and crops have drunk their fill, estuaries may hold water long enough for dead zones to form. This is the case, sadly, for rivers like the Loire in France and Po in Italy, where lack of oxygen chokes life from thousands of square kilometres of seabed every summer.

Stagnation is also a problem in enclosed seas like the Baltic, Adriatic and Black seas. Shaped like a deep bowl that is cut off from the Mediterranean by a shallow sill in the Bosporus Strait, the Black Sea was a freshwater lake when sea levels went down fairly dramatically during the last glaciation. Eventually, it refilled with salt water in a catastrophic flood 7,000 years ago, when the Mediterranean rose high enough to break through the Bosporus. The massive floods triggered a tremendous human migration and the event may be recorded in the ancient tale of Gilgamesh and perhaps Noah's flood.[12] Today only a warm, less saline surface layer is well oxygenated and able to support abundant life. This low-density layer sits over the cooler, more saline waters of the deep basin like a lid and has suffocated life below. Deeper than about 150 metres, the Black Sea is devoid of oxygen.

This lack of oxygen means the Black Sea conceals some spectacular shipwrecks, as the *National Geographic* explorer Bob Ballard has revealed. One wreck he discovered off Sinop, the ancient tuna fishing port, dates back 1,500 years. Spared the jaws of shipworms, the hull and deck are intact and even the mast still stands, a ghostly relic of a vanished age. The ship offers an extraordinary peek into the seafaring life of the ancient world, but at 320 metres deep it remains tantalizingly beyond the reach of archaeologists, its secrets, like its timbers, untouched.

The Black Sea owes its stagnant depths to nature,[13] but other enclosed seas have us to blame. Northern Europe's Baltic Sea connects to the North Sea only by a shallow, tortuous channel that curls around Denmark. The rivers that feed it drain from highly populous countries and run through croplands and intensive pig farms. They receive effluents from heavy industry and paper pulp mills and sewage from cities great and small. By the time they reach the coast, they are loaded with nutrients and organic waste. Sediment samples from one Swedish bay tell a similar story to those from the Mississippi dead zone.[14] Land was cleared and trees felled for agriculture and industry at increasing rates from 1800 onwards, which led to an ever-growing rise in nutrient runoff. Problems from plankton blooms and anoxic bottom water emerged after the 1950s, when artificial fertilizers came into widespread use. Since then, the number of people in Baltic watersheds has more than doubled and sewage has compounded the effects of fertilizers.

The Baltic is a huge brackish water lake that occasionally gulps salt water from the North Sea. When conditions are right, the North Sea pours in over the sill around Denmark, hugging the seabed as it creeps along. Brackish water (a mix of salt and freshwater) is less dense than salt water and thus flows out at the same time along the surface. This density difference restricts mixing between the salty deep and fresher surface layers and so promotes a loss of oxygen when dead plankton sink to the bottom of the sea. Species that need salt water to thrive are thus confined to the deep layers with dwindling oxygen. One reason why Baltic cod stocks fell so catastrophically in recent years was that their eggs could not survive in these deep pockets of low-oxygen salt water, although overfishing was the main reason for their downfall.

Intriguingly, there seems to be a link between the collapse of cod and other predatory fish and the vigour of seaweed growth. In shallow regions of the Baltic, thick choking mats of filamentous seaweed have overgrown the bottom like tufts of wool, blanketing areas where you could once gaze from the surface through crystal water at the waving fronds of bladder wracks and kelp below. This filamentous weed spells trouble for fishers whose nets clog with drifting clumps. The decline of big fish and seals from hunting and disease led to an increase in their prey, which in turn were predators of animals that grazed on the mat-forming seaweeds that now smother the seabed. After these smaller predators decimated the grazers, the algae spread.

Denmark and many other nations bordering the Baltic have made strenuous efforts to reduce nitrogen runoff to the sea to tackle excess nutrients and the problems they cause. Nitrogen inputs have halved since their peak in the early 1980s.[15] Unfortunately, it seems the Baltic is stuck in a vicious cycle. Low oxygen at the seabed releases phosphorus that stimulates plankton blooms, which leads in turn to worse oxygen depletion. This happens because, when oxygen declines, life on the bottom shifts from larger species that live deep in the sediments, like burrowing worms and clams, to smaller, more transient species like threadworms that live on or near its surface. In well-oxygenated conditions, the larger sediment dwellers suck particles of food from the water and deposit their wastes deep in the mud as faeces. They draw nutrients from the open water so they are no longer available to fuel further plankton blooms. By contrast, the smaller species that feed near the surface when oxygen is scarce release nutrients back into the water, where they further boost plankton growth, increasing the likelihood of more severe oxygen loss as the plankton decay. Once again, we have come to see the value to us of an intact, functioning marine ecosystem only after it has been damaged.

The Chesapeake Bay offers another sad example as we rue the loss of the vast maze of filter-feeding oysters whose reefs so challenged the navigational skills of Virginia's first settlers.[16] The reefs were scraped and nibbled away to feed America's appetite for oysters. Over the last century nutrient inputs to the Chesapeake have soared from sewage, industrial effluent and pig farms. Dead zones were first noticed in the 1930s, but their size and frequency has increased with time. By the

1960s, the declining health of the bay could no longer be ignored. But it was not until the late 1980s that a programme was begun to reduce nutrients washing into the bay. It was supposed to restore good water quality by 2010, but it hasn't worked.[17] Forecasts made in early 2011 predicted for that year the fifth largest volume of anoxic water seen in any of the previous twenty-six summers of monitoring. While excessive nutrients set the scene for anoxia, abundant filter-feeding oysters might still have prevented it from happening had we not removed them.

The nutrients that fertilize coastal and enclosed seas create another unwelcome problem, which is that plankton blooms can be toxic. Some marine plants produce potent poisons. Collectively they are known as harmful algal blooms, but more commonly they are called red or brown tides for the reddish slicks that appear at the surface when a bloom is underway. Red tides cause filter-feeding shellfish such as oysters, clams and mussels to accumulate toxins that cause paralytic and amnesic shellfish poisoning, sometimes killing people who eat them. They also poison fish and mammals, causing mass die-offs.

Florida is one of the more afflicted regions in the world. It has been troubled on and off by red tides since at least the middle of the twentieth century when scientists first began to study the phenomenon – and far longer according to the accounts of sailors and visitors. For most of the last twenty years there has been at least one red tide a year. Sometimes, red tides have persisted for more than a year. Nutrient-rich runoff from intensive agriculture seems to trigger blooms offshore, especially in the southwest after a heavy rain flushes fertilizers out to sea.[18] The limited circulation of Florida Bay concentrates the blooms.

The dinoflagellate responsible for Florida's red tides is called *Karenia brevis* and it produces a cocktail of potent toxins – brevetoxins – that act on the nervous system. When these algae are churned up by the wind and the waves, the toxins can be inhaled. Manatees, docile sea cows that graze Florida's shallow seagrass meadows, are especially sensitive, and red tides kill more of them than any other cause. A particularly severe red tide in 1996 killed 149 of them, while over 100 bottlenose dolphins succumbed in 2004. Unsurprisingly, people on land suffer from sore throats and runny eyes, and some develop more serious breathing problems when red tides strike. Hospital admissions for respiratory and stomach problems surge during red tides, and those who live close to the coast suffer the worst.[19] Inhaled toxins caused DNA damage, a precursor to cancer, in tests on laboratory rats. Bottlenose dolphins washed up dead on beaches often contain high levels of brevetoxins. In Florida, life in red-tide hotspots such as Tampa Bay, Naples and Fort Myers has become bad enough that there is concern it could drive away tourists and depress coastal property values.

There are many other toxic phytoplankton out there able to make trouble for us in places that are polluted by excess nutrients. In Australia's Moreton Bay, on the Queensland coast, a different kind grows in fibrous tufts attached to the blades of seaweeds and seagrasses. Like Florida's red tides, the problem is most severe when storms churn the sea into spray that can be inhaled, causing lung inflammation and itchy skin. The same algae can be found in Hawaii, where it is eaten by a type of mullet that has become known locally as the 'nightmare weke' – when eaten, toxic fish cause hallucinations and nightmares. The toxins aren't just responsible for short-term health problems. Some cause cancers, promote tumour growth or produce birth defects. It may be no coincidence that Hawaiians who eat seaweeds have high rates of stomach and intestinal cancers.

The tumour-promoting chemicals in harmful algae have been blamed for the growth of horrendous tumours on green turtles, although the tumours are triggered by viruses. These animals feed on seagrasses, the blades of which support toxic algae. I once lived in the US Virgin Islands, where I worked as a coral reef researcher, and most of the green turtles gliding around me when I took my daily swim suffered tumours that ranged from apple to melon size. Moreton Bay's turtles have similar problems. Could coastal pollution cause cancer? Now there's an unpleasant thought that might actually get our attention.

One group of animals has done better than most from the triple combination of nutrient enrichment, low oxygen and overfishing: jellyfish. To most of us, jellyfish are a nuisance at the beach – half glimpsed but sorely felt. I remember well my childish fascination with the amorphous lumps that sometimes littered the shore. I would prod and chop them nervously with my plastic spade, half expecting they would slide over to sting my feet. When, as a new diver, I first came upon a tiny comb jelly adrift in open water it dumbfounded me. Comb jellies are about the size of figs and are little more than blobs of seawater wrapped in a transparent glaze. But, over that skin, rows of tiny brush-like cilia beat in rainbow waves. Their beauty is fragile, as I soon discovered, for the slightest touch tears them asunder.

Since the Monterey Bay Aquarium in California figured out how to raise them in tanks, the splendours of jellyfish have been enjoyed by millions of people (the trick is no sharp corners to the tank and to keep the water moving in gentle loops). It is a testament to their loveliness that jellies can hold the limelight against the more obvious wonders of tuna, turtle and shark. Their rhythmic pulses have a hypnotic quality, like beating hearts. But the allure of jellyfish conceals a darker side. They are ferocious predators.

Jellyfish feed on zooplankton; and some hunt other jellies. When abundant they compete for zooplankton with other animals, such as herring, sardines or menhaden. Not much that is edible is off limits. They hunt eggs and larvae of fish and shellfish and, when jellyfish 'bloom', they can have major impacts on the young stages of commercially important species. A bloom of comb jellies in the Black Sea was implicated in the collapse of several fisheries.[20] Blooms can happen quickly because jellyfish grow fast (it helps that 95 per cent of their bodyweight is water – by way of comparison, only 55–60 per cent of our bodies are water). They multiply with incredible rapidity when the conditions are right, blossoming from a handful to an infestation in a matter of weeks.

One trick that the familiar moon jellies and their relatives have is a dual life cycle. We know them for their free-swimming stages, but they also live attached to the bottom as polyps a few millimetres across, much like sea anemones. When the time comes, these polyps release dozens of tiny jellies into the water, which can grow from fingernail size to adulthood in a few months. The largest jellyfish in the world by mass – Nomura's jelly – is found in Asian seas; it grows to a couple of metres across the disc and can weigh over 200 kilograms. Perhaps inevitably, Nomura are caught and eaten, but even so they have caused great nuisance in Japanese waters. They clog fishing nets, even of the boats that target them. In 2010, one boat sank under the weight of its catch. For boats that target other fish, jellyfish just get in the way and ruin the fishing. Set nets and traps are torn away by walls of jellies pressed against them by waves and currents. They also choke the cooling water intake pipes of power stations, forcing them to reduce output or sometimes even to shut down.

Mediterranean resorts have been plagued by jellyfish outbreaks in the last twenty years. The main problem species there is the mauve stinger, whose tentacles inflict slashing welts on the tender bodies of bathers. In the summer of 2004, an estimated 45,000 swimmers were treated for stings in Monaco alone. Things got so bad in Spain's Balearic Islands in 2009 that the local government issued the fishing fleet with bespoke jellyfish nets and put them on notice, should a bloom occur, to sweep up jellyfish before they hit the tourist beaches. In 2007, Irish salmon farms were overwhelmed by hordes of mauve stingers which slaughtered tens of thousands of salmon in their deathly embrace. Similar mass killings have been reported in Japan, India and Maryland.

Despite their fragility, jellyfish are remarkably resilient, perhaps a legacy of their ancient origins in the pre-Cambrian when they commanded the waters virtually unchallenged. When nutrients over-enrich coastal waters and phytoplankton prosper, there is a shift in the zooplankton from large to small species. Fish hunt by sight and suffer from the loss of easily spotted prey, whereas jellyfish, which hunt by touch, do fine. When the bodies of dead phytoplankton pile up and oxygen levels plummet, most jellyfish are unperturbed. They can survive in water with scarcely a gasp of oxygen, conditions that would kill most fish. These two attributes mean that jellyfish thrive in the low oxygen nutrient soup of seriously polluted coasts. They bloom even amid the toxic belch of the Benguela upwelling. But there is more. Most jellyfish seem to enjoy a little warming, and experiments suggest they can tolerate the ocean acidification levels predicted for 100 years hence. The concrete and timber footings of seawalls, harbours and marinas provide excellent habitat for the bottom-living polyp stages of many jellies. It is probably no coincidence that most recent explosions of jellyfish have taken place in relatively enclosed seas surrounded by densely populated coasts. Overfishing and bycatch have taken a toll on animals that eat jellyfish, like chum salmon, spiny dogfish and leatherback turtles. In short, the changes afoot in the oceans today could not have been designed better by some great jellyfish megalomaniac intent on global domination. Looking ahead to a future that is gelatinous, some scientists predict a jellyfish joyride all the way through the twenty-first century.[21]

9

Unwholesome Waters

'Filthy water cannot be washed.'

West African proverb

As a boy I lived in a small town set within a deep bay on the North Sea coast of Scotland. I could breathe the sea from my bedroom window, at least on the few days it was warm enough to open. Harbour seals basked on the shore near my front door. It was the 1970s and there was another smell on the sea breeze in those days – oil. Through my teens I remember measuring the spread of the oil industry across the North Sea in the rising count of bird carcasses that I found piled up on the shore. The tide of bodies swelled over the years: guillemots, razorbills, fulmars, and once even a great northern diver. On land, the quayside that a hundred years before had thronged with the bustle of the herring fleet was now cluttered with pipes, steel cables and other industrial paraphernalia. At fourteen, my biggest regret over the burgeoning oil industry was to see wildflowers buried beneath pipes in an abandoned coastal quarry where I used to entertain myself amid the orchids and insects, dreaming of a future life as a naturalist I am now lucky enough to have.

Oil has fuelled industry from the late nineteenth century, and oil spills have been with us since then. But they only really grabbed the headlines when they happened at sea, an inevitable consequence of offshore drilling and shipping. Much of the oil taken from the North Sea when I was young came from rigs in less than a hundred metres of water. As these wells were drained, companies looked further offshore and deeper. Drilling has now moved off the shallow continental shelves into the deep sea. If you push the limits of technology, problems are inevitable, as we learned on 20 April 2010, when disaster struck BP's Deepwater Horizon rig in the Gulf of Mexico while boring a well in 1,500 metres of water. Human error and slapdash attitudes to safety led to a blowout that turned the rig into a fireball. Eleven people were killed and many more injured.[1] For BP, this was only the beginning of their troubles.

Capping a well at 1,500 metres, they discovered, is not like capping one at 100 metres. Early on, a camera was placed on the seabed beside the wellhead. Over the following months, as BP executives writhed in the limelight, the camera beamed images into homes across the world of a violent geyser of oil and gas spewing into clear Gulf waters. While BP tried to play down the size of the spill, the camera showed the well gushing 68,000 barrels per day,[2] exposing their mendacity. For months, the oil poured out relentlessly as BP tried one trick after another, at one point even stuffing the well full of old golf balls (some wits suggested that oil executives would have been a better choice). Spring turned to summer and the spreading deep-sea oil plume burped slicks to the surface that washed ashore on beaches and marshes from Texas to Florida. In another BP gaffe, turtles trapped behind containment berms were immolated in the fires that were set to prevent oil from coming ashore. Eventually, something worked and the gush was choked off on 15 July, eighty-six days and an estimated ten tanker-loads[3] of spilled oil later (roughly 4.4 million barrels, after an estimate of oil retrieved is subtracted).

Deepwater Horizon taught us several things. We now know how much harder it is to contain a spill at depth than in the shallows. And we know that we are extraordinarily unprepared for problems at the deepening frontier of oil and mineral exploitation. BP's risk assessments for the rig detailed how walrus and sealions would be protected in the event of a spill, an obvious cut-and-paste job from an Arctic plan! In fact, the company was shockingly unprepared. It didn't possess any specialized equipment to control a deep-sea blowout. The plan might as well have said, 'If something bad happens, scratch your head for a while and wonder what to do.' Or, as a friend once said to me, 'When in danger or in doubt, run around and scream and shout.'[4]

The disaster also taught us that spilled oil doesn't always bubble to the surface, as everyone expected. Plenty of oil did appear, and an area the size of Cuba was covered by sheen for months; but most of it remained out of sight as it spread in a giant plume far below. For more than a month BP executives and some government scientists denied its existence, but repeated studies showed it was there, and that it remained there long after the well was capped, together with 700,000 gallons of chemical dispersant that BP injected at the leaking wellhead. I first discovered that not all oil floats when, as a newly minted PhD, I took a job as a reef scientist in the Egyptian town of Sharm el-Sheikh. Soon after I arrived a cargo ship hit a reef and spilled hundreds of tonnes of fuel which washed ashore in stinking globs. Apparently, the technical term for this kind of oil is 'chocolate mousse', and it certainly had that look. Beneath the waves, puddles of thick oil darkened the seabed, where they slopped around for the next year or two until they became so covered in sand and bits of seagrass they vanished. Some is probably still there today.

In 1979, there was another massive well failure in the Gulf of Mexico – Ixtoc 1. It gushed oil near the Mexican coast for over nine months. With an estimated half million tonnes spilled, Ixtoc held the record for the world's largest oil disaster until Saddam Hussein's deliberate environmental sabotage in the Arabian Gulf during the first Gulf War. Thirty years on, you can still dig lumps of oil from mud around the stems of mangrove trees near the site of the Ixtoc spill. The oysters that used to crust the trunks and roots of the trees disappeared soon after the accident and have never recovered.[5] We don't know much about the long-term impact of this spill because, once the oil was no longer visible, money for research dried up. If we had kept up our interest, we would know a lot more now about what to expect from the Deepwater Horizon spill in years to come.

There is a limit to what we can learn from past spills about what to expect from Deepwater Horizon. Until now, our experience has been of surface spills in which floating oil either evaporates, breaks down at sea, or ends up on beaches. There is no precedent for oil injected into dark, cold deep sea water at high pressure. This oil could have profound effects on deep sea life. Soon after the spill, the seabed nearby was covered by a thick flush of microbes that feasted on the carbon bonanza. Later measurements showed a virtually dead bottom, which suggests either that it was too much of a good thing or that dispersants killed them. At the time of writing, we still don't know. While the 2-million-gallon total of dispersants applied almost certainly helped limit oil impacts on Gulf shores, in combination with the oil they might also trigger a breakdown of the food web. By breaking the oil into tiny droplets, dispersant increased the exposure of mid-water creatures to oil. Some scientists fear that poisoning of microscopic plankton and microbes will interrupt nutrient cycles, perhaps leading to the eventual starvation of top predators like sperm whales and dolphins. We know that the bodies of tiny plankton are already contaminated with dispersants in the Gulf of Mexico and it is worrying that there was a spike in bottlenose dolphin miscarriages in the aftermath of the spill.

Oil spills are the starkest of all human impacts on the sea: dark, suffocating, inexorable. Images of seabirds struggling through blackened waves are shocking, but in most cases oil is not the worst of their concerns. The toll of seabirds and marine mammals tangled by fishing gear and drowned in Europe and elsewhere is far greater than from oil spills.[6] The Gulf of Mexico's fishing fleets kill more marine life in a day than Deepwater Horizon did in months. After the spill, fishing was banned in many parts of the Gulf. When the fleets were allowed to return, several months later, catches were plentiful after the brief and salutary respite from fishing.[7] The environmentalist Carl Safina wrote of how many commentators called Deepwater Horizon 'the worst environmental catastrophe in American history'. But to his mind the spill was far less disastrous than the fifty to one hundred square kilometres of marshland eaten from the Mississippi Delta every year by sea level rise, subsidence and engineered defences that starve the delta of life-sustaining mud.[8]

Some of the most iconic oil spills of the past have been from ships, such as the *Exxon Valdez* in Prince William Sound, or the *Amoco Cadiz* off Brittany. But today the main risks of large spills are from drilling and pipelines. Pipeline spills have increased more than four-fold since the 1970s, partly because so much more oil is pumped today, but also because many pipelines are ageing.[9] But as Deepwater Horizon has underscored, the real problem is our expansion of the oil industry out into the inky deep. Companies have developed their capabilities to tap oil well beyond their capacity to contain a spill when things go wrong.

Control measures for oil spills lag far behind the revolution in drilling capacity. Back in Egypt in 1987, I asked an expert what he thought the best way was to clean up the brown gloop that smothered the shore in front of us. 'A bucket and spade,' he answered. Low-tech methods may still be the best we have, but the leaky booms and fried turtles from the Gulf of Mexico remind us that they aren't very effective. The best thing for the planet, of course, would be to wean ourselves off oil, but that is unlikely to happen any time soon. In the meantime, governments must demand of oil companies much higher standards of safety and greater investment in prevention.

Oil companies are easy to demonize, but the biggest source of oil pollution in the sea is not from tanker spills or careless drilling, but from people like you and me. Two-thirds of the oil spilled into the sea around North America is carried with runoff from land (from dumped engine oil, fuel and industrial leaks) or introduced directly by pleasure boats and jet skis.[10] The great majority of recreational boats have two-stroke engines. They are cheap, light and powerful, but these assets come at a high cost to the environment. A quarter of the fuel–oil mix blows straight through the engine into the sea. (Many scooters use two-stroke engines, which is why they leave you choking on a cloud of oil and gas as they pass.) The floating fuel and oil concentrate at the sea surface, wrapping around and poisoning the floating eggs and hungry larvae of hundreds of species. I find it extraordinary that we continue to tolerate such profligate polluters when alternatives exist. Four-stroke engines have a closed system so they emit ten times less pollution than the cleanest two-stroke motors. It is time to call time on two-strokes.

Oil has actually had a few benefits for wildlife. The highly visible impacts of spills helped galvanize the creation of some of the first marine parks, much as logging spurred the establishment of national parks on land in the late nineteenth century. The Great Barrier Reef Marine Park in Australia and Monterey Bay National Marine Sanctuary in the US came into being this way. But, away from hellish scenes of burning sea and turtles dripping tar, pollution's most insidious effects in the ocean come from chemicals we can't see.

In the late nineteenth century, chemists managed to synthesize organic compounds – chemicals that contain the element carbon – with some very useful properties. (All organic compounds contain carbon, but not all carbon-containing compounds are organic. Diamonds and carbonates are examples of inorganic compounds with carbon in them.) These compounds were extremely durable, would not burn easily, did not conduct electricity, and were resistant to sun and weather. Chemists called them polychlorinated biphenyls, but most people know them as PCBs. Naturally, people quickly thought of many things they could do with these new wonders and industrial-scale production began in the 1920s. They were put in glue and hydraulic fluids, used as plasticizers in paint and as heat and fire retardants in furniture, and formed the plastic coating of electrical wires, among other things. There was just one drawback, which by the 1970s could no longer be denied: they were also highly toxic, causing cancer, liver damage, skin lesions and a catalogue of other horrors.

The problem began when manufacturers discharged PCB-laced effluents into rivers, lakes and estuaries, and chimneys disgorged them into the skies. Once inhaled or eaten, PCBs are not broken down and they tend not to be excreted. They are highly lipophilic, or fat-loving, which means that they are stored in the fat of animals that ingest them. When big animals eat smaller ones, they take on much of their load of toxins; so PCBs bio-accumulate through the food web from prey to predators. In the US Great Lakes, for example, PCB concentrations in the eggs of predatory herring gulls reached 50,000 times higher than in lake plankton at the base of the food web. In the 1970s, the carcasses of emaciated seabirds laden with PCBs began to wash ashore.

PCBs are just one class of chemicals among a wide range that are collectively known as persistent organic pollutants, or POPs for short. They include compounds like DDT, introduced to the world as a wonder pesticide in the 1940s but which lost its gleam when it was found to cause hatching failure in birds and alligators. DDT had decimated birds of prey and fish-lovers like the brown pelican before it was banned in the US in 1972. Other countries soon followed.

In the 1960s and 1970s chemists were shocked to discover traces of DDT, PCBs and other POPs in polar ice. The poles seemed so remote and pure, their landscapes carved from crisp, clean snow and ice, their skies sharp and clear. How was it possible? Pretty easily, it turned out.

Sea and air know few boundaries and can carry chemicals to the most inaccessible corners of the globe. Smoke from Chinese coal plants crosses the Pacific to North America in less than a week. Currents also move pollution, although at a more sedate pace. They transport chemicals thousands of miles in months or years. Unfortunately, certain properties in the make-up of the oceans mean that persistent chemicals find a much faster route to the poles.

Between the ocean and the air is a thin layer of water with properties very different from that of the sea beneath.[11] At around 1/16th of a millimetre, this layer is not much thicker than a piece of kitchen clingfilm. This thin membrane is stabilized by surface tension and is rich in fats, fatty acids and proteins; it is the reason why the surface of the sea sometimes has a glassy smoothness. Its stability attracts and concentrates micro-organisms, floating eggs, particles of dust and other materials. Unfortunately, water repelling, fat-loving compounds like POPs accumulate in this 'surface microlayer'. There they reach concentrations much higher than in the underlying water, often tens or hundreds of times higher.

When storms tear the sea into driven spray, contaminants race across the oceans and can bluster inland to drench coasts and people. Pollution in one region, such as the South China Sea, can leapfrog its way in a series of storm-whipped aerosols to far-flung regions. In this way, and with the help of wind, remote regions like the central Pacific can become polluted, even if they seem to be beyond the reach of serious contamination.[12] Storm spray can leap barriers like the south polar current that circles the Antarctic and separates it from the rest of the ocean. Once pollutants have reached the poles, they are swiftly taken up by animals there and work their way up the food chain.

The poles also accumulate POPs through a second kind of hopping, known to chemists as the grasshopper effect. Many POPs are semi-volatile, which means they evaporate and condense within the range of seasonal temperature variation. In summer, the chemicals evaporate off sea and land and blow around in the atmosphere until temperatures drop sufficiently for them to condense and they get redeposited. This happens either as the season changes or they reach somewhere cold at higher latitudes or on mountain tops. At the poles, chemicals are trapped within or under ice and the year-round low temperatures mean little is lost by further evaporation, so contaminants accumulate.

The surface microlayer is critical for microscopic life. Fish eggs and larvae concentrate in the top millimetre of the sea, where they benefit from better feeding conditions or escape from predators like jellyfish. Invertebrates, such as crab or sea urchin larvae, can be up to ten times more abundant in the microlayer than in underlying water, while microalgae may be ten to a hundred times more common and bacteria a hundred to ten thousand times.[13] There they come into intimate contact with toxic contaminants. Animals feeding in the microlayer will take up pollutants and pass them up the food chain.

Persistent organic pollutants differ in an important way from oil pollution. Although they are invisible, in the long run they are far more dangerous to marine life. Oil and gas have been around since time immemorial. In places like the Gulf of Mexico, natural seeps have leaked from the seabed for millions of years. What to us is toxic and unpleasant is food to vast numbers of microbes. Deepwater Horizon jetted an enormous plume of methane into the deep sea. Before long, a bloom of methane-degrading microbes (our ancient friends) had formed around the leak, and within a few months degraded much of it to harmless carbon compounds and water.[14] Toxins like DDT, PCBs and many pesticides are highly complex and the pathways of breakdown are much longer, which is why they hang around causing trouble for so long. In the extreme North and South Atlantic, these pollutants are sucked into the deep sea on the global ocean conveyor current. At the icy polar extremes of the planet, where the weather is cold and the sun weak, they resist breakdown far longer than in warmer climes.

The daily adventures of a group of bottlenose dolphins in Florida's Sarasota Bay have been followed by scientists since the 1970s. Their world closely intertwines with that of thousands of people who use this seaway every day for business and leisure. To people who live or vacation in Sarasota, the dolphins enjoy a carefree life frolicking in the waves. In reality they experience as close to an urban lifestyle as any animal in the sea, and that is stressful. Florida is one of the most populous states in America. It supports its share of heavy industry and chemical production, as well as agriculture dependent on lavish applications of agro-chemicals. The emerald gleam of Sarasota Bay is not as pure as it seems.

Tissue samples taken from the bay's dolphins show that they carry a heavy burden of contaminants.[15] They are at the top of the food web and toxins passed up from below stop with them. There is a troubling difference, however, between males and females. The toxic load carried by males increases over their lives. The most contaminated animal found in the bay was a dead forty-three-year-old male. His flesh would probably have been condemned as hazardous waste had officials known its chemical content when they disposed of the body. In females, toxins peak in adolescence and then plunge to much lower levels that remain stable until later life. The difference between the sexes is explained by pregnancy and breast-feeding.

Pregnant dolphins (and any other mammal for that matter) experience high energy demands from their growing babies. If food is short, they must mobilize fat reserves to cope, which means that accumulated toxins are liberated into the bloodstream and passed to the developing foetus. After giving birth, they continue to transfer toxins by lactation. Around 80 per cent of toxic contaminants in female dolphins get passed to their first-born calf. Not surprisingly, these calves fare badly. Only half survive their first year, compared to 70 per cent of calves born subsequently. We can't blame all of this on the chemical brew that mothers feed their offspring. By definition, first-born calves have inexperienced mothers that may also be smaller than older mothers. But there is a link. In a group of captive dolphins held by the US Navy, mothers whose calves died within twelve days of birth had about two and a half times more PCBs in their bodies than those whose calves survived.[16]

The thing about marine mammals that disturbs me is that they aren't very different from us. Early on in this book I raised the possibility that modern humans long ago developed adaptations to accommodate a semi-aquatic life, such as the fact that we have a similar body fat content to fin whales. (Actually, given the rise in obesity, some of us now rival seals and walrus!) All that fat predisposes us to concentrate a nasty cocktail of pollutants, just like dolphin, seal and whale. And, like Sarasota dolphins, we can pass them to our children in utero and through breast milk. I once told a lady I met at a conference the story of how first-born dolphin calves receive most of the chemical load carried by their mothers. She said, 'You know, that explains so much about my older brother!' Joking aside, these toxins could lead to all kinds of problems in later life, from learning difficulties to cancer.[17]

PCBs, together with a variety of other industrial chemicals including heavy metals, are hormone mimics that can interfere with the regulation of development and other bodily functions. Highly contaminated animals can therefore be sterilized by these 'endocrine disruptors', as they are called, or their offspring affected in subtle and insidious ways. It takes only tiny traces to alter our hormone systems in ways that are detrimental to health.

In the far north, indigenous peoples often carry heavy loads of toxins in their bodies. This is not because Arctic regions are more contaminated than lower latitudes; actually they are less so. It is because of what people eat. Traditional diets often contain a great deal of blubber from marine mammals, exactly the food you would want to avoid to minimize exposure to toxins. The most contaminated flesh comes from animals that feed high in the food chain, like narwhal and beluga.[18] Animals like bowhead whales, which are still hunted traditionally on Alaska's Northern Slope, feed lower down the food web and their flesh is safer. The highest levels of PCBs, DDT and other contaminants are found in the Inuit of Greenland. Tests of the mental abilities of children in some Arctic communities have revealed that those who come from mothers with high toxin loads are more likely to have problems, a pattern also found in children from industrialized countries. PCBs and other hormone mimics depress the production of thyroid hormones critical to foetal brain development.[19] In the US and Europe such relationships have led people to ask whether our increasingly contaminated world is the cause of the rising incidence of conditions such as autism and attention deficit disorders in children. The Arctic Council, an intergovernmental body that oversees the interests of Arctic peoples, recommends that women there continue to breast feed their children because they reckon the health benefits still outweigh the possible harm done by pollutants.[20] But it is a close call.

The good news is that time has been called on PCBs in much of the world. The US banned them in 1976, and many other countries have followed. The Stockholm Convention on Persistent Organic Pollutants, which was drafted in 2001, prohibits the manufacture or use of a suite of problem chemicals (but, incredibly, DDT is still permitted for pest control in some developing countries). By 2008, 151 countries had agreed to abide by the terms of the Stockholm Convention (fortunately including big emerging polluters like India, China and Brazil). The PCBs and other similar toxins already at large in the sea will cause problems for decades to come, as their release continues from the breakdown of contaminated waste and fixtures in buildings, but levels are falling gradually. A survey of the Arctic between 2001 and 2008 found that surface waters contain in total less than half a tonne of the banned PCBs, just 1/1,000th of 1 per cent of the world production of these chemicals. Levels in seabed sediments were not tested, and toxins there could keep coming back as they leach from sediments, are resuspended by bottom trawls, or are released by the crumbling remains of contaminated waste.

Other toxins have proved far more difficult to control than PCBs. Heavy metals like copper, lead, zinc and mercury also contaminate the oceans. One of the most toxic is methyl mercury because it is easily absorbed and can cross the blood–brain barrier. Much of the mercury comes from emissions by coal-fired power stations that countries seem hell bent on building more of, despite their contribution to global warming. Asian power stations today produce over half of the world's mercury pollution and much of it blows straight over the Pacific.[21] There it combines with particles of organic matter and is converted to methyl mercury by microbes. Mercury concentrations in the Pacific have increased by 1–3 per cent per year in recent decades. Mercury is highly poisonous and accumulates through food webs so top predators like dolphins, tuna and swordfish can be heavily contaminated. Tests of the feathers of museum specimens of Pacific black-footed albatrosses collected between 1880 and 2002 show an increasing burden of methyl mercury over time, with a surge after 1990, mirroring the growth of the Chinese economy.[22]

Like POPs, heavy metals can act as hormone mimics that disrupt the endocrine system. A much-publicized study of white ibises in

Florida found that methyl mercury causes male birds to pair up with other males.[23] It was a gift to headline writers: 'pollution turns birds gay'! White ibises feed in the marshes of south Florida, stalking through the swamp picking up animals like frogs and insects with their slender curved bills. Scientists took birds from wild ibis colonies and fed some uncontaminated food and others a diet laced with methyl mercury at concentrations found in the wild. Male birds given mercury-tainted food were less attractive to females and up to half of them ended up building nests with other males. Not surprisingly, mercury contamination reduced the number of chicks fledged by a third.

The world loves tuna. Canned tuna is virtually ubiquitous and the wealthy world has developed a passion for sushi, which is where many of us pick up our mercury. One estimate suggests 40 per cent of bodily mercury in Americans comes from tuna.[24] Tuna sampled in restaurants and fish markets regularly exceeds mercury levels considered harmful by the World Health Organization, the European Union and countries of North America. Large tunas, like bigeye and bluefin, are worst affected.[25] Perhaps it is a good thing bluefin tuna is far too expensive to be eaten in anything other than sliver-size portions! Samples of swordfish from California supermarkets were on average one and a half times above Federal guidelines on mercury consumption. While there are obvious health benefits from eating fish, the US Food and Drug Administration is so concerned about the problem, it advises women and children, who are more at risk of harm, not to eat shark, swordfish, king mackerel and tilefish (a bottom-living fish common off the Carolinas).[26]

Another notorious hormone mimic is tributyltin, which was widely applied to ships' hulls in the 1970s and 1980s as an anti-fouling compound. Female dog whelks – a type of snail – grew penises when exposed to this compound, oysters developed shell deformities, and some shellfisheries near ports and marinas collapsed. In recognition of its potent toxicity, the International Maritime Organization imposed a global ban in 2008.

Despite efforts to rid the world of persistent chemical toxins, our environment isn't getting cleaner – in fact just the opposite. Chemical industries are constantly creating new products, and phasing out one leads to demand for substitutes. There are about 8.4 million commercially available substances worldwide. Over 30,000 organic chemicals are produced by industry in quantities bigger than a tonne a year and most have never been formally tested for toxicity.[27] The majority probably pose little risk, but the trouble is that we don't find the ones that do until they have been in circulation for years.

Another group of chemicals – brominated flame retardants, or BFRs – was drafted in from the 1970s to replace some of the roles filled by PCBs.[28] They have found their way into every sphere of modern life. They are slathered over our furniture and packed into circuit boards and plastics used in consumer electronics. Clothes made of artificial fibres are steeped in them and they are in plastic food packaging and Styrofoam cups. Many governments demand high standards of fire prevention and compel manufacturers to use the best chemicals for the job. BFRs can make up a quarter of the weight of plastic in electronic goods and foam fillers of furniture. Although less toxic than PCBs, BFRs share their propensity to build up in our bodies and concentrate their way up food chains. Mothers pass them to their children during pregnancy and breast feeding. Just like PCBs, they also seem to be endocrine disruptors that interfere with child brain development.[29] There is now much anxiety about their safety and some countries, like Canada, have already banned the most pernicious compounds.

Pharmaceuticals are an emerging class of pollutants that is racing up the ranks of chemicals to worry about.[30] Throughout the world human populations are growing, ageing and increasingly dosing themselves with drugs to combat health problems. In developing countries, lax controls on pharmaceutical manufacture mean increased potential for releases to the environment. Pharmaceuticals are a worry because they are designed to have large biological effects at very low concentrations, and, like PCBs, some of them accumulate through food webs. Many are excreted and remain stable in the environment for long periods, during which they can cause considerable mischief. Given the many unexplained declines in wildlife around the world, attention is turning to understanding how trace contaminants like medicines could affect their populations.

One of the best known drug effects is that of synthetic oestrogen from contraceptive pills and hormone replacement therapy. Downstream of sewage treatment works, male fish have been feminized by this chemical, which means they cannot effectively reproduce. In a lab experiment, ibuprofen, a common anti-inflammatory drug, reduced reproduction of freshwater fleas.[31] The good news is that lab tests often produce effects only at higher chemical concentrations than are normally seen in the environment. The bad news is that subtle effects may only manifest themselves over the long term so cannot be detected in short experiments. Many drugs, like antidepressants, are designed to affect mood and behaviour. Fluoxetine, the active ingredient of Prozac™, can cause symptoms in fish such as erratic swimming, unresponsiveness, and decreased aggression and feeding.[32] Happy shrimps live dangerously, according to another study, swimming away from shelter and into the waiting jaws of predators. Any of these effects could potentially disrupt the life cycle, increase mortality or reduce breeding success.

Another class of emerging pollutants are nano-particles, extremely small particles that may be just billionths of a metre across. One familiar use of nano-materials is the silver particles incorporated as antibacterials in underwear and socks. Nano-materials are used to make better cooking oils, improve the efficiency of solar panels and enhance drug absorption of pharmaceuticals, among almost countless other possibilities. Because they are so small, nano-particles might easily pass into emissions and effluents that end up in the sea. By virtue of their size they can become incorporated into body cells. Research is still in its infancy, but there is evidence that such minute particles can do more damage than larger particles of the same substances.[33] There is currently enormous investment in the research and development of nano-technologies, but much less in exploring their possible environmental effects. We should be concerned that this revolution in materials science doesn't outpace caution. For example, filter-feeding mussels exposed to nano-particles in glass wool developed to clean up oil spills accumulated particles in their gills and vital organs where they caused cell damage and death.[34] There is much interest in engineering nano-particle-sized pesticides to increase their potency. Although this might decrease the quantities applied, the potential for new forms of pollution damage is obvious.

I don't want to leave the impression that the oceans are toxic. In much of the sea, especially far from human habitation, pollution is very low. Dangerous chemicals are concentrated where inputs are high, around estuaries and cities, ports and shipping lanes. Only a few places are so contaminated as to be dangerous, like New Bedford Harbour in New England. This site received industrial wastes for many years such that sediments are loaded with a cocktail of chemical nasties. Fishing is banned there, and there is a clean-up programme underway to remove contaminated mud. The wider problem is that chemicals at large concentrate from trace amounts in the water to harmful levels in animals. Tens of thousands of tonnes of BFRs, for example, are produced per year and they are already well embedded within marine food webs. Tissue samples taken from a group of false killer whales around the main Hawaiian Islands contained high levels of flame retardants, while southern California sealions had levels forty-five times higher still.[35] The bodies of both species were also laden with PCBs and DDT.

It is easy to get indignant about the ubiquity of chemical pollution and rail at the rapacious corporations that peddle these products. But we shouldn't forget the thousands, perhaps millions of lives they save. Flame retardants have spared many of us from being torched in our beds; DDT prevented countless deaths from malaria. But a balance must be struck between one kind of safety and another of danger. Once chemicals get into the sea it is very difficult to remove them. And we have only recently begun to realize that toxins like mercury and PCBs combine with another kind of waste that is on the increase in the world's oceans: plastic.

10

The Age of Plastic

One tempestuous night in 1992, a container ship from China ran into heavy seas in the Western Pacific. Huge swells breaking over the ship washed a container-load of plastic bath toys overboard. They set some 29,000 plastic ducks, turtles, beavers and frogs free on a voyage that for some has not yet ended. When bath toys began to appear on North American beaches, Curtis Ebbesmeyer, a Seattle-based oceanographer, latched on to the importance of this event as a way to track the movements of the great ocean currents.[1] The toys hitched rides on several currents circling the Pacific and over the years have made landfall in Hawaii, Alaska and Washington State. Some even threaded their way through the Bering Strait into the Arctic Ocean, where they froze into pack ice. That ice then bumped and scraped its way around the pole, pushed by winds and underlying currents, before it disgorged the toys into the North Atlantic. These travellers have since been picked up on beaches in Maine and Scotland.

More than two millennia have passed since the great philosopher and naturalist Aristotle walked the shores of the Mediterranean island of Lesbos, deep in contemplation. It was on Lesbos that he laid the foundations for his masterwork of natural history whose influence would endure until the Age of Enlightenment. Those beaches would have been strewn with natural flotsam, the strand line drawn in palm fronds, seaweed and seeds. Here and there a fragment of worn plank or the sole of a leather shoe or decayed rope would betray the work of man. Fast forward to a hundred years ago and beaches had by this time become littered with the flotsam and jetsam of human societies, but the main difference from Aristotle's time was in the quantity, rather than in the type, of rubbish. Fragments of nets, strands of hemp rope and storm-shattered wooden boat spars would still have been there, but this time with glass fishing floats, barrel staves and bigger heaps of rotting organic refuse carried to sea by rivers. Today, a beach-comber is faced with a very different kind of garbage.

Two thousand years from now, archaeologists excavating the remains of our world will call this the 'Age of Plastic'. Victor Yarsley, an English chemist, helped usher it in. Yarsley was born in 1901 and early on became interested in the industrial potential of synthetic poly-mers. He spent his early career trying to develop non-inflammable celluloid film, but eventually abandoned the search, declaring it impossible. He then struggled for decades to find ways to realize the potential of plastics, working long hours from a lab jerry-built in his garden shed. His daughter later recalled having to tape over holes and cracks in her bedroom to keep out the awful smells.[2] Success eventu-ally came as he discovered ways to make plastics without air bubbles, and perfected the mixtures for new products like dentures and pros-thetic limbs. By 1941 Yarsley felt confident enough to set down his vision for a future world built around the miracle of plastic:

> This plastic man will come into a world of colour and bright shining surfaces where childish hands find nothing to break, no sharp edges, or corners to cut or graze, no crevices to harbour dirt or germs ... The walls of his nursery, his bath ... all his toys, his cot, the moulded light perambulator in which he takes the air, the teething ring he bites, the unbreakable bottle he feeds from. As he grows he cleans his teeth and brushes his hair with plastic brushes, clothes himself within plastic clothes, writes his first lesson with a plastic pen and does his lesson in a book bound with plastic. The windows of his school curtained with plastic cloth entirely grease- and dirt-proof ... and the frames, like those of his house are of moulded plastic, light and easy to open never requiring any paint.[3]

Predictions about life in the future are often funny and usually wildly off. According to the books I read as a ten-year-old, by now we should all be skimming the skies in airborne cars and relaxing in the garden as robots cook our dinner and clean the house. What is remarkable about Yarsley's world is that it has come true and we are living it. By 2008, the latest year for which I have a figure, 260 million tonnes of plastics were produced using 8 per cent of global oil production in raw materials and energy.[4] The curve of production over time bends upwards like a cliff face, increasing by 9 per cent per year. The stark reality of this ever-steepening upward climb is that more plastic was made in the first ten years of this century than all of the plastic created in history up to the year 2000. The world is awash with plastic – most of us are rarely out of contact with something made of the stuff. We are literally and figuratively swimming in it. Yarsley's powers of foresight deserted him on one key matter, however. His future was a brighter and better world: 'a world free from moth and rust and full of colour, a world largely made up of synthetic materials ... a world in which man, like a magician, makes what he wants for almost every need out of what is around and beneath him'.

Actually, what is around and beneath us today is an endless vista of discarded plastic – a world full of colour indeed. The wildflower-filled roadsides of old have for many of us been replaced by a near-continuous succession of plastic food wrappers, bags and lost or abandoned junk. By weight, about a tenth of our waste is plastic, and a much bigger proportion by volume. A third of all plastic is turned into disposable packaging, used once and thrown away. We pack our discarded plastic into landfills and recycle some, but a significant amount finds its way to the sea. There is a river at the bottom of my garden. When floods rush by, I sometimes watch as a stream of rubbish bobs past. One fairly ordinary day I counted them: twenty-seven pieces of plastic in an hour. It is a story repeated in every one of the world's streams and rivers that drains inhabited land. When it rains hard over the Caribbean island of St Lucia, for instance, the sea turns brown with sediment washed from exposed soil and the bays fill with bobbing plastic bottles and drifting shopping bags. I once opened my Sunday newspaper to a double-page photo that showed a solid mass of plastic bottles, bags, buckets, shoes and other refuse. So choked was the scene it was hard to tell whether this was land or water. Eventually I spotted the face of a little Filipino boy amid the rubbish, swimming in Manila Bay. I was stunned. Years later the image still haunts me.

Rivers launch their garbage into the sea, where some is picked up by offshore currents to join bath toys, trainers, cans and stuff lost from boats on transoceanic voyages. Curtis Ebbesmeyer called his book about this floating world *Flotsametrics*, but its subtitle, to paraphrase Daniel Defoe, might be *The Life and Strange Surprizing Adventures of a Plastic Duck*. Actually, there is another book about these toys with the marvellous title *Moby Duck*![5] The great currents of the oceans move in counter-rotating circles that interlock with one another. Due to the Earth's rotation, winds blow to the right in the northern hemisphere and to the left in the southern, giving them an easterly or westerly heading (Coriolis again). Near the equator in both hemispheres, the wind tends to come from the east – these are called the Trade Winds – while at mid- to high-latitudes they blow from the west – the Westerlies. Winds move the sea beneath, but again the Earth's rotation gives them a twist and continents block their movement. The result is that air movements spin up vast rotating currents known as gyres.

Sailors have recognized these winds and currents for thousands of years, but they weren't mapped until the nineteenth century, when a US naval captain, Matthew Maury, gathered together all of the oceanographic measurements available in a pioneering book, *The Physical Geography of the Sea*.[6] Maury recognized at least three different gyres – a number that has increased to nine as modern oceanographers have added detail. Curtis Ebbesmeyer added two more small loop currents in the Arctic to bring the total to eleven.

In general one finds two main gyres in each ocean, one to the north and another south of the equator. The Pacific Ocean is so vast it has two in the north. The Indian Ocean has just one to the south, because monsoon winds dominate the north and the Indian sub-continent blocks water movement there. An unbroken loop of sea circles Antarctica so the gyre flows all the way around this continent in an endless circle. If you have ever wondered why round-the-world yachtsmen often run into difficulties in the mountainous seas of the Southern Ocean, it's because this is the shortest and fastest way to circumnavigate the world.

Thanks to the Earth's rotation, revolving currents in the gyres push water towards the centre, where it piles up into hills. You might have thought the sea was perfectly flat but there are hills and valleys in the ocean just as on land. Where the Earth's crust is thin, such as

Figure 4: Curtis Ebbesmeyer's map of the world's ocean gyres: giant surface loop currents. Grey spots inside the gyres show areas of rubbish accumulation, the so-called ocean garbage patches. Redrawn from C. C. Ebbesmeyer and E. Scigliano (2009) *Flotsametrics and the Floating World*. HarperCollins, New York.

over ocean trenches, the pull of gravity from the dense mantle beneath is strong and it creates hollows in the sea. Admittedly, these are very gentle valleys and hills; the difference between a peak and a valley floor amounts to less than a metre of height over hundreds or thousands of kilometres. In the middle of each gyre, the piled up water is pulled down by gravity and there is a downwelling to the deep sea like water flowing down a gigantic bath plughole. A nautical almanac published in 1845 described these currents thus:

> It has been observed that the waters of the Atlantic have a greater tendency toward the middle of the ocean than otherwise, and this seems to indicate a reduced level, forming a kind of hollow space or depressed surface . . . in the middle part, although within the region of the Tradewind, the currents are not regular, but indicate a kind of vortex.[7]

This author mistakenly believed the currents flowed into a hollow in the ocean, when in fact they were pushed uphill by the combined force of the gyre and the Coriolis effect. But he rightly spotted the vortex. The central parts of some gyres – those with high atmospheric pressure air cells above them – gather everything that floats into enormous rafts, which are held there by the unseen hand of downwelling water. In the Atlantic gyre there is a place so distinctive it has a name of its own, the Sargasso Sea. Maury called all of the central regions of the world's gyres 'Sargasso Seas' due to their accumulation of the seaweed *Sargassum*, a floating weed buoyed up by shiny green clusters of floats that to early Portuguese navigators looked like grapes. But the name has stuck only in the Atlantic. Maury wrote:

> Columbus passed through [the Sargasso Sea] on his first voyage of discovery to America. His crew were alarmed at it, for in places it is so thickly matted over with brownish weeds, as to hide the face of the sea, presenting somewhat the appearance of a drowned meadow, and looking as though you might walk on it, without sinking more than shoe deep in water. The sargassos of the southern hemisphere are not so well marked as this, nor are they as rich in drift or floating matter.
>
> You observe that in both the North Atlantic and the North Pacific, the sargasso is in the middle of the pool as it were, or between the warmer current to the west, that flows north, and the cooler one to the east, which returns south. Into this sargasso of the Atlantic are carried the trees and plants and creepers that are brought down from the Andes by the Amazon, and from the Rocky Mountains by the Mississippi and its tributaries. These, with the wrecks of steam-boats that are blown up or snagged, sunk and lost on the Western waters of America, are all drifted

out to the sea, and placed within the influence of the Gulf-stream. In the course of time they find their way into this sargasso. On their voyage to it, – their sea of rest, – they become loaded with barnacles and shellfish; arrived there, these die, or, loading their floating domicile too heavily, with increase of inhabitants, it finally sinks, conveying them down with it into the depths below.

When Curtis Ebbesmeyer visited the Sargasso Sea in 1977, he was taken aback not so much by the rafts of weed, but by the number of Styrofoam cups that blossomed from it like water lilies.[8] His meticulous tracking of ocean garbage has revealed the oceans' remarkable interconnectedness. The great current that circles the planet, the global ocean conveyor, is powered at the surface by many gyres that act together, like the cogs in a clock. Those cogs are of different sizes and spin at different speeds. Slowest of all are the Arctic gyres that rotate about once in thirteen years. The vast gyres of the Pacific Ocean – Ebbesmeyer calls them the Turtle and Heyerdahl gyres, the latter after Thor Heyerdahl who used it to cross from South America to the Marquesas on a balsa raft – turn once every six and a half years. The loop that circles Antarctica is driven faster by ferocious winds and turns once in three and a third years.

At each cycle, some of the floating garbage escapes from the outside edge of the gyre and finds its way to the coast where it is tossed up onto beaches. Once again, Ebbesmeyer's extraordinary obsession with flotsam gives us a number: about half of the garbage carried is freed at every turn. There are still bath toys out there from the 1992 spill, but not many. At the time of writing, millions of tonnes of junk that were washed offshore by the devastating Japanese tsunami of 2011 are headed towards Hawaii and North America. The first wave is expected to make landfall by 2012, but, as the bath toys show, beaches will continue to receive debris from this disaster for years to come.

For the last several decades, beach clean-ups have been a regular activity for marine conservation groups, and an essential task for resort towns. The majority of that junk is plastic. Beach surveys from America to the Southern Ocean, and from the coast of China to remote islands of the central Pacific, show that the ratio of plastics to other rubbish typically lies in the range of 2:1 to 9:1. Averaged across the European Union there are approximately 500 to 600 pieces of litter per hundred metres of beach. This only includes pieces large enough to bother picking up. A survey of Orange County beaches in California that included all visible plastic produced a figure of over 100 million items of trash for the whole county, most of it tiny pieces of plastic.[9] This is equivalent to more than 150,000 items per hundred metres of beach!

Despite the oceans' capacity to carry garbage halfway around the world, plastic is most abundant in populous places and close to cities. The northern hemisphere is much worse affected than the south, although right now rates of accumulation of beach rubbish are increasing faster in south polar latitudes than anywhere else.[10] Each year, about 2,000 items of human-generated rubbish (most of it plastic) washes ashore for every kilometre of beach in the northern hemisphere compared to about 500 in the southern.[11] Enclosed seas like the Mediterranean are worse affected than open ocean coasts. There, junk can wash in at 1,800 pieces per hundred metres of shore per year.

Most plastic is easily overlooked, too small to be seen by a casual glance, and much too small to be included in these surveys. Next time you are at the seaside, draw a square on the beach near the shoreline about the length of your forearm. Then pick from inside every fragment of plastic you can find. You will need to look very closely, since much of it will be less than five millimetres across. Many of the pieces will be rounded drops of white or yellowed plastic known as nurdles or mermaid's tears. They are the raw material of the plastics industry – pellets ready to be moulded into finished products. Over the years, vast quantities have been lost at sea and in spills on land. Give yourself plenty of time, because you will probably find hundreds, perhaps thousands of them. A survey of some New Zealand beaches found up to 300,000 per linear metre of shore! Ninety-eight point five per cent of the plastic found in the Orange County beach survey were nurdles.

Plastic stuff gets thrown away in such abundance partly because it doesn't last very long. I am often annoyed at how something made out of decent materials like wood or metal has to be junked because some critical bit of plastic in it has broken. But that fragility belies an incredible durability in the substance itself. Plastics break down very slowly but over time they fragment into ever-smaller pieces. When you get bored picking them from your plot of beach, why not relax with a swim? As you enjoy the caress of waves, consider this: the water around you probably contains a multicoloured soup of tiny plastic fragments. While there is plenty of the stuff near coasts, it has built up in offshore waters, and spectacularly so in the Kuroshio Current that flows past Japan and East Asia. There densities of plastics were reported up to 3.5 million per square kilometre of sea.[12]

Not all plastic floats. Roughly half is denser than water and sinks at various rates, while the other half is what we are familiar with from ocean garbage patches, as Curtis Ebbesmeyer dubbed the middle regions of the gyres. Few surveys have been made of seabed litter so there is little to go on, but it is clear there is another hidden problem which we will soon have to reckon with. In some parts of Europe, contamination can reach levels of a piece of plastic for every square metre of bottom.[13] Plastics and other rubbish have made it to the deep sea as well. Scientists who deploy instruments on the deep sea bed often return to find them clogged with drifting plastic bags. One submersible pilot reported that plastic bags passed him by like a succession of ghosts. Heavier rubbish thrown from boats reaches the bottom of the sea quickly. The deep sea is not so remote in distance, even if to us it is a profoundly alien environment.

I mentioned that ocean gyres lose about half of their flotsam at every cycle. Some of it comes off the outer edges and escapes towards the coast, but the rest goes inward. Just like the mats of floating seaweed that so amazed the sailors of old, great rafts of plastic, fishing line, nets, ropes and a thousand and one other bits of junk have accumulated within the gyres. In a story often retold, a yacht captain, Charles Moore, took a different route from Hawaii to Long Beach, California in 1997 and crossed through the centre of the gyre in the north-east Pacific. He was so stunned by the amount of garbage he found there that it changed the course of his life. Since then he has dedicated himself to the study of plastics at sea, and is now a relentless campaigner for something to be done about them. One of his most famous findings was that, by the turn of the millennium, there were six times more floating plastic fragments in the surface water by weight than zooplankton.[14] The gyre Captain Moore traversed has gained notoriety as the Great Eastern Garbage Patch; at about the size of Texas it is one of the world's largest. Surveys across the Sargasso Sea show the scale of contamination. On average, plankton tows there revealed densities of floating plastic fragments in the order of tens to hundreds of thousands of pieces per square kilometre.[15]

Plastics are completely foreign to life, so animals have been unable to mount any evolutionary defence against them. The Laysan albatross, which nests on Kure Atoll in the central Pacific Ocean, has the misfortune to live close to the Great Western Garbage Patch, as the centre of the gyre on the western side of the Pacific is called. Albatrosses feed by picking live and dead prey from the sea surface. Unfortunately, these ones can't distinguish between plastic and flesh. Researchers recently found that adults returning from long-distance foraging trips fed their chicks an average of seventy pieces of plastic per meal.[16] As the flesh rotted from chicks that had starved to death, it left piles of fishing line, golf balls,[17] pens, bottle caps and sundry other fragments, each framed by a mute halo of feathers. Some chicks contained more than 500 plastic items. Plastic content of Laysan chicks has increased ten-fold in the last forty years, presumably in line with increases in the mass of rubbish circling the Pacific. As early as 1965, three-quarters of Laysan chicks found dead had plastics in them.[18] There is a heartrending possibility that these albatross deliberately fly to areas of rubbish concentration for the rich 'feeding' opportunities they provide.

Autopsies of leatherback turtles in the Atlantic Ocean show this beast has similar difficulties telling food from plastic. Leatherbacks eat jellyfish and other gelatinous zooplankton. To a dewy-eyed leatherback, plastic bags and balloons look much the same as prey. A study of dead leatherbacks washed ashore since the 1880s found plastics present from 1968 onwards.[19] Some contained lumps of tangled plastic bags and Mylar balloons (the helium-filled variety we give to kids) the size of a football. Today, over a third of dead leatherbacks examined have plastic in their guts. The strain of hauling itself ashore to nest obviously got to one leatherback in French Guiana. It pooped a huge plug of plastic bags and sacks that weighed over two and a half kilos.[20] Normally turtles fast during the breeding season and don't shit on the beach. How many other bags, I wonder, have been recycled through the guts of turtles far out to sea, and how many turtles' lives are threatened by the plastics packed in their bellies? As jellyfish are 95 per cent water, they make a pretty meagre meal. To break even, adult leatherbacks must consume hundreds of kilos of jellyfish every day, over half their bodyweight. Even if plastics don't kill outright by blocking the gut, they could interfere with digestion and cause animals to starve by tipping their caloric balance into the red.

Hundreds of other species ingest plastics, either directly or by eating something else that has. In the North Sea, nineteen out of every twenty fulmars – a smaller relative of the albatross – washed up dead on beaches contain plastic. On the Dutch coast, four out of five plastic fragments had peck marks from birds.[21] Plastics have reached some of the remotest places on the planet. At sub-Antarctic Macquarie Island, fur seal scats contain many plastic fragments that were eaten by the deep water fish they prey on.[22] Even whales have eaten plastics. A pygmy sperm whale stranded on a New Jersey beach in 1993 had its stomach blocked by fragments of plastic bags. The whale recovered after they were removed and it was released into the Gulf Stream.[23] Others have been less fortunate. A dead pygmy sperm whale stranded in Texas had a plastic rubbish bin liner, a bread wrapper, a crisp packet and two other pieces of plastic sheeting choking off its stomach.[24]

These whales may have taken in plastics accidentally while feeding, but there are signs that whales target pieces of plastic junk deliberately. Two sperm whales that died on the California coast between them contained over a hundred kilos of net fragments, line and plastic bags.[25] The pieces packed into their stomachs came from all kinds of nets that may have been up to twenty years old, and seem to have been tossed overboard by fishermen mending their gear.

I once came across a photograph taken far out to sea in one of the oceans' garbage patches. A huge ball of fishing net suspended by a couple of hundred floats stretched across the frame for at least ten metres. At one end, a sad-looking and exhausted turtle stared at the camera, helplessly entangled. This poor beast had dragged the net with it across how many kilometres of ocean nobody could tell. Like Sisyphus in Greek mythology, it was condemned to a life of immense hardship undertaking a task that could never be completed. Some of the most disturbing images of the harm done by human detritus are of sealions that have put their heads through loops of plastic or net as pups. As they grow, the loops tighten and cut a deep, bloody slash into their neck. When the sealion finally dies from this slow-motion throat cut, it decomposes and leaves the loop of plastic to be picked up by another. On California's Farallon Islands, the number of seals entangled in plastic and fishing line rose from a handful in the 1970s, to more than sixty a year by 2000.[26]

Plastics came into widespread use after the Second World War, so pretty much all of the plastic at large on the high seas got there after 1945. Since then, the tide of plastic pollution has grown relentlessly, and it is fragmenting into ever-smaller pieces. It is often said that plastics take hundreds, or even thousands of years to degrade. Japanese researchers found plastics that had been floating in the North Pacific for decades were degrading much faster than expected.[27] This apparently good news story has a poisonous twist. These particles are not just harmless roughage. They concentrate toxic compounds on their surfaces, sometimes to levels a million times or more above concentrations in the seawater around them.[28] Recall from the last chapter that the surface microlayer of the sea, in which many plastics can be found, concentrates toxins. They attach to plastics. In one Japanese experiment, polystyrene beads soaked in seawater for several days picked up PCBs. When they break down, plastic particles release toxic compounds like flame retardants, styrene, phthalates and bisphenol A into the sea. These toxins exert their insidious effects on wildlife, and us, at minute concentrations.

Bisphenol A and phthalates are ingredients of a large variety of plastics. Phthalates are added to increase flexibility and transparency, but they don't bond with the plastic so are easily leached out. Bisphenol A is added to the resins that coat the inside of food tins among many other products. Both have become highly controversial in recent years because of health concerns about their endocrine-disrupting properties, and both have been banned or restricted in a number of countries, including Canada, Australia, the US, Japan and the European Union. Some organizations, like the UK's Food Standards Agency, insist that such substances are safe, but research is ongoing and many subtle developmental effects have been detected in lab studies. Phthalates are attracted to fats like those they would come into contact with in the surface microlayer. Given the ubiquity of these compounds in plastics, the possibility of harm to marine life (and us) must be taken seriously.

We have only recently been confronted with the plastic confetti that is the legacy of decades of pollution. In the late 1990s, Hawaiian beaches were inundated with a multicoloured snowstorm of plastic fragments. People there were used to bits of plastic bobbing around, but had seen nothing like this before. Periodically the vast ocean currents that circle our oceans twist off rings of water that escape from the pool trapped inside the gyre. In some cases this might happen but once in half a century. It seemed that the ocean gyre near Hawaii had belched a gutful of plastic that had been building within it since plastic came into widespread use.[29]

An alarming new twist is that most cosmetics manufacturers now add sub-millimetre-sized plastic granules to hand lotions and face creams as exfoliants. They are too small to be filtered out by sewage works, and most of the particles are washed to sea, where they can be ingested by tiny plankton, which mistake them for food like copepods or fish eggs. Like other plastic fragments, these granules have a high surface area with which to attract and concentrate poisons like PCBs and mercury. Since plankton are the foundation of almost all ocean food webs, the problems this could create are clear. It is nearly impossible to buy an exfoliating face or hand cream today that doesn't include plastics. Take a look at the labels on yours. If you find polyethylene in the list of ingredients, you are washing yourself with plastic.

Many of the particles at large in the oceans range from millimetres to micrometres in size. They now fall into the size range of planktonic food for a huge variety of life at the bottom of the food web. A third of plankton-feeding fish sampled in the North Pacific Central Gyre had plastics in their guts.[30] Experiments with captive animals show that microplastic particles are eaten by almost any species that filters water, picks drifting particles of food or slurps deposits from the seabed.

Few people have looked, but, in places where they have, beach and bottom sediments are stuffed full of microplastic particles. These microplastics have concentrated chemicals to higher levels and could pass into animals that eat them, providing a fast track for the accumulation of toxins all the way up to top predators that we like to eat. Scarcely any research has been done on the transfer of chemicals from plastic fragments to animals that eat them, but this is certain to change in the next few years because the world's attention has turned to plastics at sea. At this point, what we do know is that plastic particles fed to mussels end up in their circulation system and stay there for long periods. Chicks of a scavenging seabird, the great shearwater, that carried higher loads of ingested plastic had higher concentrations of PCBs in their tissues.[31]

Remote regions of the ocean, like the Sargasso Sea and north-east Pacific, have become slowly rotating graveyards of plastic junk, some of it decades old; the tangled remains of lost and abandoned fishing nets drift past golf balls, toothbrushes, gas lighters and plastic bags. For how many hundreds or thousands of years will they continue their aimless voyages? In one telling incident, a dead albatross chick from Midway Island in the North Pacific had been fed a piece of plastic engraved with a serial number. It was traced to a US bomber that had crashed into the sea in 1944.[32] The amount of plastic at large is still growing exponentially year by year. You come face to face with this truth on virtually any beach in the world. In South Africa, for instance, bottle lids on beaches, used as an indicator of the prevalence of smaller plastic rubbish, increased by over fifty times between 1984 and 2005.[33] That is not 50 per cent, but 5,000 per cent. The genie is already out of its plastic bottle. The oceans are choking on plastics and will continue to do so for hundreds of years, even if we were to stop dumping plastics today. But it is never too late to start the clean-up.

Problems from plastics, sewage, oil and toxic chemicals are all too familiar, but there are some things we don't normally think of as pollution, at least in the sea. Noise and 'biological' pollution – that is, the spread of species beyond their native haunts – are growing concerns and I turn to them in the next two chapters.

11

The Not So Silent World

When I was a graduate student I spent three unforgettable summers exploring Saudi Arabia's coral reefs. I was part of a team there to map its biological riches for the first time and we were allowed to visit places long closed to foreigners. I remember well the keen thrill of slipping into the water, knowing that I was the first person ever to scuba-dive there and wondering what I might discover. In 1984, we reached Tiran Island by sailboat after days beating up the Red Sea against bone-jarring waves raised by powerful northerlies. It was bliss when at last we found shelter in the lee of Tiran's mountainous flank. We had hardly seen a soul for weeks and Tiran felt like the end of the world. That night, far across the sea, a few lights from the Egyptian coast suggested we were not alone. But the peace was palpable.

Next day I dived one of the coral reefs that defend the entrance to the Gulf of Aqaba. Below water the bottom was hard to make out at first through a blizzard of plankton-feeding damselfish, basslets and glassfish that swayed to the silent pulse of unseen currents. Purple seafans and chocolate seawhips bent and wafted on the liquid breeze amid a tapestry of corals of marvellous shapes and colours. Surrounded by such tranquil beauty, I lost all sense of time.

Little more than a decade later, I returned to dive the same spot, this time setting forth by day-boat from the Egyptian resort of Sharm el-Sheikh with fifteen other divers. We charged north with ten other boats, all hoping to be the first to tie up to one of the moorings that marked a string of dive sites. The Egyptian coast behind me had been utterly transformed; whereas before there had been only a few dwellings, now resorts and hotels studded the clifftops, coves and beachfronts. At the dive site I had to pick my moment carefully to leap into the sea, to avoid the propeller blades of two latecomers jostling for space.

I expected to find immediate relief from the confusion and fumes above water, but even ten metres below the surface I was overwhelmed by the engine roar from boats overhead and others coming and going from more distant sites. The near-constant bombardment made this dive feel like the underwater equivalent of standing in the middle of a motorway. I tried to find a quiet corner to while away the dive but the noise was everywhere. From a rocky ledge a stonefish looked at me with an unblinking scowl. It seemed an appropriate expression. I wondered whether the fish on this reef were relieved when the last boats left in the evening and a semblance of peace was restored.

Jacques Cousteau published his first book in 1953. He called it *The Silent World* but his title was misleading. As well as the timeless notes of whale song, Cousteau's seas were filled with the scratch and rasp of foraging animals, the whistles and clicks of dolphins, the low groans of whales and the rumble of breaking waves. By that time the thump and growl of boat engines had also invaded the oceans. Noise levels have grown to a roar since the 1950s, as the world's globalized economy has launched tens of thousands of new merchant ships. The sound of a supertanker is ear-splitting, and because sound travels further underwater than in air it can be heard by a whale a day before the ship arrives. Background noise levels in the sea have not been measured systematically, but in the few places we know about they have increased by about three decibels per decade since the 1950s.[1] The loudness of sounds underwater doubles for every six decibels. So today's seas could be over eight times noisier than those that Cousteau taught us to love in the early 1950s, and in some hotspots are over a hundred times as loud.

A few years ago I had a haunting experience off the tiny volcanic island of Saba in the Eastern Caribbean. I had begun a dive to count fish when I heard a loud drawn-out rumbling growl or rather I felt it as it thrummed through me. I spun round expecting to see a leviathan bearing down, but none was there. Then came the unmistakeable melodious moan of a humpback whale. For the rest of the dive I listened to a song as beautiful as any by Verdi. I never saw whales on this or many subsequent whale-song dives in Saba. They were probably far away, perhaps tens of kilometres, for their songs have evolved to carry enormous distances in the sea.

The much-greater density of seawater compared to air carries sounds further and faster. Noise travels about five times more quickly in the sea, reaching 1,500 metres per second in temperate waters. Sound drops off less quickly in water than in air, so noises can be heard at far greater distances.[2] High frequencies attenuate more quickly than low frequencies, so deep notes go further. Some of the great whales communicate over distances of hundreds or even thousands of kilometres using the kind of low-frequency rumbles I heard in Saba. Sounds go furthest of all if animals take advantage of a strange property of the sea. Between the warm, lighter layer at the surface and the cool, dense water below there is a narrow region where sounds travel in a similar way to light in a fibre-optic cable. Differences in density across this boundary – the thermocline – keep soundwaves produced there in a narrow channel where they lose little energy as they travel. Before the days of satellite beacons, the US Navy issued pilots with an explosive charge. If they had to ditch in the sea, they were to lower the charge on a string and explode it in this acoustic pipeline. The sound would carry to sensors perhaps thousands of kilometres away in places like California, Hawaii and Panama and rescuers could then work out where the pilot was by triangulation.

Several species of whales are thought to use this aquatic telephone to broadcast calls over hundreds of thousands of square kilometres of ocean. Their voices are well adapted to be heard from far away. Blue whales are the largest and loudest animals on the planet and they can bellow at over 190 decibels. Sperm whales are also very loud. Doug Anderson, a cameraman on the BBC's *Planet Earth* series, once came across a newborn sperm whale calf in the Azores that he began to film while the mother hung around below with an older calf. The baby played with him for a while before the camera housing caught in one of its foetal folds. This obviously hurt because it gave out a stream of clicks and blew hard. Anderson turned to find the mother's head filling his entire field of vision. He thought she would charge but instead there was a crack like the splintering of a falling tree, so loud it shook his body and left him stunned.[3]

The peace and quiet of the oceans is today broken by far more energetic sounds than ships. All over the world, seismic bangs from oil and
gas exploration race through the sea to probe the rock beneath. Survey ships have air guns that produce pulses of sound above 200
decibels to send shock waves through the seabed. The boom of military sonars used to track submarines can reach 235 decibels. Things
sound about 61.5 decibels quieter in the sea compared to air.* Sonar
noises and seismic blasts therefore come in at an above-water equivalent of around 150 to 175 decibels. For comparison, a rock concert
can reach 110 decibels, and people feel pain from 120 decibels, or
about the noise of a big chain saw close up. The sound of a seismic
gun is over thirty times louder still and the sound of the most powerful military sonar more than a thousand times louder.

It isn't quite this simple and I must admit to shameless anthropo-
morphism in my description of the disgruntled Red Sea stonefish
earlier on. There is a second and equally important dimension to
sound, which is frequency. High-pitched noises have a high frequency,
whereas low-pitched sounds have a low frequency. Most animals can
only hear over a limited frequency range. Bats hear higher frequencies
than we can. My stonefish would have been stone deaf to many of the
sounds that you and I hear since its hearing range is set lower than
ours (although boats produce loud noise at low frequencies so the reef
would probably have been as noisy for the fish as it seemed to me).
What would constitute a gut-churning blast of sound for a human
might register as only a whisper or not be heard at all by some other
creature.

Beaked whales are especially susceptible to sudden loud noises in
the sea. These odd animals look like a whale crossed with a porpoise
or dolphin. Some have blunt foreheads and thick fleshy lips, while
others have drawn-out snouts like bottlenose dolphins. They are shy,

* The decibel scale provides a measure of sound levels, or sound pressure levels more
precisely, that is referenced to the sensitivity of the human ear, where 0 decibels cor-
responds to something at the lowest limit of our hearing in air. All decibel measures
must be referenced to a specific sound pressure level, which is conventionally taken to
be 20 μpa (micropascals) at 1m from the source in air and 1 μpa at 1m from the source
underwater. What this means is that, for a given level of sound pressure, the loudness
underwater would be equivalent to hearing the sound 1m away, whereas the loudness
of the same sound on land would be equivalent to hearing the sound 20m away.

can dive for more than an hour, and rarely spend more than a few moments at the surface so they are hardly ever seen. We think there are twenty-one species, but there could be more. These whales are so enigmatic they have been called 'the least understood group of large animals on Earth'.[4] Some species have never been seen alive and are known only from their dead bodies that have washed ashore. They range in size from four metres long and a tonne in weight to thirteen metres and fifteen tonnes. It is hard to imagine that we could overlook animals the size of elephants or bigger so easily, but it is their lifestyle that makes them elusive.

Peter Tyack, from Woods Hole Oceanographic Institution in the US, has devoted his life to understanding how whales use sound to communicate. I once spent a lively bus journey with him crossing Santa Cruz Island in the Galapagos. In just an hour, he completely turned around my notion of what it means to be a whale. Until that time, I had thought of whales as animals that lived at or near the surface, but dived into the deep blue to find food. Peter said that we should think of beaked whales as deep-living animals that occasionally pop to the surface for breath. The abyss is their home. It is this deep-lifestyle that makes intense sonar noise a real problem for them.

In 1996 there was an unusual mass stranding of a species called Cuvier's beaked whale in Greece's Kyparissiakos Gulf. When whales strand together, they usually all come ashore at once in the same place. This time the whales came ashore separately along nearly forty kilometres of coast. No wounds or other signs of harm were found on the bodies. Most of the whales had stomachs full of squid, which suggested they had fed just before their death. For Alexandros Frantzis, then a young biologist at the University of Athens, the smoking gun was a NATO research vessel that had been in the gulf to test a naval sonar system at exactly the time the whales died.[5] The system used extremely loud sounds at low frequencies – some well within the range of whale hearing – to detect ultra-quiet submarines. He became convinced that the sounds had somehow either injured the whales directly or disoriented them so badly they beached themselves inadvertently.

Beaked whales of many different species have since stranded in similar ways, and almost always the military are in the vicinity, testing some extreme noise-making device.[6] Closer scrutiny reveals these victims to have suffered serious internal injury. Some even come ashore bleeding from the ears. Like human divers that have spent too long underwater, many whales had suffered decompression sickness.[7] When divers breathe at depth, nitrogen gas dissolves in their blood. As any diver knows, spend too long too deep, or ascend too quickly, and the nitrogen will fizz out into your blood to cause the bends. Severe cases can kill you. As the whales surfaced, bubbles of nitrogen formed in their blood, fat and other organs, literally tearing them apart internally and blocking their circulation. How these whales get decompression sickness is still a bit of a mystery. Initially, people thought it was because they had been frightened to the surface too quickly. But Peter Tyack doesn't think this would cause injuries of the kind he had seen.[8] Instead, he believes that when sonars scare whales they spend too much time 'bounce diving' between thirty and eighty metres deep, snatching short breaths at the surface, and so they accumulate dangerous amounts of nitrogen in their blood.

There are other sources of intense sound in the sea. In the 1990s, physicists working with US defence agencies came up with the idea of using sound transmission to measure the temperature of the sea and thereby track global warming. The idea called for 195-decibel pulses of sound to be produced a kilometre down between Hawaii and California for twenty minutes every four hours ... continuously. Test transmissions were made between 1996 and 2006. Whose brilliant idea was this? If noise travelled as far above the sea as below, and with as little attenuation, we would have dismissed the scheme after little more than a moment's contemplation. But physicists are sometimes inclined to forget that there is a living world. In one spectacular example, they sunk an array of highly sensitive hydrophones – instruments that record underwater sounds – into the Mediterranean off the coast of Sicily in 2005.[9] They hoped to listen for the sounds that tiny sub-atomic particles called neutrinos might make as they smacked into water molecules. Because neutrinos are vanishingly small, the chances of that happening are low, even in a huge volume of sea. More than a hundred physicists collaborated on the project, but the possibility that background noise might interfere was not even mentioned until late in the day. Everyone assumed that the deep sea would be virtually silent. Of course, there was far too much noise, even before boat traffic was factored in, to detect the sound of a colliding neutrino. Instead, the hydrophones proved adept at pinpointing sperm whales using echolocation clicks to hunt squid in the liquid darkness. At least the physicists discovered something: there were more whales around than anyone had suspected.

The tests proved the viability of using sound as an acoustic ocean thermometer but the method remains controversial. Research showed that humpback whales seemed to avoid the sound sources, but nobody can say for certain what the long-term impacts of these sounds might be.[10] Since physicists still have plans to create a web of sound throughout the world's oceans to measure water temperature, this question needs an answer.

The loudest noises in the sea cause the most dramatic effects and have attracted most concern. But the rise in background noise from boat engines (and now wind farms) is probably more significant. Sound is one of the most important means of communication for creatures underwater. In the clear water of a coral reef, you can see fifty or sixty metres at best. Usually, visibility is far less; in the dark and muddy waters of estuaries it can shrink to under a metre. Because sound travels so far and fast, it is the best way to communicate. Many marine animals therefore have excellent hearing and use sounds to visualize their surroundings, to avoid predators, attract mates and find prey. Background noise can mask important sounds and make life more difficult and dangerous.

Whales and dolphins, as their amazing vocalizations suggest, will likely number among those worst affected by noise. The din of boats is shrinking their world. Before we invaded the sea with our noise, fin whales could hear each other over distances of at least 1,000 kilometres. It seems that only male fin whales produce very loud calls, perhaps to find and attract mates over these vast distances.[11] Given the spectacular decline in whale populations caused by two centuries of commercial whaling, it must be much harder now to find mates. Many whale populations have nowhere near recovered from whaling, so long-distance calls could be a vital way to reach out to receptive mates. Today, noise from ships and wind farms fills the sea with a low frequency rumble that masks their calls beyond distances of just ten kilometres. Because area scales as the square of distance, the space over which they can contact one another has shrunk 10,000 times.

Sarasota's dolphins are famous for more than just the chemical toxins in their bodies. They know one another by name. Each dolphin has a distinctive whistle that others recognize.[12] When boats pass by, the dolphins whistle to each other more often and move closer together, which suggests they worry about traffic. Several have been badly injured by boat strikes, so they have good cause. Since boats pass within a hundred metres of a dolphin roughly once every six minutes during daylight hours, they could interfere with the dolphins' ability to feed, breed and look after their young.[13] Another explanation for the more frequent calls when boats are about is that the deafening sound of engines makes it harder to stay in touch.

Fish exposed to boat noise or other loud sounds suffer stress. Smallmouth bass live in the lakes and streams of North America. The hearts of captive fish beat faster when they were exposed to the sounds of boats, and engine noise produced a stronger and more prolonged response than the sound of a paddled canoe. Several kinds of freshwater fish exposed to boat noise have been found to react by producing more cortisol.[14] This is the 'fight or flight' hormone that increases blood sugar levels. It is found in a remarkable array of species throughout the animal kingdom, from fish and birds to ourselves. Animals release it in response to acute danger to mobilize energy, sharpen thinking and dull pain. But they also produce it in response to longer-lasting stress. Wired executives and harassed mothers have raised cortisol levels. This is when benefits turn to costs. Cortisol can wreak havoc within our bodies. It raises blood pressure, increases risk of heart attacks and strokes, and suppresses the immune system. Fish and other animals in noisy, stressful places may also suffer health problems.

In an experiment that entertained me mightily, Greek researchers played Mozart to captive carp for hours every day over several months.[15] Fish grew faster in the presence of Mozart than with the usual background noise of pumps and filters. The researchers thought that Mozart might have soothed away their stress. Whether or not you believe this, I am sure the composer himself would have had a laugh.

Wind farms are a relatively new source of sound but, given the push for green energy, one that is set to become ubiquitous throughout large regions of shallow sea. During construction, the noise from pile-driving is loud enough to damage the hearing of dolphins or other marine mammals that stray within a few hundred metres. During operation, wind farms produce a low frequency underwater rumble, the volume of which depends on wind speed. These noise levels are probably similar to those produced by ships, but the difference is that they will affect the same area continuously. There has been very little research on how marine life might respond. On the one hand, it might render large areas of habitat unsuitable for sensitive animals. On the other hand, the footprints of wind farms often exclude fishing so might benefit some fish and shellfish. It is too soon to tell for sure.

Fish and many other forms of marine life use sounds to visualize the world around them. Think of the way that sound enriches your own appreciation of where you are. The tapping on the window tells you it is raining outside. The whirr from the kitchen lets you know the dishes will soon be clean. The sound of footsteps upstairs shows the kids are in their bedrooms, while the light patter from the corridor tells you to expect a hungry cat to come and chance its luck. We may not think of ourselves as being particularly aural creatures compared to bats or owls, but our ears give us a good sense of the world around us.

The noisy underwater world sets the scene for marine animals in much the same manner, but with one important difference. The far greater density of water means that sounds move particles in ways that can be felt. Fish have evolved the lateral line organ along their flanks to sense this particle displacement. Like us and other verte-brates, fish have ears with internal 'bones' made of crystalline calcium carbonate to hear sounds too. They share with us the ability to pick out specific sounds from background noise, just as we can hear the voice of a friend through the buzz of a party, or tell which instruments are being played in an orchestra. For a time, it was a mystery why goldfish, which don't make noises themselves, had such finely tuned hearing.[16] Trapped within their paradigms, scientists sometimes fail to notice the obvious. Clearly, there are benefits from listening to sounds other than your own voice!

The soundscape underwater has been called 'acoustic daylight' for the way it reveals things about the world that cannot easily be seen. Larvae of fish from coral reefs must escape the reef for life in open water, where they grow for a period of days or months before they are ready to return as fully formed juveniles.[17] If they cannot find a reef at this point, they die. Many larvae follow sounds they might have heard before they hatched from the egg to guide them back when the time is right – the whisper of breaking waves, the crackle and rasp of shrimp and crab, the scrape of parrotfish beak on coral. Even some coral larvae possess the ability to hear reef sounds.[18] It is easy for us to picture how animals make sense of these sounds, but particle motions from underwater sound are harder for us to comprehend. Different sources probably produce distinctive particle motions that paint the surrounding scene with details we can only imagine.

Most marine animals use sound to tell when predators are about: the whoosh of a school of baitfish fleeing the thunderclap of a predator's strike, the crack of a snail being crushed, the pistol shot of a snapping shrimp (actually their claws move so fast to smash prey they produce a sonic boom[19]). Some also call for mates, or use sounds to weigh the risk of a challenge for territory. Drums, as their name implies, are fish that form noisy spawning aggregations in estuaries, thumping out the rhythm of procreation on their swim-bladder instruments. In Puerto Rico, red hind groupers surprised researchers by shouting 'whoo hooo' as they rushed to spawn.[20] Background noise might make it harder for larvae like those of coral reef fish to follow sounds to the right habitat. Already, the number of offspring produced by many fish and invertebrates has fallen precipitously because of overfishing. Noise is yet one more way in which the viability of their populations may be compromised. Few other species have been tested to see how important noises are to navigation, but it seems unlikely that only coral reef animals have this ability so the problem could be widespread. There can be more immediately harmful consequences of background noise. Hermit crabs in the Caribbean, for instance, became less attentive to possible predators when they were distracted by the sound of passing boats.

The underwater world is now awash with noises that evolution has not prepared marine animals for. The maigre is a kind of drum, a lithe fish of metallic grey that lives in Mediterranean estuaries. In the days before intense fishing, it grew nearly as big as the average man, but these days few reach more than a metre. Maigres exposed to the noise of a powerboat in experiments were a hundred times less responsive to the sound of other maigres. Like right whales, the seascape over which they could communicate would shrink 10,000 times when boats pass nearby.

Male toadfish 'sing' by vibrating their swim bladders with 'sonic muscles' to entice choosy females to their nests and warn off rivals. Lusitanian toadfish are flat-headed with wide rubber-lips that make them look like their heads have been run over by something heavy. They live along coasts and estuaries of the east Atlantic and Mediterranean where males set up territories among rock piles from which they glower at one another. There they display to female fish with a buzz called a 'boatwhistle', though a better description would be an extravagant fart. The most appealing males persuade females to leave behind a hoard of perfectly round golden eggs for them to fertilize and protect. When captive toadfish were played the sound of ferryboats in a 2005 experiment, they found it hard to hear one another.[21] Someone has turned up the stereo at the party and now it is impossible to hold a conversation any more. Just as revellers can switch to sign language there may be other ways that animals can communicate, but at the expense of subtlety and with an increased risk of misinterpretation.

The Pacific coast of North America is home to a relative of the toadfish whose vocal dexterity throws light on the evolution of sound communication in backboned animals such as ourselves. Midshipmen are fish that look a bit like overgrown tadpoles, with round flat heads and narrow tails. Some are drab, but others blaze as if sprinkled with gold dust. Lines of glowing photophores decorate their bodies with points of light, like Christmas lights. These fish live in the intertidal zone where males set up mating territories in pools beneath rocks. From these muddy grottos they sing love songs to passing ladies and grunt warnings to other males. Their songs are short on melody and make

Tuvan throat songs seem operatic by comparison. Mostly they just buzz, like lawnmowers drifting through summer suburbs, sometimes interspersed with 'boinnngggs' reminiscent of twanged rubber bands. But, inside their heads, the part of the brain that controls their songs turns out to be the same part that frogs, reptiles, birds and people use, suggesting that the ability to 'speak' to one another evolved some 400 million years ago.[22] These fish can get so boisterous they keep houseboat owners in San Francisco Bay awake at night.[23] One resident described the sound as akin to the noise of a generator; another said it was like ten electric razors run at once.

The clutter of low frequency hum we create in the sea overlaps pretty closely with the hearing ranges of fish and whales. There are signs that some try to compensate. Northern right whales call louder when there is background boat noise.[24] The more noise there is, the louder they yell. They also call about an octave higher than they did in the 1950s, probably in an effort to be heard above the low frequency drone of engines. But there is a limit to how loud you can shout. As noise levels continue to rise, whales will become increasingly isolated. Since there are fewer than 400 northern right whales left, isolation could seriously threaten their survival. Alternatively, they might be able to compensate by calling during quiet periods or by making longer calls. Killer whales in three British Columbia pods made longer calls when surrounded by boats of whale watchers than when left in peace, which suggests that they struggle to communicate above the clamour.[25]

Birds try to compensate for noise in similar ways. Great tits in cities sing shorter, faster songs at a higher pitch than those in the countryside.[26] Birds next to busy roads often have lower breeding success and are less fit than birds in quiet places. Oven birds nesting in Canada close to a noisy generator couldn't get their messages across and attracted fewer mates than those further away from the generator.

Noise pollution interferes with feeding. If every time a boat bears down upon an animal it stops foraging, it may soon go hungry. It also affects species more directly if they use sound to hunt. Beaked whales hunt by echolocation in the deep sea. They produce a stream of clicks, made by blowing air through a structure in their blowpipe affectionately known as 'monkey lips'. The noise is beamed through enlarged lobes of acoustic fat in their heads, and they receive echoes back in the flesh beneath their jaws. Beaked whales are particularly fond of squid. Aside from a thin horny internal sheath, squid have a similar consistency to water, which makes them hard to detect. I imagine that you would need to listen closely to find a squid in the dark. A Cuvier's beaked whale recorded hunting when there was background noise had less than half the success others had with no sound interference.[27]

Some whales and dolphins use echolocation to find their way around. If their hearing is damaged they may be prone to stranding in shallow water, or entanglement in fishing nets. Experiments with captive bottlenose dolphins exposed to high-intensity sonar showed short-term hearing loss.[28] Longer exposure may cause permanent damage. When stranded animals rescued in North America and the Caribbean were tested, a third of rough-toothed dolphins and over half of bottlenose dolphins had levels of hearing loss that would have been judged severe or profound in humans.[29] Endangered northern right whales thread their way through some of the busiest shipping lanes on the planet as they migrate up America's east coast to summer feeding grounds. If noise impairs their hearing, they may fall victim more easily to a ship strike. This is now one of the two main causes of death for these whales, the other being entanglement in fishing gear. Even whales with normal hearing find it impossible to hear one another when ships are closer than a kilometre or two, which is the case much of the time.

Human noise in the oceans is a relatively recent phenomenon. For most of history, the only sounds of any note would have come from the clatter of oars against wood and the hiss and slap of boat hulls as they scythed through waves. That changed in the middle of the nineteenth century when the first paddle steamers were launched. But it was in the last two decades of the nineteenth century that the peace was broken in a major way, as tens of thousands of steamboats were built. The first offshore oil wells were drilled off Santa Barbara in California in the 1890s, starting a raucous rush for oil that shows no sign of letting up. Since then, background noise has grown inexorably, tracking global economic growth. Eighty per cent of goods traded internationally are now shipped by sea. As trade volumes increase, so does noise.[30]

It would be a mistake to suggest that all creatures in the sea are adversely affected by noise. Some like to gather beneath floating objects, even if they are noisy ships. And although sounds travel much further in the sea than on land, away from shipping lanes, ports, wind farms and oil and gas fields, there are still plenty of peaceful places in the oceans. But it is hard to escape the conclusion that the rising clamour is but one more stress piled onto life in the sea. If nothing changes in the way we build or run ships, the din beneath the sea will only get louder.

The great majority of noise in the sea is from boats, and there seems little prospect of hushing them for the sake of marine life alone. Motives for developing quieter boats are more likely to be fuel economy or comfort of crew and passengers. There are simple ways to reduce ship noise. The loudest sounds come from propeller cavitation. As a propeller spins it creates a low pressure zone behind each blade. The faster it rotates the lower the pressure, until a point is reached where a gap opens up between the water and the blade. As these 'cavities' spin off they slam shut with an audible slap. When propellers turn at low speed there is no cavitation, so boats run more quietly. Therefore a simple way to reduce noise is just to slow down. That is exactly what the world's merchant shipping fleet has done in recent years to combat the soaring cost of fuel, because cruising slowly is also cheaper. A 20 per cent cut in speed can save 40 per cent in fuel costs. In fact, many shipping lines have dropped steaming speeds from twenty-five to twelve knots, about the speed reached by sailing clippers, the workhorses of international trade and communication in the late nineteenth century.

Slower ship speeds have also been prompted by a growing awareness of the greenhouse gas emissions by the global fleet. The biggest ships emit more carbon dioxide pollution than the world's smallest countries, and collectively the world's merchant fleet releases about 4.5 per cent of global carbon dioxide pollution. Critics argue that speeds will rise again as soon as the price of oil goes down, but there are ways to reduce both noise and fuel emissions through better design. The military have developed propellers that minimize cavitation at high speeds to make ships harder to detect, and cruise ships have also found ways to minimize noise. There is no reason why the merchant fleet shouldn't adopt similar designs, and many ships could be retrofitted. Ships could also reduce the need for noisy engines and fuel by using kites or skysails to catch the wind, which has the added benefit of being free.

The world's shipping fleet is governed by the International Maritime Organization, which has overseen a huge improvement in the safety of oil shipments over the last forty years. Oil spillage has fallen considerably as a result of the imposition of double hulls, the compartmentalization of oil in the hold, pinpoint satellite navigation and a raft of better sailing practices. It is high time for them to tackle the twin problems of noise and fuel efficiency.

12

Aliens, Invaders and the Homogenization of Life

Charles Darwin and Alfred Russel Wallace discovered the principles of evolution by natural selection around the same time, each working entirely alone, by obsessing over patterns in the distribution of species. As both of these astute naturalists travelled the world, they noticed that some species came and went, while the characteristics of others changed from place to place. Eventually it became clear to them that species were not fixed by some godly act of creation. Rather, life's forms were constantly reworked by evolution. It is Darwin we celebrate for this theory, but it was Wallace who spurred him to publish it when he later came upon the same idea. Geographic history has left some places with more species than others, and some with more unique species. Other places, by virtue of environmental stress, isolation, or lack of habitat, are less rich.

The geographic ranges of species are circumscribed by many factors, most notably physical barriers such as landmasses, currents or insurmountable gaps in suitable habitat, like ocean basins. When natural barriers divide regions, as the Isthmus of Panama did when it separated the Atlantic from the Pacific 3.5 million years ago, evolution may take life in different directions.[1] Over great stretches of time new species arise and others go extinct. Some natural barriers to the dispersal of species rise or fall slowly, others more quickly in relative terms, like shifts in current patterns, or the appearance of a new island from a volcano under the sea.

Humankind has lowered or entirely removed these barriers. For the last several thousand years, but especially the last several hundred, we have made long-distance voyages with animal and plant passengers in our cargo and attached to our ships' hulls or carried in ballast. These journeys opened up new areas for colonization. The distances species could be carried underwent a step change in the fifteenth and sixteenth centuries as the voyages of Dias, da Gama, Magellan, Columbus and others began the age of global exploration by sea. When Columbus sailed to America, he unwittingly took a bunch of freeloaders with him. In the Indian and Pacific oceans, early Arab and Chinese voyages would also have carried species long distances.

Getting there is only the first hurdle for a would-be colonist. Conditions must be right for survival and a sufficient concentration of colonists have to be introduced to establish a self-sustaining foothold in the new territory. In centuries past, ships often anchored or tied up for many weeks at a time on foreign shores (as opposed to only hours or days today), allowing ample time for attached fouling organisms to shed eggs and larvae. In addition, vessels had their bottoms scraped every few months to keep them clean, and countless such voyagers would find themselves marooned in alien waters. Plants and animals could also be carried from place to place in water that sloshed around the ships' bilges, or attached to rocks used as ballast to increase the draft of lightly laden ships. Sailors carried rocks (together with their hitchhikers) from the Old World to the New, only to dump them on arrival to load cargo. The genes of introduced periwinkles and serrated wrack seaweeds in Nova Scotia give away origins in Scotland and Ireland, from whence most of the ship traffic arrived.[2] Ballast became an even more significant means of emigration from the late nineteenth century onwards, as large vessels began to take in water as ballast in one port and discharge it in another. Ballast water now ranks as the foremost means by which marine species are introduced to new environments. Many marine organisms begin their lives as tiny eggs or larvae that can be transported in ballast. Even big animals can get around in ballast – fish 30cm long have been found in ballast tanks.

We call displaced species 'alien' or 'non-indigenous' to distinguish them from the natives. Some become troublemakers. These 'invasive' species run wild, freed from the control of their natural predators, parasites and competitors in their native haunts. Invasive species are like nature on steroids. They often multiply with extraordinary rapidity and overconsume or out-compete indigenous species for space, food and other vital resources. The results can devastate ecosystems.

Such invaders can eradicate local species and reduce other once-dominant life forms to the role of bit players. One such invasion, underway in the Caribbean right now, will change reef life there for ever.

Red lionfish are natives of the tropical Western Pacific and are popular aquarium fish. They have broad pectoral and dorsal fins that they hold open like delicately patterned Chinese fans. Their beguiling beauty is deceptive for they are ruthless predators. They use their fins to usher small fish, ever so gently, into awkward corners of a reef, where they dispatch them with mouths that expand in an instant to engulf their prey. In the late twentieth century, enough lionfish were released or escaped from aquaria in south Florida that a breeding population became established.[3] Lionfish have few Caribbean predators. Their lacy fins conceal sharp spines that deliver a potent shot of venom to the unwary. The Caribbean was ripe for invasion.

Early this century, surprised divers began to report sightings of lionfish off the coast of North Carolina. Reports followed from the Bahamas, Turks and Caicos, and Bermuda. The Caribbean has not seen anything like the lionfish for millions of years, if ever. Never having encountered such a belligerent and aggressively defended predator, the local fish don't know what to make of them and show little fear. Consequently, lionfish are slurping them up by the millions. Indeed, speared lionfish have been brought up with their bellies packed so tightly they seem scarcely able to move. With food so easy to come by, these lionfish are growing faster and larger than they do in their native waters, and, with their populations increasing exponentially, they are laying waste to Caribbean reef fish.[4]

In less than twenty years, lionfish have exploded across the Caribbean. They can now be found from Rhode Island south to Colombia, and from Belize east to the Windward Islands, and may ultimately range along the entire coast of Brazil. Although divers and conservationists soon rallied to challenge the threat, their efforts are futile, except for divers keeping the invader off specific reefs by means of frequent removals. Lionfish are there to stay. Mark Hixon, a biologist at Oregon State University who has studied the lionfish in the Bahamas and Caymans, thinks 'this could well become the most devastating marine invasion in history'.[5] Although lionfish are not large – the biggest invader has been recorded at just under half a metre – they target small fish and invertebrates that include the young of large species. Such animals not only are valuable as food for people, like groupers and snappers, but they also provide important ecological services, such as parrotfish that graze seaweeds that would otherwise overgrow corals. Imagine if some predatory beast invaded Europe and began to slay millions of birds, to snatch dogs in parks off their leads, and to indiscriminately kill cats in gardens and deer in the country. To imagine this is to think of what life has become on Caribbean reefs. The lionfish are particularly ruthless, but they are only one of many pernicious invasive species that have begun to cause mayhem in marine ecosystems. One small window of hope may be that lionfish are becoming popular menu items in restaurants in the US and the Caribbean, but whether harvesting this edible fish will suffice for control remains to be seen.

Ships have long been the most convenient route to a new home. At the last count, over 50,000 vessels grind their way around the world carrying goods.[6] They move about twelve cubic kilometres of ballast water from place to place every year, and at any moment some 10,000 species are thought to be in transit.[7] Being slopped around inside the dark belly of a ship where temperatures can swing widely and food is at a premium is not the surest way to survive a long journey. But it does select for the most tenacious and resilient, characteristics that serve alien species well in new places. Estuaries, which support many of the largest ports, have become hotspots for invasions, and are places from where animals and plants may spread like plague. Chinese mitten crabs and Japanese seaweeds are invading European and North American seas, while Asian kelp is spreading in California. In Germany's Elbe River, where mitten crabs have been established for a century, at certain times of year writhing legions of crabs swarm downriver like a locust plague on their way to spawn. A West Atlantic comb jelly called *Mnemiopsis* devastated planktonic life and fisheries in the Black Sea after it invaded in the 1980s, and it has now reached the Mediterranean, Baltic and North seas.[8] European green crabs are crunching their way through naïve intertidal natives on North America's west coast and the coasts of Japan, South Africa and Argentina. Throughout North America, dense stands of phragmites reeds have begun to block coastal views and choke out native wetland plants. Intriguingly, the reed is native to North America, so scientists were puzzled as to why it had turned bad. Genetic analysis gave them their answer.[9] A variety introduced from Europe is causing the nuisance and in New England has entirely replaced the native variety.

San Francisco Bay is one of the most invaded ports in the world.[10] Over the last 140 years, nearly 300[11] alien species have settled in the bay and its rivers, and the rate of arrival has increased. From 1851 to 1960, a new species arrived just over once a year on average, but between 1961 and 1995, there was a new arrival every fourteen weeks. These new species have transformed the bay, much as San Francisco itself has become a cultural cornucopia. Hardly any corner has less than a third of its species count made up of aliens, and in many places they have complete control, having ousted all the natives. Atlantic smooth cordgrass has formed dense stands over areas of open mud that once were rich feeding grounds for birds. Beneath the tranquil surface, three kinds of gribbles, tiny isopod crustaceans related to woodlice, are busy boring their way through wooden docks and pilings. In Seattle, they have inflicted $700 million worth of damage.[12] Clam diggers around San Francisco Bay are now much more likely to unearth Asian clams than native ones. So enthusiastic is the feeding of these clams that they are sucking the water clean of plankton, which could deprive other species of essential food and cause declines of animals whose eggs and larvae drift in the bay. The rise in ship traffic is part of the reason why San Francisco has been so comprehensively invaded, but there are other ways to get from place to place. Aquaculture, the farming of marine life for our consumption, is a major source. Rather than farm local species, we like to move the fish and shellfish around that we enjoy most and are best able to raise. In the 1800s, vast numbers of eastern oysters were transplanted from the Atlantic to the Pacific coast of North America, and the same happened in the 1900s with Japanese oysters. Atlantic salmon are grown along cool stretches of the Pacific coast of North and South America. East coast oysters are cultured on the west coast. California's red abalone is raised in Chile. But it isn't just the cultured species themselves that get moved: along with the oysters came scores of unwanted hitchhikers, many of which are now abundant in San Francisco Bay.

Fish feeds can also bring unintended guests, and are often responsible for outbreaks of parasites and diseases that I will come to in the next chapter. The expansion of aquaculture across the globe has been extremely rapid and will continue, which means this is a problem that will run and run.

Marsh grasses, like the Atlantic smooth cordgrass, were introduced to San Francisco Bay intentionally by the US Army Corps of Engineers to 'restore' wetlands and prevent erosion. It was spectacularly successful, but now state officials want it out and have launched an eradication campaign. Perhaps as much as 6 per cent of marine invasions worldwide have come from unintentional introductions from aquaria. After lionfish, the most notorious is a tropical seaweed, *Caulerpa taxifolia*, a dark-green weed that scrambles over the seabed in mats of twisted stems and has been dubbed the 'killer algae'.[13] *Caulerpa* doesn't normally grow in cool places like the Mediterranean, but keepers at the Monaco Aquarium[14] heard of a strain that German aquarists had discovered could survive in similar conditions and imported it for their own tanks. Almost immediately, *Caulerpa* escaped into the Ligurian Sea through discharge pipes from the circulating seawater system.

In 1984, divers noticed a patch of the strange new weed, smaller than a tablecloth, growing wild.[15] Curiously, it was not removed. Five years later, it covered a hectare of seabed. Ten years after its release people knew they had a serious problem. More than fifteen square kilometres of the bottom were blanketed in the weed, which blocked the light and suffocated all beneath; where meadows of seagrass once waved in sunlight, now there was only *Caulerpa*, like a cancerous growth. And, like cancer, this invasion soon metastasized. Fishing boats picked up the weed in their nets and pleasure craft snagged it in their anchors. Soon it flourished in hundreds of new places. Just fifteen years after its release, *Caulerpa* covered 97 per cent of available surfaces between Toulon in France and Genoa in Italy, a distance of nearly 300 kilometres. It isn't over yet. Much more of the Mediterranean could soon fall victim to the blight. The alga has now secured bridgeheads in Spain, Croatia and Tunisia.

Caulerpa racemosa is a related species that is also highly invasive and has settled in the Mediterranean. It was first reported in 1990 and, if anything, is an even more vigorous competitor for light and space than its namesake. It even seems to be out-competing it! Within six months of arrival, this plant had gained complete dominance in some places, indiscriminately laying waste to seagrass beds, seaweeds, sponges and all manner of other creatures that make a living attached to the bottom of the sea. *Caulerpa racemosa* came all the way from southwest Australia and has now spread from Cyprus to Spain and beyond to the Canary Islands.

Like many invasive species, both these algae have far-reaching effects within the communities they have entered. In the space of just a few decades, their arrival has brushed aside relationships formed among Mediterranean plants and animals over tens of thousands of years. Seagrass beds are rich in fish, crustaceans, starfish, urchins and hundreds of other species that call them home or use these meadows for a time before moving on. Green turtles graze contentedly on the flexible green blades beneath shafts of light that dance and dapple over the lush pasture. They aren't keen on *Caulerpa*. Most fish also fare badly where *Caulerpa* mats occlude the bottom. The hunters can no longer reach invertebrates living in the sand. Those that once sheltered among blades of grass have found their lives brought to a standstill as the thickets close around them, much as the palace and its dozing inhabitants were overgrown by a dense tangle of briar in the fairy tale, *Sleeping Beauty*.

So what factors facilitate a successful invasion? The bad news is that many of the problems highlighted elsewhere in this book increase the chances that invaders will establish and make trouble. It turns out that, where overexploitation or pollution have depleted local life, competition for space and other resources is less intense and invaders can establish a beachhead from which to spread. Some pollutants may also render conditions more amenable for invaders. Nutrient pollution in the Black Sea, for instance, is thought to have favoured the establishment of the *Mnemiopsis* comb jelly. The simplification of marine ecosystems by multiple stresses opens up opportunities for invaders. And the creation of structures such as seawalls, rigs and wind farms can facilitate the spread of invasive species that require hard sea bottom across expanses of otherwise unfavourable habitat.

One of the great paradoxes of invasive species is that they can spring to abundant life from just one or a handful of individuals, like a forest fire caught by a lone spark. Yet we struggle so hard to keep life's spark alive when endangered species have been reduced to the last few dozen individuals. What makes invaders different is freedom, opportunity and, of course, luck. We forget that some invasive species may have failed many times to settle successfully before they eventually do. We only see the winners. When a species is plucked from its homeland and thrust into a new world, it can find itself reborn, freed from the shackles of traditional enemies. The burden of disease and parasites may be lifted. Likewise, predators and competitors can be thin on the ground for a stranger. Life is good.

The contrast between old haunts and new can sometimes create conservation ironies. Sea lamprey invaded North America's Great Lakes after a canal was opened around the natural barrier of Niagara Falls in 1919. These primitive fish are slippery, slim as eels and mottled like leopards. They have a ring of conical teeth around a jawless mouth with which they attach themselves to bigger animals. Lampreys rasp holes in the flesh of their hosts, leaving their flanks pocked with bleeding sores. So many now live in Lake Superior that there is hardly a big fish left that has not been attacked. For lamprey, the Great Lakes have been one long lunch. Back in their native Europe, sea lamprey numbers have plunged in the last half century, a decline usually blamed on poor water quality in the rivers they use to spawn. But I suspect the real cause is loss by overfishing of the big animals they use as hosts while at sea. In Europe the sea lamprey is protected; in the Great Lakes it is vermin.

Of course, there are qualities that predispose a species to be a successful invader. Weedy species the world over have been crafted by evolution into skilled opportunists. They often have resistant seeds or spores that can wait as long as it takes for the right moment to germinate. It helps if you grow fast, produce lots of offspring quickly and don't mind mating with relatives. It is also good not to be too picky about where you live and to enjoy novelty and variety in your diet. I wouldn't have singled out lionfish as likely invaders of the Caribbean from my encounters with them in the Red Sea. They seemed too shy and delicate and were never particularly common. Surrounded by animals that know well to avoid their dangerous beauty, they have to work hard for a meal. There was no hint of the profligate gluttons they would become in the all-you-can-eat buffet of the Caribbean.

Few places on this planet have not been reached by some marine invader or other. Not surprisingly, some of the worst-affected places have the highest ship traffic. The US west coast, Northern Europe and the Mediterranean are all hotspots of invasion. But some places are more vulnerable than others. Hawaiian seas have been invaded time and time again.[16] Like the islands above them, Hawaii's seas are rich in species found nowhere else. On land, invaders have devastated plants and animals ill-equipped to deal with aggressive conquerors. In the sea around Oahu, many areas are now dominated by invaders: mangroves from Florida line Pearl Harbor; one of the most common shore barnacles is newly arrived from the Caribbean; and blacktail snappers from Moorea and bluestripe snappers from the Marquesas are among the most abundant large fish in shallow waters. Over 300 alien plants and animals are now known from Hawaiian seas, and these numbers do not take into consideration potentially hundreds more that may be alien but whose histories are less well known. On the face of it, Asian seas have been less affected by aliens, with reported numbers less than a fifth the number of new arrivals into the US west coast and Hawaii. But the long-term presence of Southeast Asia at the epicentre of trade in consumer goods tells us to distrust these statistics. Asian waters are probably crawling with invasive species, but surprisingly few people have cared to look so far. A Caribbean variety of zebra mussel, the black striped mussel, has established at high densities all around Singapore and part of Malaysia, for example.[17]

The least invaded seas in the world are at the poles. At the time of writing there was just one marine invader known in the Southern Ocean and nine in the Arctic. The poles have been spared for several reasons. Not many people live there so boat traffic is sparse, and polar conditions are extreme and few species from anywhere but the opposite pole could cope with them (and it is hardly likely that they could find a way between the poles). Times are changing though, and it is very probable that polar seas will soon find themselves home to many more alien species. In the Arctic Ocean, loss of sea ice due to climate change has led to the mixing of seawater from the North

Pacific with the North Atlantic for the first time in 800,000 years. We know this from the arrival in the Atlantic of a species of North Pacific planktonic diatom that has been absent this long.[18] A far bigger Pacific traveller reached the Mediterranean in 2010 where astonished Israeli scientists spotted a grey whale.[19] The last Atlantic grey whales were slaughtered by New England whalers in the seventeenth or eighteenth century. This one must have lost its way and inadvertently navigated the newly ice-free route to the Atlantic.

Global climate change is warming polar seas enough to make life there possible for species from lower latitudes. On a changing planet, these range expansions will come to make up a large fraction of new introductions. They might also gain a helping hand from us. The prospect of an ice-free Arctic in summer has once more excited great interest in a north-west passage for ships between the Atlantic and Pacific. Dozens of explorers lost their lives between the sixteenth and nineteenth centuries in search of this elusive route, their efforts doomed by savage ice. As the lost grey whale shows, it might soon be a reality, and without great care ships could take many new species into the Arctic and provide a fast-track route between the Atlantic and Pacific for countless others. The increasing quantities of floating plastic and fishing debris to which fouling species can attach themselves may also carry aliens to polar seas, although there is little firm evidence yet.[20] Ocean garbage has multiplied the means of hopping from place to place in the sea, and can go where few ships travel.

Not all alien species cause problems. Some may blend into their new homes with little fuss. They add colour and variety, like a new picture hung on the living-room wall. But significant numbers, perhaps the majority, are harmful, as judged by their impacts on native wildlife or to human interests. An estimate made in the year 2000 suggests that invasive species cost us $1.4 trillion every year in trying to undo the damage caused, eradicate the invaders, or in economic losses sustained from the degradation of ecosystem services such as clean water provision or soil stabilization.[21] Those troublesome Chinese mitten crabs have so filled Germany's levees with burrows that they are crumbling. The vast majority of this damages estimate can be attributed to terrestrial invaders. So far little attention has been given to alien species in the sea, but we can be certain that the costs of aggressive new colonizers, like Seattle's gribbles, are quietly building beneath the waves.

While much of the attention around invasive species has concentrated on costs, some create new opportunities. When the French engineer Ferdinand de Lesseps achieved his life's goal to join the Mediterranean and Red Sea by digging a canal between them, he managed what countless others had dreamed of since pharaonic times. The Suez Canal runs at sea level. As soon as it opened in 1869, water began to flow between seas that had been parted for over 11 million years, and has carried tide upon tide of emigrants. Since Mediterranean Sea levels are lower than those of the Red Sea by 1.2m, most of the flow has been northward, and invasions were much facilitated by the lower biodiversity of the Mediterranean. To begin with, the invaders' path to virgin territory was blocked by a series of briny lakes – the Bitter Lakes – in the middle of the canal. Over the course of a century, flushed by seawater, the lakes became less inhospitable. The rate of colonization of the Mediterranean increased from drip, to trickle, to steady stream.

Early colonists burst forth into a place with few predators or competitors to challenge them. The Eastern Mediterranean had long been thought of as impoverished, with low productivity and few species compared to the richer west. The contrast from north to south through the canal was even greater. The Red Sea hosts an extravagant diversity of life that thrums around vibrant coral reefs. While few corals can tolerate the cool Mediterranean, many reef inhabitants were more cosmopolitan and soon settled in. Several of them now sustain valuable commercial fisheries in Israel, Egypt and along the North African coast as far as Tunisia.[22] They include rabbitfish, lizardfish, goatfish (only the lizardfish looks anything like its terrestrial counterpart; imagination seems often to have deserted the namers of fish) and a clutch of prawns, three of which were fat and juicy enough to be welcomed.

It is premature to say this without fear of contradiction, but places with few species seem to be easier to invade than richer communities.[23] The idea is that resources are less fully used so would-be invaders can gain a foothold more readily. According to this logic, the Eastern Mediterranean was ripe for the taking. The problem is, by extending this logic one has to conclude that places where human or other impacts have picked off the natives will be highly vulnerable to alien invaders. Mass mortality of corals and sponges from disease and heat stress has left vast areas of free space, for instance. Given the ubiquity of human sources of stress and their multiplicity of effects, it is a near certainty that most ecosystems in the sea have lost at least some of their resistance to invaders. Their defences are down just when they need them most. The onslaught has only just begun.

In some ways, the changes to nature that we have unleashed through travel and trade mirror the effects of globalization on human culture. Some cultural influences are like aggressive invaders that, within the space of half a century, have subsumed and erased cultures that flourished for millennia in isolation. Explorers and filmmakers today struggle for days or weeks through sweltering jungle, across torrents and treacherous ravines only to find on arrival people sporting ragged T-shirts emblazoned with Manchester United or Nike. So it is that in countryside and town, lake and ocean, we find familiar faces in unusual places. What we often miss are the plants and animals they have displaced that in the past made these environments exotic and strange.

Invasive species are one of the five horsemen of the apocalyptic loss of biodiversity that is sweeping the planet (the others are habitat loss, pollution, climate change and overkill by hunting or fishing). Yet, strangely, invasives have been neglected relative to the others, particularly at sea. Although they have purged habitats and ecosystems of their rightful owners – just as with habitat loss, pollution or climate change – they do not yet seem to have caused many outright extinctions in the sea.[24] Native species often hang on in pockets of poor quality habitat, just as peoples quashed by conquering armies find themselves marginalized and browbeaten. In the Mediterranean, native mullet have been pushed deeper by invasive goatfish. But local population crashes can eventually lead to disappearance, and there are many scientists who believe that invasive species in the sea may yet drive natives to extinction given enough time. The first may be closer than we think.

The Derwent River estuary of Tasmania is home to a little fish with the bizarre habit of walking on its fins. Its pectorals – the fins on either side of the body behind the gills – look like arms and each ends in a webbed 'hand', hence its name: the handfish. Spotted handfish, as this one is called, have flesh-coloured bodies speckled red, fins edged in lemon and dark eyes lined with gold. They display to one another in a semaphore of fin flashes and showy standoffs. Nobody knows why they are found here and nowhere else, but such a limited geographic range means local population loss and global extinction are one and the same. In the 1990s, their universe was turned upside down by the arrival of the North Pacific starfish, an undiscriminating predator from Japan and East Asia whose numbers have since swelled to millions. Among their predilections are spotted handfish eggs, laid in nests on the seabed. There could be as few as a couple of hundred to a couple of thousand spotted handfish left in the wild, but it is hard to know since they are difficult to count. While all species have a right to exist, the loss of one so charming seems especially heartbreaking.

Geologists recognize distinct epochs in Earth history according to the fossils and rocks present, and give them names. The most recent, the Quaternary, covers the last 2.6 million years and is subdivided into the Pleistocene and Holocene. The Pleistocene spans the time of the ice ages, while the Holocene covers the 12,000 years from the end of the last ice age to the present. Some scientists with a humorous bent have suggested that the present era should be called the 'Homogecene' because it is a time of unprecedented mixing of animals and plants from one part of the world to another. While efforts are being made to control the spread of invasive species, I wonder about their efficacy in the long term. This is an uncontrolled experiment with life. Who knows what will come of it?

When I was a student in the 1980s, my professors made much of the problems of alien species and how they could lay waste to the native fauna and flora in their new homes. Rabbits ate their way through Australia, water hyacinths choked lakes from shore to horizon, killer bees swarmed into the southern US, and Nile perch chomped their way through an irreplaceable evolutionary showcase of brightly coloured African cichlid fish. The solution, in those days, usually involved introducing something else, following the logic of the old nursery rhyme about the woman who swallowed a fly and then a spider to catch the fly, and a bird to catch the spider, and so on, until she had eaten her way up the food chain. Like this hapless lady, who dies in the end as you may recall, the risk is that further introductions will do more harm than good.

There was a dramatic case of such an effect in French Polynesia. Giant African land snails were accidentally introduced and began to damage crops, so a voracious predator was brought from its homeland to control it. This snail, called the rosy wolf snail, developed a taste instead for the colourful tree snails that lived deep in moist valleys. Every valley had its own species of tree snail that had evolved in isolation, separated from the others by craggy mountain ridges they could not cross. The rosy wolf snail soon jumped these barriers to munch its way through an evolutionary cornucopia whose remnants now hang on only in zoos scattered across the world. As far as I know, there haven't yet been any tit-for-tat introductions in the sea, so dreadful mistakes like this haven't been made, but the temptation is there.

Introduced species – biological pollution – is one of the hardest kinds to clean up. People have spent lavishly on efforts to rid tiny scraps of land from the scourge of invasive rats so as to spare rare seabirds or unique species like lizards. The biggest islands on which they have been successful to date are little more than a few square kilometres in area, and they are unlikely to get much bigger. If you can catch an invader early and hit it hard, success is possible.[25] The black striped mussel, a native of the Caribbean now established in Singapore and Fiji, was detected in Darwin Harbour in Australia in 1999. Within nine days, the harbour was quarantined and pumped full of bleach and copper sulphate, which killed pretty much everything, including the mussel. In California, a parasitic worm of the commercially valuable abalone was introduced with animals brought from South Africa. It established at Cayucos and was successfully eradicated by hand removal of infected abalone and a reduction in overall abalone densities below the level needed to sustain the parasite. This was easy to do for an animal that lives in shallow water stuck to rocks and is the coveted target of skilled fishermen. Once an invader has spread beyond a few sites, however, it is pretty much hopeless to try to eliminate it. Local control is the only option, such as that by the enthusiastic bands of spearfishers attempting to rid Caribbean reefs of lionfish, and the chefs who encourage us to eat them.

Until recently, it was almost heresy in conservation circles to admit that introduced species were here to stay and should be treated as valid members of their new ecosystems. They are persecuted and discriminated against and treated as the most reprehensible forms of life. In the UK, some conservationists still treat species with contempt that arrived 500 years ago! When do you draw the line? When I was small I loved beech trees, with their muscular grey trunks, smooth coppery leaves and funny triangular nuts. It was only years later that I realized what a fool I had been to be taken in by its shallow appeal. This tree was an alien! An impostor that had only made the hop from mainland Europe in the Middle Ages. I am joking of course; I still love beeches. People mourn the loss of the familiar, but there are good reasons to resist the blending of the world's fauna and flora as the planet shrinks beneath our globalizing influence. This mixing represents a real loss of diversity and variety, and sometimes it takes generations to realize how much intruders have altered their new environments. When I look at a picture of a coral reef, I can tell from the species present where the photo was taken, give or take a few thousand kilometres. Each place has its own distinctive collection of species.

Even if only a small fraction of introduced species cause trouble, their effects on native ecosystems can be hugely disruptive and expensive. Like pollution by toxic chemicals or plastic, the best means of control is prevention. Since the two main vectors for the spread of invasive species are ships and aquaculture, the onus is on these industries to deal with the problem. One simple solution for aquaculture would be to stop raising species outside their native ranges. With nearly 400 species in culture right now, that shouldn't restrict options much, and would benefit consumers by giving them a wider range of ingredients to choose from. Ships could solve the problem of stowaways relatively easily by installing purification systems that use spare heat from the engines or UV light to kill unwanted passengers. Another simple measure that can reduce harm is exchanging ballast water offshore rather than in coastal seas. The idea is to avoid exchanging coastal species since these are the ones that cause most harm. Jim Carlton from Williams College in Connecticut thinks he has seen fewer invasions of US seas due to ballast water since open-ocean exchange became mandatory in 2004. Similarly, he has heard far fewer reports from Australia or New Zealand, where similar regulations were introduced a decade ago.

On the outside of boats, the usual approach to invasive prevention has been to slather toxic chemicals over the hull, but some of these, like tributyltin, proved terribly toxic to other species. An alternative approach involves coating the hull with a non-toxic layer of vinyl resin reinforced with glass[26] which reduces drag and limits the ability of fouling species to get a grip. It can be maintained by high-pressure tools that kill any species that do manage to establish themselves on the ship. Like agreements on toxic chemicals, these measures can only succeed with international regulation.

13

Pestilence and Plague

Soon after North America slammed into South America and the Isthmus of Panama blocked the exchange of water, animals and plants between the tropical Atlantic and eastern Pacific, a wave of extinctions convulsed the Caribbean and wiped out three-quarters of all coral species and many other forms of life. Evolution has since gone its own way in each region and the two oceans now have few species in common. The result is that Caribbean reefs look different from those everywhere else. One feature that distinguishes them is their great abundance of seafans.

Seafans are corals with a flexible horny skeleton over which is stretched a thin layer of tissue sprinkled with tiny polyps. These polyps are like miniature sea anemones. By day they are mostly retracted but at night they expand to catch drifting food with a crown of stinging tentacles. As their name implies, seafans are flat and fan-shaped, with a maze-like filigree of minor branches stretched between upright ribs. They are almost ubiquitous on Caribbean reefs, their supple forms bending to the movement of wave and current. One day in 1995, on a dive in Puerto Rico, I came across a seafan that didn't look well. The tissue between its upright ribs was blotched with dark purple stains and several ribs stood bare where parts had died. It was the first of many such seafans I saw.

Soon after I got home I was contacted by another scientist who had spotted sick seafans on reefs in Curaçao, some 800 kilometres to the south. At the time, I was based at the University of the Virgin Islands in St Thomas, and he was keen to know if I had come across anything like it. By the year's end, reports from other countries confirmed that the Caribbean was in the midst of a full-scale epidemic. It would reduce the population of seafans by as much as 75 per cent in the next few years.[1] So what was killing these seafans and why did it spread so quickly? The pathogen turned out to be a land-living fungus, and so attention swung rapidly to the red soil bleeding onto reefs from deforested island landscapes. But others thought the source could have come from somewhere far more unexpected – Africa.

In their idle moments, shuttle astronauts sometimes contemplate the Earth's great convulsions as they unfold in graceful slow motion far below. The fury of a hurricane is reduced to a twist of bright cloud; the flood of a mighty river a delicate discolouration of blue sea; dust whipped by desert winds a fan of brown streaks above the ocean. It is hard at first to accept that a Saharan storm might affect Caribbean reefs but jetstreams of air far above the sea carry millions of tonnes of African dust across the Atlantic every year. When I lived in St Thomas, sometimes the rain would fall in dirty red drops that left a thin layer of Africa over everything it touched. Incredible as it may seem, the island of Bermuda owes its red soils to thousands of years of such atmospheric dustfall. Charles Darwin remarked on African dust near the Cape Verde islands when he stopped there on the *Beagle*:

> The dust falls in such quantities as to dirty everything on board, and to hurt people's eyes; vessels even have run on shore owing to the obscurity of the atmosphere. It has often fallen on ships when several hundred, and even more than a thousand miles from the coast of Africa, and at points sixteen hundred miles distant in a north and south direction.[2]

As Darwin soon discovered, this windblown dust carried the spores of land plants and fungi. Although it is difficult to prove conclusively, some scientists believe that African dustfall was responsible for infecting Caribbean seafans with the fungus that started the epidemic.[3]

Seafans are not the only organisms to have been felled by disease on Caribbean reefs. In the 1980s, waves of disease wiped out nearly all the staghorn and elkhorn corals, both signature species for the Caribbean and ubiquitous in underwater pictures and resort brochures from the 1950s to 1970s. Elkhorn coral formed dense arbours of robust branches as thick as human limbs whose tips broke the surface at the lowest tides. As the bottom fell away to the seaward side of these ramparts, they blended into bushy thickets of more delicate staghorn coral whose branches sheltered flamboyant troupes of shimmering fish. The cause of the elkhorn epidemic was later discovered to be a human gut bacterium, carried in sewage effluent.[4] This outbreak was one of a collection of pathogenic diseases that has contributed to the loss of 80 per cent of living coral cover in the Caribbean since 1977.[5] Another disease destroyed almost every long-spined sea urchin in the Caribbean. Before the epidemic, waving forests of sea urchin spines blackened the bottom and often obscured the corals below. Their loss left reefs vulnerable to rampant seaweed overgrowth.

As I mentioned in an earlier chapter, green turtles around the world have in recent decades blossomed grotesque disfiguring viral tumours that bulge from armpits and groin, face and throat. Badly afflicted animals even have tumours that overgrow their eyes and mouth. This is just what you can see. They often have more tumours in their stomachs, throats, lungs, kidneys – pretty much anywhere there is soft tissue. The disease is caused by a papilloma virus and was first reported in the 1930s, but it has reached epidemic proportions since the 1990s in densely populated areas such as Hawaii, Florida and Barbados.[6] Since then it has been seen in other turtle species, such as the critically endangered leatherback, but green turtles remain worst affected. This disease, as I said, appears to be linked to tumour-promoting chemicals released by harmful algal blooms, which in turn are worsened by coastal pollution.

Other marine habitats have also experienced disease epidemics. Since the 1930s, a wasting disease has destroyed seagrass meadows from the Gulf of Maine to Florida. Wild and cultivated oysters in Europe succumbed to disease epidemics in the early to mid-twentieth century. In the 1980s, harbour and grey seals in the North Sea were decimated by a distemper virus similar to one carried by dogs, and pilchards in Australia were wiped out by a virus in the 1990s.

Diseases certainly seem more visible today than in the past, but is that just because there are more people looking now? The problem is that we lack a baseline from the distant past against which to measure the prevalence of diseases and parasites today. Although there were reports of disease outbreaks in commercial sponges and oysters in the nineteenth century that destroyed fisheries, people didn't tend to get overexcited about marine diseases because they couldn't see them. If birds fall from the sky, dying dogs litter the streets or stands of forest trees wither, they are obvious and worthy of attention. But animals and plants in the sea can sicken and die by the millions unnoticed.

Jessica Ward, from Cornell University and Kevin Lafferty, from the US Geological Survey in California, made an attempt to build a disease baseline of sorts by trawling through all the reports of marine diseases made since 1970.[7] The total number of reports of anything to do with marine life has climbed steadily since then – there are more scientists at work today and more are interested in the sea. So they corrected for this increase by finding the proportion of reports about disease in any given year. This increased over time for turtles, corals, mammals, urchins and molluscs. There was no upward or downward trend for sharks and rays, seagrasses, crabs, lobsters or prawns. Fish appeared less afflicted by disease today than in the past. Of course, this approach has its problems. Science is as amenable to fashion as any human endeavour and people get excited about different subjects as time passes. Professors train students in the subjects of their own passions and those students often pursue the same interests in their own careers. But the results have a ring of truth that chimes with experience for those of us who have dived the oceans over this period. We see a lot more pestilence and plague today than we did when we first jumped in the water.

I have said a lot about the downside of parasites and diseases so far, but we must remember that they are a natural and important part of the way the world works. Their presence indicates that an ecosystem is complex and functioning well enough for them to make a living, particularly where parasites or diseases have complex life cycles, perhaps with multiple hosts. One of the richest places Kevin Lafferty has seen for reef fish parasites is the near-pristine Palmyra Atoll in the mid-Pacific, where he discovered a particular abundance of parasites that use sharks to complete their life cycles.[8] He thinks that having parasites in the system is akin to having top predators in the sense that they help maintain checks and balances so that dominant species don't overwhelm their rivals. In other words, while we anguish over coral disease, to a certain degree, these diseases probably help maintain the coral diversity that fascinates us.

But something has clearly changed. What is responsible for the rise of illness in the sea? For outbreaks of disease or parasites to occur, you need a source of virulent pathogens, a population of susceptible individuals, and a means of transferring infection from one to another. The many ways in which we are changing the oceans offer fertile opportunities for pathogens to get established and spread. People who are stressed tend to be less healthy than their more relaxed colleagues because stress compromises our immune systems. In an analogous way, multiple stresses leave animals and plants more susceptible to illness. So illnesses endemic to a region and those inadvertently brought in by people will often result in an outbreak. Disease outbreaks illustrate the intersecting effects of many different drivers that increase susceptibility. They show how cumulative stresses sum together to compromise life in the sea. The epidemic that wiped out Caribbean long-spined sea urchins began in Panama, close to the canal that connects the Atlantic and Pacific. Although marine species cannot pass through the canal unaided because of its freshwater lakes, they can hitch rides on ships. Ballast water not only carries invasive species, it also carries disease. When diseases find susceptible animals in abundance, they can cause catastrophic epidemics. As Jared Diamond pointed out convincingly in his book *Guns, Germs and Steel*, disease was a greater force than weapons in the suppression of indigenous peoples as Europeans colonized the world. Epidemics of influenza, smallpox and cholera swept through tribes, islands and entire continents, destroying societies and cultures along the way. Every time contact was made, whether hostile or in friendship, the result was massacre by disease. In some places, disease left only one survivor for every ten people originally present.[9] The Caribbean island of Hispaniola (today's Haiti and Dominican Republic) lost 240,000 of its 300,000 inhabitants within twenty years of Columbus's arrival, probably to smallpox.

European diseases had such profound effects on people in the New World and the Pacific Islands because they had never encountered them before and had no immunity. They spread like the wind and felled almost everyone they touched. Much the same is true for life at sea. The ancient disconnect between the Caribbean and the Pacific means that long-spined sea urchins in each ocean developed their own diseases. Pass one to another and the result is an epidemic on a massive scale.

The wave of death from bacterial infection that engulfed long-spined sea urchins in the 1980s sped round the Caribbean so fast that it was over almost before anyone noticed, the disease having run out of hosts. A later effort to trace the timing of this outbreak suggested it swept across 3,000–4,800km of reefs in a year.[10] Similarly, a herpes virus that infected Australian pilchards in 1995 raced around Australia and New Zealand at 10,000km per year. It was probably brought in with unquarantined shipments of assorted frozen bycatch fish destined to feed captive southern bluefin tuna being grown for Japanese markets. Anyone who has been to Australia and been doused with insecticides on arrival will be surprised that such shipments are still let into Australia without any precautions.

Epidemics in the sea seem to spread faster than almost any on land. The only land outbreaks that have come close are myxomatosis and calicivirus in rabbits and West Nile virus in birds. All these are carried by flying insects that quickly spread the virus through susceptible populations. There are fewer barriers to dispersal at sea than on land – no mountain ranges, deserts or inland seas to halt the progress of outbreaks. Ocean currents can move pathogens hundreds or thousands of kilometres in a matter of days or weeks. But epidemics like the ones that attacked long-spined sea urchins or Australian pilchards also spread against the current, which is puzzling until you remember the sea-surface microlayer. This millimetre-thick skin over the sea can be churned into an aerosol by storm winds, together with any floating particles, and whipped long distances across the ocean surface regardless of the direction of underlying currents. During the peak of the pilchard epidemic, entire shoals floated belly up in kilometres-long rafts, so it is easy to see how viruses could have found their way into sea spray.

Across the world, populations of many species of marine mammals have struggled back from the brink over the last fifty years or more. Historic hunting of whales, dolphins, porpoises and seals pushed some to the edge of extinction long ago – animals like the Guadalupe fur seal in the south-western US and Mexico, or the Mediterranean monk seal. Others did succumb and are no longer with us, like Steller's sea cow from the North Pacific and the Caribbean monk seal. Porpoise meat was greatly admired in places like the Baltic countries. The name itself derives from the French *porc-poisson*, or 'pig fish'. In England, it was served at banquets and declared a 'royal fish' in the early fourteenth century, presumably to protect stocks from overexploitation by common folk. These animals and others like them were among the first to benefit in the twentieth century from conservation efforts in the sea.

Over the course of decades, mammals like elephant seals, harbour seals and sealions have gradually repopulated their former haunts. Once again they fill breeding beaches, laze on drying rocks at low tide or frolic in the waves. This change in fortune has cost some populations because it has made them more susceptible to epidemics. Diseases thrive where there are many susceptible hosts. To trigger an epidemic there must be a sufficient density of vulnerable animals to sustain transmission otherwise a disease would fizzle out after felling a handful of animals. In 1988, harbour seals began to sicken and die across northern Europe.[11] After much head scratching, the culprit was finally isolated. It was a virus very close in identity and nature to canine distemper virus and was named phocine distemper virus. This disease depresses the immune system of seals. Victims become lethargic and their eyes runny, but usually it is a secondary infection like pneumonia that kills.

Phocine distemper swept through the harbour seal population and took a few hundred grey seals along the way. By the time it was over a year later, 18,000 were dead, half the harbour seal population. Epidemics end when disease transmission rates fall below a critical threshold. Mass deaths of seals caused the population size to plummet and transmission rates fell. Those left behind probably have some resistance so it is harder for the pathogen to find susceptible victims and the disease disappears ... for a time. After the outbreak, seal numbers began to recover. By 2002, they had risen sufficiently for distemper to take hold once again. Like all pathogens, distemper is able to mutate over time into novel strains to which its host has little or no immunity. The epidemic in 2002 was a near exact replay of the one which preceded it, but this time it cut down over 20,000.

The similarity of phocine distemper to canine distemper gives a clue to its possible origin. Other animals provide one of the main sources of emerging diseases. The same is true for pathogens that affect us. Ebola is one of the most feared illnesses and leads to near certain death through internal bleeding as blood vessels disintegrate. It is passed to people through contact with primates in Africa, perhaps animals that have been hunted for bushmeat. West Nile virus is passed to people from birds via mosquitoes, while swine flu came from pigs. About three-quarters of emerging infectious diseases of people have their origins in other animals. Phocine distemper was probably passed to seals from dogs, which carry the virus without suffering harm. When an epidemic dies down, dogs continue to harbour the illness until conditions are ripe for another outbreak. In the late 1980s, 80–100,000 seals were killed in Lake Baikal, this time by canine distemper virus itself. In Antarctica, crab-eater seals were affected by a distemper virus that was probably passed to them by sled dogs.

The much-heralded recovery of sea otters in California has recently been stalled by disease.[12] Otters have been dying from a parasite of domestic cats, *Toxoplasma gondii*. The parasite causes brain inflammation, and, while it does not always kill directly, it makes sick otters much more likely to be taken by sharks. Otters living near urban centres with rivers flowing into the sea were three times more likely to be infected than those further away, which leads some scientists to believe that rivers carry parasites from cat faeces in rainwater runoff. Once in the sea, they think that parasites could be concentrated from the water by filter-feeding clams and mussels that the otters eat.

Transfer of disease from land to sea is a common theme of recent outbreaks. The African dust suspected to have inoculated Caribbean seafans with fungus is laden with other spores and micro-organisms, many of which remain viable even after their high altitude journeys. Runoff also carries soil and a mix of pathogens to the sea. California sealions have been infected by a soil fungus that causes the occasionally lethal Valley Fever in people of the southern US and Mexico. The toxoplasmosis that culled sea otters has also recently been discovered in a spinner dolphin and beluga whales. Spread of disease agents from land to sea is hastened by land clearing, which promotes soil erosion, and by increasing human population densities in coastal areas.

It seems likely that these trends will lead to many new instances of disease and parasite transfer in future.

As I mentioned, diseases thrive where susceptible animals or plants are common. In this time of rapid global change, several forces have come together to increase the numbers vulnerable to infection. Earlier I touched on how chemical contaminants like PCBs and methyl mercury can suppress the immune system. Mass mortality of harbour seals in the Baltic raised the possibility that pollution in this highly contaminated sea had weakened their ability to fight off infection. An experiment in which one group of seals was fed herring from the Baltic and another from the less contaminated Atlantic confirmed this hunch.[13] After two and a half years of dining on Baltic or Atlantic herring, those from the Baltic group had lowered immunity, an effect blamed on PCBs, while the Atlantic group remained healthy.

Those long-suffering dolphins of Sarasota show much the same thing. Animals carrying a heavier burden of toxic PCBs and DDT in their bodies had less responsive immune systems than less contaminated beasts. Concern is growing that immune suppression by pollutants contributed to mass mortalities of dolphins from morbilliviruses (viruses related to measles) in the Mediterranean and around North America in the 1990s. In the Indian River Lagoon in Florida, there have been horrible cases of dolphins whose living bodies are being engulfed by fungus, a result of immune suppression by a cocktail of pollutants. Similar concerns have also been raised about the role of immune suppression by toxic chemicals in people.[14] Perhaps they have contributed to the emergence of several new diseases that have appeared since the 1970s. The jury is still out.

There are other ways in which susceptibility to disease can increase. Physical injury can weaken animals and plants and wounds provide relatively easy access to the body for infectious agents. In St Lucia one of my graduate students, Maggy Nugues, noticed that diseased corals were more common in places that were heavily dived than ones that were only occasionally dived.[15] No matter how careful they are, scuba divers bump into, scratch and scrape the delicate tissues that cover coral skeletons. These injuries could provide sites for infection. Sure enough, experiments show that injured tissues are much more readily infected than healthy ones. Resorts like Eilat and Sharm el-Sheikh in the Red Sea are visited by millions of snorkelers and scuba divers every year, each one of whom knocks, tramples or crushes some delicate creature or other. In some sites, it is hard to find any coral that hasn't been injured. Vast swathes of the northern Red Sea, and many other places with coral reefs, have been turned over to scuba diving.

There is another form of injury to marine life that in scale and severity dwarfs the impacts from scuba divers: bottom trawling and dredging. When a trawl net is dragged across the seabed it causes immense collateral damage to anything that lives on or near the ocean floor. The footrope of a trawl is designed to run close to the seabed. Sometimes it is weighted and other times it has rollers to enable it to work over areas of rougher bottom. In many beam trawls, where the net is held open by a heavy steel beam that drags above the bottom on metal runners, there are several 'tickler' chains set ahead of the net mouth to scare up fish that hug the bottom like plaice, sole and flounder. Footropes and chains slice off marine life such as corals, sponges, seafans and seaweeds. Ahead of the net, they catch boulders that can be metres across, sometimes entire chunks of living reef, and then roll them along the bottom as the net is towed.

Dredgers are even more destructive, although generally smaller than trawls. Those used to catch scallops are made in the form of a steel frame with vertical teeth set beneath to dig up the sea bed. Behind is a bag of chain mail and net to hold the catch. Scallop dredges look much like the harrows farmers use to break up clods of earth after a field has been ploughed, and their effects below water are not far different. Dredges weave trails of devastation across the bottom. Within these tracks, the bottom is littered with the bodies of dead, dying and injured animals: three-legged crabs with punctured carapaces struggle over ripped fragments of sponge; chopped pieces of seafan lie across corals torn by ragged wounds; scallops with chunks of missing shell wait to be eaten by one of the many scavengers that this kind of fishing encourages.

The carnage wrought by bottom trawling and dredging is multiplied from coast to horizon and beyond, from sea to sea and ocean to ocean. Virtually nowhere above a kilometre deep is spared. Some places get hit once every five or ten years, while places where trawling is unusually intense can be trawled five times in a year. One widely regarded estimate says that an area of 15 million square kilometres gets hit by trawls and dredges every year.[16] That is more than 40,000 square kilometres of damaged, dead and dying bottom life every day. For the sake of comparison, that equals an area about one and a half times the size of Europe or America hit every year. It isn't hard to see how this could increase the number of individuals susceptible to disease.

The hidden hand of climate change also plays its role in the emergence of disease. Raised water temperatures stress host animals and pathogens thrive where stress prevails. It is often said that a warmer world will be a sicker world, although some people are less sure. However, there are good reasons to think that diseases might flourish in the future. It won't just be fish or plankton on the move as the climate changes. Diseases and parasites will also find opportunities as they spread with their hosts or colonize places where new hosts are found. Such shifts have already been seen. In the early 1990s, an oyster parasite moved 500 kilometres north into the waters of New England as the sea warmed enough for it to survive over winter.[17] This single-celled parasite proliferates within oyster tissues and can take several years to kill its host.

Mass coral bleaching is one of the most obvious sentinels of severe warming stress. In some places, disease outbreaks have been reported following sub-lethal bleaching.[18] Diseases like the fungus that wiped out Caribbean seafans seem to grow faster in warmer seas,[19] and, in the Mediterranean, sponge mass mortality has been linked closely with unusually warm water.[20]

Acidification will weaken many life forms that use calcium carbonate to make shells. Thinner, more brittle shells and skeletons will mean greater chance of injury and therefore sites for infection. Nutrient pollution from terrestrial fertilizers and sewage can also encourage disease. Diseases on Caribbean seafans and corals spread faster in an experiment when nutrients were added, which suggests fertilizer boosts growth of pathogens as well as crops.[21] There is a possibility that the general background increase in nutrient pollution of coastal seas has helped to trigger some disease epidemics, but there isn't yet much support for this and its role probably isn't decisive. Instead, nutrients are just one of a combination of conditions that, on their own, might be shrugged off easily, but together have been lethal.

There is one transformation of the sea which, from the point of view of diseases and parasites, makes it a healthier place. As I said earlier, diseases have not increased in all groups looked at. There was no upward trend for disease in sharks and rays, or crabs, lobsters and prawns, and fish seem to have less disease now than they did in the 1970s. All of these groups are intensively targeted by fishing, which means that the densities of most have nosedived in the last hundred years. We can see in ourselves how a change in transmission rates can influence the development of illness. Colds and flu peak in winter in temperate and cold climates because we spend more time indoors and huddled together than in summer. Viruses may be assisted in a few other ways, such as our increased susceptibility in winter, or by greater survival outside the body in weak light. But the basic element is crowding. At low host densities diseases find it harder to get a foothold and spread, so exploited species may be less at risk of epidemics than they were before we fished them so intensely. At last, something positive about overfishing, but not really a cause for celebration.

The effect fishing exerts on disease and parasites goes beyond the simple reduction of host densities.[22] You will remember that fishing tends to selectively remove the big, old and toothsome beasts, while leaving the young and supple behind. By virtue of age, or diminishing resistance, old animals carry the greatest load of parasites. As we have fished down populations and felled their oldest members, we have inadvertently driven parasites below critical density thresholds that dictate whether they persist or disappear. Put another way, we can fish out parasites well before we fish out their hosts.

When we stop fishing a place, we play the tape of decline backwards and the area fills up again with life. Corals, sponges and seaweeds sprout from the seabed and coalesce into glades and thickets. Small fish grow into great stalking predators. Diminutive plankton feeders multiply from twinkling points into constellations of fish. As densities rise, diseases and parasites could make a comeback. I don't know of any research that demonstrates such an effect as yet, but it is a possibility. On the other hand, there are studies that demonstrate the opposite effect in prey or other species not targeted by fisheries. In a California marine reserve, sea urchins were less abundant but more healthy because population explosions were prevented by the recovery of their lobster predators.[23] And, in the Philippines, coral diseases were less common in marine protected areas with healthy fish populations, probably due to prevention of seaweed overgrowth by herbivorous fish.

Disease epidemics are reshaping ecosystems in profound ways. Coral reefs give us a rare opportunity to compare present conditions with what has gone before. Corals constantly rebuild reefs on foundations laid by previous generations. Just as archaeologists piece together the course of history from their excavations, cores drilled through reefs tell us what they looked like tens, hundreds and thousands of years ago. Cores through Caribbean reefs show that recent branching coral losses from disease are unprecedented on a timescale of several thousand years, as far back in time as the cores were able to show. As coral skeletons crumble in the aftermath of mass mortality, habitable space for fish, invertebrates and hosts of others dwindles. The reef haemorrhages life and, as it empties, its ability to support human needs and livelihoods declines. Dense palisades of branching corals once protected islands and coasts from Florida to Panama and South America. Where once divers had to thread their way gingerly through close forests of sharp coral to reach the open water beyond, today they pass over gently undulating hummocks of seaweed and flattened mounds of sponge, sea squirt and encrusting coral. As coral ramparts tumbled, they left shores open to attack by wave and storm. Beaches in front of Caribbean resorts and condos that reefs once protected have washed away, forcing developers to replace them with ugly concrete defences.

Diseases can cause other economic losses. The seagrass wasting disease I mentioned earlier destroyed critical nursery habitats for commercial fish all the way up the US eastern seaboard. It also caused the extinction of a tiny limpet whose only habitat was stolen from beneath it. As outbreaks of disease in the oceans multiply and spread, we can expect worse to come. They warn us that life there is increasingly stressed.

14

Mare Incognitum

Like children the world over, my daughters love turtles. At once incongruous and graceful, they connect us to the world of 15 million years ago when very similar turtles swam alongside megatooth sharks, or 75 million years ago when they rubbed shoulders with dinosaurs. Only eight species of marine turtle remain from a lineage that stretches back little changed deep into the age of the dinosaurs. The largest living reptile is the leatherback turtle, a barnacle-encrusted eminence that can reach three metres long and weigh two tonnes. Today we confront the stark possibility that people will drive the leatherback turtle to extinction within the next human generation. Already there is just one leatherback left in the Pacific for every twenty in 1962, the year I was born.

I find this prospect awful to contemplate. What will the world seem like to our grandchildren if the only leatherbacks left are those that inhabit books or computer screens? At the rate we're going, by that time the world could have been robbed of the majesty of whale sharks, the frivolity of sea otters and the frenzy of bluefin tuna, beating the sea to foam as they hunt down prey. These animals are not just icons of children's books and television channels; they are creatures we can pursue with our imaginations far into unknown depths. Without them, life will lose some of its grandeur.

The last thousand years have seen human influences gather like a green wave, scarcely perceptible at first but slowly lifting and steepening over the centuries to burst across the globe in the last sixty years. Our planetary remodelling did not stop at the shore; it just came a little later to the sea. There is no precedent for the speed and variety of changes underway today, save perhaps the asteroid impact that

ended the reign of dinosaurs 65 million years ago. Even the cata-
clysms of the other great extinction events seem sedate by comparison.
There is no sign of any let up. The rate of change continues to accelerate in step with human population and economic growth. Our impacts
have intensified with time, and where once habitats and species were
subject to just one or two influences, like fishing or siltation, now they
are caught amid a morass of stresses whose effects accumulate and
reverberate through every layer of the living world.

While life's players differ from place to place, how the play of
human impact has unfolded is much the same. The abundance and
sheer physical mass of life has declined over time. There has been a
shift from dominance by large-bodied animals and plants to small
ones. Populations have been truncated so that age and wisdom have
given way to youth and inexperience. The web of life has been simplified as predators have yielded to prey and some species have vanished.
Habitats built over millennia by the exertions of countless generations of creatures have crumbled from three-dimensional magnificence
to two-dimensional monotony. And for our own species, perhaps
uniquely gifted with a sense of aesthetics, in too many places there has
been a shift from beauty to sordid poverty.

The shifting baseline syndrome has blinded us to these alterations.
A trip to the seaside appears to hold the same joy and wonder for my
children as it did for me some forty years ago, when there was more
to find amid rockpools and waves. But what disturbs me, as my own
perspective extends into middle age, is how rapid the alteration has
been. The vibrant coral reefs I dived as a youth seem less bright, less
imposing and less wild today, as indeed most of them are. In that time,
the carpets of stony corals that build reefs and give them much of their
vivid colour and complexity have shrunk by a third to a half. The fat
beady-eyed groupers that skulked beneath ledges have become rare
and walls of predatory snappers and emperors have thinned. In many
places, silt washed off the land has fogged visibility so the vistas are
less impressive and their colours dimmed.

Just fifty years ago there were thirty times more Atlantic bluefin
tuna swimming wild than now exist. The shoals that raced into the
North Sea each spring to thrash among the herring have disappeared.
Vast ice shelves have broken free in the Southern Ocean, the Arctic ice

thins, and open water near the North Pole has allowed surface waters of the North Pacific and the Atlantic to mix for the first time in 800,000 years. Seagrass meadows in the Mediterranean have been swallowed beneath choking blankets of invasive seaweed. Dead zones have grown and multiplied. Mighty currents have slowed. In sum, humanity has achieved dominion over the oceans and marine life is in difficulty.

It is hard to grasp the prospect of seas so compromised that they no longer sustain the ecological processes which we take for granted, and upon which our comfort, pleasure and perhaps even our very existence depends. What will the future look like? Prediction is difficult given that we have never been here before. In the early days of European seafaring, unexplored areas of ocean were marked on charts as *Mare Incognitum*, or 'Unknown Seas', and the truth is that we are voyaging into such seas again today. But we have a special interest in the outlook so prediction is worth a shot. After all, as the American engineer Charles Kettering once said, the future is where we will spend the rest of our lives. To see where the ocean world is going, a good place to begin is to look at the drivers of change.

Fishing is the most ancient of human influences on the sea, but it is only in the last 150 years that its effects spread beyond the local. The late nineteenth-century industrialization of fishing came when we added engines to boats. Technological innovation gathered pace through the twentieth century in a race to snag fish and beat others to profit, and has never let up. The footprint of fishing spread from traditional fishing grounds to distant seas in the early 1900s, and then in the mid-twentieth century from coasts to the high seas, and from shallow water to deep. Over time we have substituted new species as past favourites waned. The price of fish has outstripped inflation for decades, reflecting the increasing difficulty and cost of sustaining supplies. In the last third of the twentieth century, developed countries turned to the developing world to supply their fish, having exhausted their own grounds. The UN Food and Agriculture Organization (FAO)'s scorecard of the world's major fisheries shows an increasing number and rate of collapses since 1950, when it began to collect statistics.

Growing demand for fish from the rapidly expanding human population has driven these trends. In 1880, on the cusp of the industrial fishing revolution, there were 1.4 billion mouths to feed. By 2011, that had reached 7 billion, five times more. Demand is set to rise by another third in the next forty years with the addition of at least another 2.1 billion people, so pressure for fish protein isn't about to let up.[1] If we exploit the oceans as we have done for the last century, then overfishing will continue to eat away at populations of the world's big fish; some will be driven to extinction, while many more will become too rare to play any further meaningful role within their ecosystems. As they disappear, we will continue the switch from large predators like cod and hake to animals low in food webs, like prawns and anchovies. But they too will become overexploited (as some already are), and we will have to seek seafood from other sources, such as Antarctic krill, which will be processed into more palatable-looking foods like fishcakes and fish sticks.

In fact, fisheries for krill and jellyfish are already on the rise. FAO figures show jellyfish catches rose exponentially from almost none in the 1960s, to more than half a million tonnes today. If we fish as we do now, it is doubtful that catches from these animals – which based on their fleeting lifestyles could be dubbed the mice and cockroaches of the sea – will be able to substitute for the peak landings reached in the 1980s, so wild seafood will get ever scarcer. A billion people rely on seafood as their main source of animal protein today, most of them in the developing world.[2] The continued decline of wild fisheries threatens malnourishment for many more by the middle of this century.

Just as consumers in the Western world are developing scruples about the sustainability of the fish they eat, economic growth and increased disposable income in Asia, especially China, have released a surge of demand for seafood. Asian buyers have fanned out across the world in search of supplies, especially targeting poor communities in developing countries. Judging by their actions, most have little thought for sustainability. They buy while the fish last and then move on. The Asian market has forced prices for some products sky-high, turning once worthless bycatch species like blue and silky sharks into valuable targets. In 2004, a single large whale shark fin could sell for over US$50,000.[3] The long-term prospect does not look good unless Asian consumers can be persuaded to value sustainability soon.

Things don't have to get this bad. Aquaculture could help sustain supplies of fish, and better management could maintain or perhaps increase catches from wild stocks recovered to populations far larger than today's. I will come back to how we could achieve this metamorphosis later on. But let's return to the other drivers of change first.

Human population growth and technological development have also driven the rise in pollution. The Industrial Revolution opened the veins from which toxic metals and other chemicals began to bleed into the sea. While time has staunched some of these flows, others have opened as miners and heavy industry have shifted their dirtiest operations from developed to developing countries. All the while we have invented new ways to make things. Our enthusiasm for novel materials and processes seems always to outrun prudence. Highly toxic chemicals are set free long before we notice their darker sides. The legacy of industrialization is written in no-go zones too toxic for use. But many chemicals have spread so widely they cannot be avoided. They swirl around all of us. They compromise wildlife health and reproduction and contribute to the rising incidence of disease epidemics in the sea. And as people often sit at the apex of the food web, trumping consummate predators like sleek tuna and grinning sharks, we too are vulnerable. Seafood lovers the world over are accumulating in their bodies toxins drawn from waters where we thought they had been laid to rest for good.

Population growth has driven up demand for food. The world avoided the much-trailed population crash expected in the 1980s and 1990s by the green revolution that industrialized agriculture. We have squeezed ever-greater production from the land by lavish applications of artificial fertilizers and other agrochemicals. They have bought us time, but, like fisheries, farmers have won greater production at the cost of the natural capital of their soils.[4] Rains and irrigation water carry soils and their chemical burdens to the sea through estuaries and deltas whose filter feeders have all but disappeared and whose wetlands have been stripped and converted to other uses. Once in the ocean, nitrogen and phosphorus fire plankton and microbial blooms that overwhelm the assimilative capacity of food webs decapitated by overfishing, and exceed the physical water mixing needed to keep up with demand for oxygen as dead plankton accumulate.

22. Climate refugee in Bangladesh. The coastal belt of Bangladesh is eroding at an unprecedented rate, aggravated by a rising sea level and extreme weather events induced by changing climate. Rising seas could displace 20 million people in Bangladesh by 2050. This woman lost her home to storm surge brought on by a cyclone in 2010.

23. *above left* Healthy Papua New Guinean coral reefs bathed in seawater of normal pH.

24. *above centre* Reefs exposed to carbon dioxide bubbling from underwater vents are subject to a water acidity equivalent to levels that could be reached by the late twenty-first century under a business-as-usual scenario of carbon dioxide build up in the atmosphere.

25. *above right* Reefs exposed to intense carbon dioxide venting equivalent to levels that could be reached in the twenty-second century if we do nothing to combat carbon dioxide emissions. Only a few corals can survive this level of water acidity.

26. *above* These microscopic coccolithophores are phytoplankton (plant plankton) that produce exquisitely ornamented shells made from calcium carbonate. Each ball is a single cell that measures about $\frac{1}{250}$th of a millimetre across. Fossilized skeletons of coccolithophores are a major ingredient of chalk. They are expected to be highly susceptible to rising acidity as dissolved carbon dioxide accumulates in the sea.

27. *above right* A die-off of menhaden fish in Narragansett Bay, Rhode Island, USA, due to a severe low oxygen event in August 2003.

28. *right* Coastguard vessels and oil industry firefighters battle the Deepwater Horizon blowout of 2010 in the Gulf of Mexico. The subsequent oil spill became one of the largest in history, with over 4.4 million barrels discharged into Gulf waters.

29. *below right* A red tide blooms off the coast of La Jolla, California. The alga responsible – *Karenia brevis* – produces toxins that can harm wildlife and people. Such 'harmful algal blooms' are often fuelled by nutrients from sewage, industrial effluents and agricultural runoff.

30. Shrimp trawlers at work off the coast of Ecuador can here be seen from space as they churn hundreds of tonnes of mud from the seafloor into the overlying waters. Aside from the damage done to bottom living animals and plants by the trawl nets themselves, such sediments can re-release nutrients that fuel harmful algal blooms, as well as liberating toxic contaminants from polluted runoff.

31. Tiny plastic particles caught in a fine mesh plankton net alongside minute molluscs called by-the-wind sailors, *Velella velella*, in the middle of the North Pacific Gyre, otherwise known as the Great Eastern Garbage Patch.

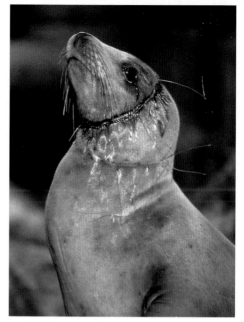

32. Plastic contamination of the Santos estuary, São Paulo, Brazil. Estuaries are especially prone to waste build up as they often receive garbage washed down from highly populous catchments.

33. California sealion trapped in a piece of monofilament gill net. The fishing net has gradually tightened in a slow-motion throat cut as the animal has grown in size.

34. Laysan albatross live in the North Pacific and fly thousands of miles over the open sea in search of food for their young. Unfortunately they cannot distinguish floating plastic trash from food, with heartbreaking results.

35. Remains of a Laysan albatross chick from Kure Atoll in the Northwest Hawaiian Islands. This chick starved to death on a diet of junk food.

36. Humpback whale and calf near Tonga. Whales are highly vocal animals, few of them more so than humpbacks. They use the remarkable sound-transmitting properties of water to call each other over distances of hundreds, sometimes thousands, of miles. But their ability to communicate has diminished in the last fifty years owing to the rising clamour of human-generated noise in the ocean.

37. *left* A pair of plainfin midshipman fish mating in the intertidal zone of Quadra Island, British Columbia, Canada. The overlying rock with its clutch of golden eggs has been lifted carefully to show the female (upside down and pale), and male (right way up and dark). Male plainfin midshipman produce loud buzzes and grunts to attract mates and drive away predators.

38. *right* Red lionfish on a shallow water reef in the Bahamas. In the space of less than twenty years, this native of the Pacific and Indian Oceans has successfully invaded coral reefs throughout almost the entire tropical Western Atlantic, in the process munching its way through countless indigenous fish.

39. Jellyfish possess an ethereal beauty as they drift in the open sea, which belies their ruthless efficiency as predators of the planktonic realm. They are highly effective opportunists that are beginning to benefit from overfishing, coastal pollution and climate change. As a result, dense blooms of jellyfish have reached plague proportions in some places, driving people from the water, blocking power station cooling water intakes and killing stock in fish farms.

40. Scallop dredge being lifted into a boat in the Irish Sea. This boat is fishing with four dredges either side. Each individual dredge is composed of a steel frame with vertical teeth that dig up the bottom, and a chain link bag to collect the scallops and anything else that is caught (usually a wide assortment of rocks, weed, fish and other invertebrates). Scallop dredges are notorious for the collateral damage they do to life on the seabed.

41. Migration of a large shoal of Munk's devil ray (*Mobula munkiana*) along the coast of Baja California near Cabo Pulmo. Animals such as these rays have increased in numbers as many of their natural enemies, mainly large predatory sharks, have been overfished.

42. Red mangrove in the Galapagos Islands. Many tropical coasts are protected by dense forests of mangrove. Their interlocking roots stabilize sediments and protect coasts from waves and storms. By trapping mud and growing upwards over time, healthy mangroves could help shield coasts from the worst effects of sea level rise. In many countries, however, they have been cleared to make way for shrimp ponds and other coastal developments.

43. A killer whale dives into the face of a wave in pursuit of a turtle in the Galapagos Islands. Moments later it seized the animal and then shared the spoils with an accompanying female. We can engineer replacements for some of the things that nature does for us when ecosystems falter, but there is no substitute for its raw beauty.

44. *inset right* A barge-load of retired New York subway cars destined to be dumped in the sea to create 'artificial reefs'.

45. *below* Tyres dumped into the sea off Fort Lauderdale, Florida, in the 1970s to create 'artificial reefs'; garbage by any other name. There isn't much evidence of marine life inhabiting this 'reef'. Natural habitats do the job far better than artificial ones.

Despite the proliferation of dead zones, future seas will not be life-less. We are creating winners as well as losers. Jellyfish, for example, are great opportunists, and some scientists fear that large parts of our most productive seas will transform into jellyfish empires.[5] Only a few animals eat adult jellyfish – leatherback turtles and giant ocean sunfish for instance (both in decline) – because they lack substance and you have to eat huge quantities to break even. There are many more predators that eat snack-size juvenile jellies.[6] However, with the wholesale decline of marine life from overexploitation, pollution and climate change, the ranks of these predators have thinned so more jellies survive.

Jellyfish positively thrive in pollution-enriched seas. Given unlim-ited food, they can reach adult size fast. With their stinging tentacles, they are formidable predators. Here one of the quirks of ocean food webs comes into play to seal their dominance. Most animals that might eat jellyfish go through tiny egg, larval or juvenile stages when the tables turn and they are themselves jellyfish prey. Such role revers-als of predator and prey are rare on land. In the sea, however, they are prevalent, with surprising effects. (The American oceanographer Andrew Bakun invites us to imagine a world in which zebras and antelopes are voracious predators of the young of lions or cheetahs. What would the Serengeti look like if this were so?)

The jellyfish joyride begins when high levels of nutrients combine with a fall in abundance of their predators. When plentiful, jellyfish suppress their predators further by eating more of their young, so paving the way for a full-blown population explosion. If food runs short jellyfish don't just die off; instead they shrink and wait until conditions improve (although if nutrient levels fall far enough and for long enough, jellyfish blooms can snuff out). In a future with more acidified seas, jellies won't have troublesome carbonate skeletons to handicap their chances. The altered oceans that haunt our possible future could offer them worlds of opportunity. They have been here before. Enigmatic traces in rocks from the earliest Cambrian Period, some 550 million years ago, tell of an age of jellyfish that preceded the great radiation of life that established most of the animal groups alive today. Collectively, the modern reappearance of seas dominated by gelatinous animals, microbes and algae has been dubbed 'the rise of slime'.[7] It signals a reversion towards conditions that prevailed in the earliest days of multi-cellular life.

Climate change adds a third set of drivers of change to life in the sea. Heating will thicken the lower density warm layer of water that sits like a lid over much of the ocean, to restrict oxygen transfer to underlying waters and constrain upward flow of nutrients. The result will be reduced productivity over great expanses of sea away from coasts and expanded low-oxygen zones beneath. Already there are signs that plankton productivity has fallen, probably as a consequence of warmer seas, and the volume of oxygen-starved waters has grown.[8] Nutrient starvation of surface waters may also be due to the collapse in fish stocks from overfishing, causing nutrients to disappear into the depths rather than being recycled in shallow waters. Great whales may once have played a role as a nutrient pump from deep to shallow water.[9] Many feed at depth and crap at the surface. Before whaling, when the oceans thronged with more than a million whales, that could have made a big difference to productivity.

Along western coasts, tropical heating will force faster redistribution of heat to the poles, strengthening winds that power upwellings of deep water. Upwelling could benefit fish production by raising nutrients from the deep, but intensified upwelling could push some regions into hyperproduction, creating noxious eruptions of hydrogen sulphide like those that plague Namibia's coast today. Upwellings might also pull low-oxygen water of higher acidity onto continental shelves to cause mass mortality of bottom life, like episodes seen recently along the Oregon coast.

The effects of acidification are hard to predict. At the very least life is likely to get much more difficult for species with carbonate shells, which include some of the most important primary producers in the sea, the phytoplankton that sustain food webs and release life-giving oxygen. We don't yet know whether the photosynthesis-boosting effect of extra carbon dioxide might outweigh or compensate its corrosive qualities in the balance between benefit and cost. If not, there are non-calcifying phytoplankton that might step up production. The worst-case scenario would be large-scale depression of primary production, which is the quantity of carbon fixed into plant material, which also means less oxygen produced.

Any fall in the rate of plankton production would reduce the snow of organic debris that sinks from sunlit surface layers to the deep sea. Deep-sea communities survive on meagre handouts from above and failure in supply would thin their numbers. As the sea is the ultimate sink for all of the carbon dioxide we are releasing into the atmosphere, it would also reduce the rate at which carbon gets locked away in bottom sediments. This is the 'biological pump' that takes carbon from where it can be a nuisance to where it does no harm. If phytoplankton secrete less carbonate because of acidity, their shells will be lighter and they won't sink as fast. The pump will run more slowly so the oceans will absorb less of the carbon dioxide we add to the atmosphere, and global warming will accelerate. Limited evidence from coccolithophores in deep-sea sediments suggest their mass has actually increased by 40 per cent since the Industrial Revolution. For them, the growth-boosting effect of extra carbon dioxide has so far outweighed the acidity cost.[10] Some experiments suggest the reverse is true,[11] and in the wild the skeletal weight of foraminifera in the Southern Ocean, another key plankton group, has declined by 30–35 per cent since pre-industrial times.[12] The leading edge of science is always fraught with uncertainty. We still have much to learn.

The Roman Emperor Marcus Aurelius said, 'Look back over the past, with its changing empires that rose and fell, and you can foresee the future, too.' There are more than enough portents to be able to see where the oceans are headed. Scotland's Firth of Clyde is a window to a world without fish. Italy's volcanic Ischia lets us dive in an acid sea. China's Bohai Gulf shows how nutrient pollution and overfishing can carpet a sea in weed and spread plagues of jellyfish and toxic plankton blooms. The Gulf of Mexico, Oregon, the Baltic and hundreds of other places across the world give us a taste of the oblivion spread by anoxic water. Europe's North Sea reveals life on the move as the world warms. The Gulf of Maine and San Francisco Bay are filling with alien species. Sea level rise is eating away at the Mississippi and Nile deltas and threatens densely populated coasts like those of Holland, Germany and Thailand. And the rotting bodies of Pacific albatross chicks, bloated and starved by a diet of plastics, show us how life struggles as the oceans fill with human refuse.

Each of these places illustrates nature under stress from one or two of the impacts I have described in previous chapters. But these primary stressors are not the only ones they face. The Firth of Clyde, for example, has been polluted by decades of sewage sludge dumping, the waste from a major city and heavy industry. The Gulf of Mexico's seabed is raked over year after year by shrimp trawlers. Multiple stresses have more serious effects because they can act in synergistic and often unpredictable ways.

Stresses exert their insidious influences in several ways. Species are sensitive to different stressors to different degrees. If each stress was applied separately, a snail might be noticeably affected by slight acidification, made uncomfortable with a half degree rise in water temperature, and be indifferent to heavy metal pollution. Add two or more together and the effects could be severe. In one dramatic example, brittlestars are able to survive slight acidification, but add trace amounts of the antibacterial triclosan, an ingredient of handwashes and face creams, and they collapse into fragments.[13] Stresses don't usually combine to such spectacular effect, but their collective impact can erode the fitness of individuals and populations. Their effects are best understood in the currencies of life and death. Populations endure when the birthrate equals or betters the death rate. Stresses often drive down births and increase the likelihood of death, so they upset the balance of life's equation.

The extra physiological cost of coping in low-oxygen water, for example, could be met by increasing food intake or reducing calorie demand. Extra food may not be available, so animals could compensate by laying fewer eggs or skipping a breeding season. Or they might grow more slowly, which would also mean fewer offspring since small animals usually produce less eggs. Low oxygen in itself is a powerful constraint on body size, favouring small animals over large because they can breathe more easily, and thereby also cutting egg production as big animals produce more eggs. High body burdens of toxic contaminants, many of them hormone disruptors, might further compromise reproduction through reduced fertility or miscarriage. When the tipping point is passed at which deaths exceed births, a population declines.

The cocktail of human stresses is mixed differently from place to place but whatever the combination, their collective impacts cause worse harm than each one acting alone. In other words, the kill rate from multiple stresses acting in concert is greater than the summed kill rate from each of the same stresses alone. There is a parallel here between the environment and the experience of intensive care doctors. They must jump in quickly to support patients when they have only one problem to deal with, like blood loss or stroke. When problems begin to cascade through the complex systems of the body, mortality shoots up.

When a species succumbs, the effect multiplies through the ecosystem via the web of interactions that connect species. Some creatures are strongly connected, like salmon sharks and salmon, anemonefish and anemones, or dugongs and seagrass. Others form loose associations. A crab might be happy to live among the branches of half a dozen different types of coral, or a fish might feel equally at home amid the waving fronds of several different seaweeds. Still more are linked indirectly through intermediaries: the food your prey eats is ultimately your food (a truth we too should keep in mind in our relations with the rest of the biosphere).

Stresses act on ecosystems at this higher level when they weaken or knock out links among species. When one species goes down, it takes others with it that in turn affect the animals and plants with which they are connected. The more different stresses we add to the blend, the wider the spectrum of life that is directly affected, so the consequences penetrate further through the tangle of existence. As species dwindle or disappear, ecosystems adapt and reform as the remaining inhabitants find other ways to make a living. Weak links can strengthen as animals find alternative food or living space.

Knocking out one species, inserting something else, or adjusting the abundance of another can lead to effects on species that at first seem completely unconnected with them. The nineteenth-century American national parks pioneer John Muir's famous phrase is apt here: 'When one tugs at a single thing in nature, he finds it attached to the rest of the world.'[14] The scale of our ignorance of these interconnections is breathtaking. It would take a thousand lifetimes of research to figure out all of the ways in which we are affecting the species in an ecosystem of only moderate complexity. Even within a species, effects differ. While adult lobsters may happily cope with increased acidity, the same change halts development of one of their several juvenile stages.

When species that play key structuring roles disappear, their loss has profound consequences. In the Pacific Ocean, longline fisheries have cut populations of large ocean-going sharks like the thresher to a tenth of their former abundance.[15] In the wake of their departure, smaller species like the pelagic stingray have proliferated. Off America's east coast a similar reorganization has played out. Megapredators like hammerhead, great white and tiger sharks have collapsed, leaving the field open to their prey.[16] Five species of smaller shark and ray have grown in abundance, some spectacularly so. The cownose ray – blunt-faced stingrays with dark round eyes and long whiptails – travel in groups and raise thick billows of sand as they excavate buried clams. With no big sharks to prey on them, their groups have grown to hundreds strong. Added together there could be as many as 40 million up and down the US east coast. Cownose rays are particularly partial to bay scallops, and with this many mouths to feed there is scarcely a scallop left, leaving US fishing communities empty handed and bewildered. It isn't an obvious connection – that fishing out giant sharks would cause the collapse of scallop fisheries – but that is how impacts proliferate. It does, however, illustrate the difficulty of predicting how stresses will reshape ecosystems, even when there is only a single stress involved.

Canadian cod may have fallen victim to another kind of ecological reorganization since their population crashed in the 1990s and a moratorium on cod fishing was imposed. Cod once reigned supreme in the chilly waters off Newfoundland. Their ferocity and indiscriminate tastes made them almost universally feared and avoided. Lobsters lurked and crabs cowered as throngs of cod advanced over the continental shelf to their spawning grounds. Like most fish, however, cod are vulnerable when young. Eggs and larvae are snapped up by capelin, herring, mackerel, jellyfish and hosts of others. When cod were mighty they produced so many young they overwhelmed their predators. It mattered not that tiny cod poured in endless streams into almost limitless mouths because plenty escaped. But, when cod were humbled by overfishing, the survivors could only produce a tiny fraction of the former abundance of young, few of which survive the carnivorous gauntlet today. Here is another of those role reversals in which a predator falls victim to its prey.

When stresses take out habitat-defining species like kelp, oysters or seagrass, they can set off chains of events that remodel the habitat so it becomes hostile to the original inhabitants, just as political revolutions sweep aside old regimes and those who thrived in their orbit.[17] When they go, or are pushed out, space is freed up for competitors. If strong or predatory interlopers establish themselves, they can change conditions in ways that inhibit recovery of the original species. European coasts were developed at the rate of a kilometre per day between 1960 and 1995, causing widespread wetland loss. Below water, rocky reefs carpeted in lush seaweed forests were subjected to siltation, reduced light, anchoring and dredging, and both the direct effects of destructive fishing and the indirect effects of removal of key fish species that kept grazing herbivores in check. When large, canopy-forming seaweeds disappeared from Mediterranean reefs, they were replaced by shorter turf-forming seaweeds. Beneath the old canopies, sediments were swept away by fronds that lashed back and forth under the influence of waves and currents. Turf seaweeds, by contrast, trapped sediment to create an unstable layer over the rock so that it cannot long sustain larger weeds that require firm anchors to hold them in place.

In the Baltic Sea, lush meadows of brown, red, purple and pink weeds struggled as light penetration faltered under the increasing plankton growth stimulated by excess fertilizing nutrients. The dead bodies from these blooms piled up on the seabed and drove down oxygen levels so that a thick organic soup began to churn above the bottom to cut light levels further and drive out animals that once called the weedy forest home. On coral reefs, seaweed again causes problems when mass coral mortality from high temperatures or disease opens up space. Where grazers have been overfished seaweeds rapidly take hold, especially in places where nutrient-rich water also bathes the reef. Just as in the Mediterranean, these seaweeds trap sediment, which makes the coral comeback much harder, and the weeds may themselves harm tiny corals by poison or contact abrasion to make them more susceptible to diseases, some of which the seaweeds harbour.[18]

In many cases, loss of habitat-forming species like oysters, sponge or coral opens the way for invasions by other species. Sometimes it is the invaders themselves that oust the habitat creators, such as the rampant *Caulerpa* weed that engulfed the native seagrasses in the Mediterranean. Invaders complicate the business of projecting the future for marine life still further. In part, the wave of new species advancing into habitats is a product of climate warming. Some arrivals can be predicted quite easily because they live nearby, but their effects may be hard to judge because of the complex ways in which they ripple through the web of life. Other introductions are more capricious and come about through the chance arrival and unlikely survival of creatures from far afield. The red lionfish that have so rapidly settled into Caribbean coral reefs seem certain to cause a major reorganization of life there, the outcome of which we can only guess at.

Habitats have been opened up and flattened out and the new regimes usually support fewer species and are less productive than the old. The contrast between kelp forest and turf pavement is like the difference between woodland and pasture. Think of how rich is the life of the forest. Its multiple layers and kaleidoscope of conditions produce communities within communities: the lush and open canopy; the fuzz of lichen and fern on trunk and branch; the dark, mossy wrinkles and cavities; the tangled undergrowth and deep loam beneath. Compare that to open grassland grazed by rabbits or livestock. A close look might reveal a miniature world more varied than is apparent at first glance, but the simple fact is that grasslands offer fewer ways to make a living than woods, and, while some species do well (the Serengeti's rivers of wildebeest in full migration come to mind), the abundance of life they support is less.

Complex habitats offer more ways to make a living than the simpler, flatter habitats that remain in the wake of degradation. They also sustain more big fish, who in turn keep productivity high by recycling nutrients. Just as gardeners love horse manure, so fish shit is great for seaweed and phytoplankton, the engines of production in the sea. They liberate nutrients to the legions of detritus feeders and animals like corals, molluscs and sponges that feed on floating particles. A recent study based on computer-simulated removals of fish species from food webs in tropical freshwater lakes and streams showed how some species play disproportionately important roles.[19] Inconveniently for us, the most important species are often those favoured by fishermen, because they tend to be large, or abundant, or both. When the simulated extinctions followed the sequence of importance of species in catches, the loss of nutrient-recycling function was far greater than when species were removed at random.

In these simulations, when species with similar ecological roles were allowed to increase in abundance following the removal of others, they were often able to compensate for the roles played by their vanished colleagues. Diversity is a key to resilience. Where there are lots of species and much overlap in what they do, there is more scope for some of them to disappear before losses trigger catastrophe. The trouble with fisheries is that they are often unselective and take out several species at a time. And, when one declines, fishers switch to others so the scope for compensation is limited. This is certainly what I have seen on Caribbean coral reefs, where fewer and smaller species are left behind as fishing intensifies. The dainty little parrotfish that pluck seaweed from Jamaica's ransacked reefs are all that are left after fishermen have taken their fill. They cannot possibly control the runaway seaweed growth that is overwhelming coral there, while the dense groups of bulky parrotfish that exist on lightly fished reefs easily strip seaweed back to a green baize film.

It saddens me to see the loss of beauty and wild energy of the sea. Tangled forests of mangrove that teemed with life have been replaced by the bare embankments of shrimp ponds. Verdant green meadows of seagrass have withered or been buried under blanketing mud. The labyrinthine creeks of saltmarsh, where once you could leave the modern world behind, have been drained and built over time after time. These are just the places that are easy to see, but we find similar losses in less accessible corners. Cold water coral reefs in the North Sea were already half destroyed by trawlers when they were first discovered in the 1990s. The deep reefs off Miami, where liveried grouper once held court in palaces of coral, are now mostly rubble, their trawled ruins inhabited by tiny fish and worms. At some level these losses discomfort every one of us. But the disintegration of untamed nature reaches beyond aesthetics. It is undermining the ability of the oceans to sustain human needs and well-being.

15

Ecosystems At Your Service

'To people who allege that we
Incline to overrate the sea
I answer, "We do not;
Apart from being coloured blue,
It has its uses not a few;
I cannot think what we should do
If ever 'the deep did rot'".'

Sir Owen Seaman, 1861–1936[1]

My graduate student Ruth Thurstan spent the summer of 2010 sampling seabed sediments around Britain, wrestling with coring equipment and a sometimes restive team who didn't share her enthusiasm for 5 a.m. starts. They pounded coring tubes into the bottom, raising clouds of mud that thickened as they worked until they had to communicate by touch alone. Once removed, the cores were split from top to bottom; half went to be dated using radioisotopes, while Ruth spent the next six months sifting fragments of shell and spine from the other half. These remains help trace the transformation of the bottom from an exuberance of oysters, horse mussels and clams, to silty scarcity. The change happened when trawlers and dredgers scraped the bottom clean in pursuit of fish and oysters. They sundered the layer of invertebrate life that once crusted the bottoms of estuaries and great expanses of continental shelves. Over years, decades and centuries, the trawlers pulverized and buried creatures that once enlivened the seabed. Ironically, oysters introduced to Europe from North America to replenish exhausted beds set loose a disease that wiped out the remaining natives.

British fishers immediately regretted the loss of the invertebrate crust over the bottom because it destroyed their source of bait for longline fisheries for cod, haddock, turbot and the like. Some also saw that it would undermine the productivity of the fishing itself, since a breakdown in habitat would compromise living conditions. The British surgeon and naturalist John Bellamy predicted as much in 1843:

> ... the employment of the Trawl, however, during a long series of years, must assuredly act with the greatest prejudice towards these [fishy] races. Dragged along with force over considerable areas of marine bottom, it tears away promiscuously, hosts of the inferior beings there resident, besides bringing destruction on multitudes of smaller fishes, *the whole of which, be it observed, are the appointed diet of those edible species sought after as human food* ... An interference with the economical arrangement of Creation, of such magnitude, and of such long duration, will hereafter bring its fruits in a perceptible diminution of those articles of consumption for which we at present seem to have so great necessity.[2]

Like all good Victorians, Bellamy continued in poetic vein,

> ...we might suppose the species thus injured by an abstraction of their food, to utter some such expressions as the following:-

> Nay, take my life, and all, pardon not that:
> You take my house, when you do take the prop
> That doth sustain my house; you take my life,
> When you do take the means whereby I live.

Bellamy's warning and those of many less poetic scientists after him went unheeded, and the trawlers obliterated habitats formed over thousands of years, not just around Britain, but the world. More than 150 years later, Bellamy's prediction has been proved true. A study of fish on south-west British trawling grounds showed that several species are skinny and stunted compared to those in less trawled areas because they cannot find enough food.[3] Had there been any places left that trawlers had not worked over, the contrast would have been more obvious still. Productivity has fallen not just as a result of direct removal of the plants and animals, but because simple habitats offer

fewer living spaces and sustain less life than complex ones. The cita-
dels and piazzas, avenues and apartments of these underwater worlds
have crumbled and their flattening, not just by trawlers and dredges
but by the panoply of stresses I have described, has left room for only
a fraction of their former inhabitants.

European seas are far less productive than they once were. The fact
that the UK bottom trawl fleet lands only half the fish today that it
did when records began in 1889, despite a massive increase in fishing
power, says all we need to know about how we have squandered natural capital.[4] People in wealthy nations have so far been sheltered
from the impacts of overexploitation. They are rich enough to buy
fish from parts of the world where there are still some to be caught.
But, when fish stocks collapse in the developing world, desperation
follows. Nowhere has this been more potently illustrated than in the
Philippines. This country should be the world's underwater wonder-
land. It lies at the heart of the most diverse sea on the planet, crammed
full of a dazzling variety of fish, coral and almost every other form of
life. Filipinos have lived by and from the sea for thousands of years.
They depend on it, getting some 70 per cent of their animal protein
from seafood. In recent years, explosive population growth has upped
demand far beyond the capacity of the sea to supply all these people.
The result is a cycle of overfishing that has become ever more frantic
with each passing year. The most extreme manifestation of overfishing in the world has got to be the practice of pa-aling.

When times were better, Filipinos fished using traditional methods
like traps, nets, hooks and spears. A day's fishing would easily supply
a family's needs and leave extra to sell. In a familiar pattern, as stocks
declined, people devised ways to maintain catches by upping the
power of these simple methods. The process of escalation eventually
led them to invent pa-aling – or drive net fishing – in which a team of
men and boys suffers extreme peril at the limits of human endurance
for a few bucketfuls of fish. When reefs have been emptied of life in
the shallows, there is nowhere to go but down.

A team of ten or fifteen men and boys will dive down as far as thirty
or forty metres with a huge ball of net, breathing diesel-tainted air
through a length of garden hose supplied from a compressor on board
a boat. Meanwhile another team on the boat struggles to keep the

engine going and all the hoses from turning into a lethal tangle. Underwater, the fishermen unfurl the net into a dome the size of a village church. It is weighed down at the sides and one end with rocks and buoyed at the top with air-filled plastic bottles. After a break at the surface, the dénouement comes as the team swims along the bottom with a scare line to drive fish into the net, trampling and bashing the bottom with rocks in their advance. One drive can clear a place of half its fish and leave the reef in tatters.

The human cost of pa-aling is brutal; divers frequently far exceed safe diving times and are stricken by agonizing bends that can paralyse or kill. Others drown when they are tangled in the net or air supplies are cut off. Captains often exploit child labour in their pursuit of the last fish. This is the last gasp of overfishing. When the fish have been taken from these depths, there will be none left within reach of divers. The Philippines should have the world's most spectacular coral reefs. Instead, it serves as a dreadful warning.

The world is living on borrowed time. We can't cheat nature by taking more than is produced indefinitely, no matter how fervently politicians or captains of industry might wish it. Rich nations can outsource production to poor ones, but, at some point, fish stocks will collapse there too and then there will be no fish to be had at any price. In essence, what we have done in the last few decades is to mine fish, bringing them in at rates faster than they can replace themselves. Sharks, bluefin tuna, cod, Chilean seabass, all have declined steeply as a result of excessive fishing. The price that must be paid for today's rapaciousness will be tomorrow's scarcity, or in some places, seas without fish. If we follow the trajectories of loss seen today, that point may be only forty or fifty years away. Fishermen and women of the future will have a tough time keeping their nets full.

Nature has survived planetary upheavals many times before and life will survive this one. Natural selection is a powerful force. It has brought us life in all its wonderful guises and leaves us in breathless admiration at how such exquisite adaptations as the eye, or the colour-shifting skin of a squid, or our self-aware brain could have come about. Life is endlessly adaptable and can eke an existence in some of the most hostile environments on Earth, like superheated hydrothermal vents under extreme pressure in the complete darkness of the deep sea. For over 4 billion years, the planet has remained habitable; life has endured through times when the oceans turned more acidic, when the atmosphere choked on carbon dioxide and methane, when the continents coalesced into parched desert, and when ice ages froze the sea.

But timescale is everything. A starving person will not be spared by the prospect of plentiful food three years hence. Their needs must be met now. Evolution can take a very long time. It works in the currencies of life and death, and the ticking hand of evolution's clock counts out time in generations. Each generation gives forth its young to face the vicissitudes of life. As Charles Darwin deduced, those best suited to the conditions prevail to give birth to their own young, who in turn share some of the characteristics that favoured their parents. So it is that species can adapt to changing conditions over periods of many generations.

Human stresses on the environment exert potent selective pressure on living organisms, winnowing those less able to cope. The hand of evolution can already be seen in the emergence of slower growth and smaller size at maturity in exploited fish, for example. The trouble is we are changing the environment at such breakneck pace it will far exceed most species' ability to evolve adaptations quickly enough. For an animal like a whale, or deep-sea fish, or reef coral, whose generations can span decades, there is too little time. Microbes, and many phytoplankton and zooplankton, will produce generations in hours, days or weeks. There is plenty of scope for them to evolve their way out of trouble. But evolution imposes constraints too. The loss of coral reefs for long stretches of geological history suggests that it isn't easy to find a way to precipitate solid carbonate from seawater when acidity gets too high. So there is probably no 'get out of jail' card to be had here.

Animals, plants and microbes do everything they can to survive. When conditions turn, they adopt all kinds of strategies to persist, like switching to different foods, going deeper to find cooler water, or discovering pockets of favourable habitat in otherwise hostile terrain. We can draw comfort from the fact that life has been through many crises before. Pretty much everything alive on Earth today made it through the de-glaciation that happened from about 20,000 to 10,000 years ago. This transition saw the world warm several degrees and included rapid shifts in conditions that would have tested the hardiest. Life has experienced some twenty glacial cycles in the last 2 million years so it is well versed in the business of tracking environmental change, and ecosystems have undergone many reassortments of their component species.

Financial products these days always include a caveat in the small print below their headline claims of amazing returns: 'past performance is not necessarily a guide to future performance'. There is an important difference between those past climate swings and the present day: us. Species didn't have to cope with the multitude of stresses they face today. We really are sailing into the unknown.

We are beginning to count the human cost of altered oceans in currencies other than food. In the far distant past, when Roman emperors ruled and North America was a patchwork of tribal territories, estuaries and coastal seas were clearer and bluer than they are today. In estuaries such as America's Chesapeake Bay, where oyster reefs lined channels and inlets, they could strain the entire volume of water in a few days, a rate that has fallen a hundred-fold today with the decline of these shellfish. Species on the seabed of open coasts did much the same. The massed ranks of mollusc, sponge and coral drew nutrients and mud from the water and locked them away in shell and sediment. We need them more than ever today to help deal with the wastes that pour from our industries and towns, streets and fields.

In Seattle, coastal residents have suffered growing discomfort and health problems since the 1980s due to heaps of sea lettuce rotting on their beaches. The lettuce produces the toxic gas hydrogen sulphide, which wafts from the beach to cause streaming eyes, sore throats and in some cases breathing difficulties. It may be significant that hospitalizations from asthma in Washington State are higher than the US average and have increased in the last decade.[5] However, it still isn't possible to disentangle trends in ocean pollution from increasing traffic pollution or poverty as causes. The seaweed problem is due to nutrients from sewage outfalls and lawn fertilizer washed from the city's gardens into Puget Sound. You may want to think twice next spring before liberally fertilizing your lawn. Similar problems are popping up in other parts of the world. Residents in the Western Australian town of Busselton saw their beach transform from one of the best in the country to an 'environmental nightmare' within the space of a decade as heaps of rotting seaweed piled up.

In France, the effects of intensive pig farming in Brittany hit the headlines in 2009 when Vincent Petit was exercising his retired racehorse on the beach at Saint-Michel-en-Grève. Without warning he and the horse were sucked into a pit of rotting seaweed up to the horse's shoulders; the horse died within minutes and Petit lost consciousness. He was saved by a bulldozer operator working nearby who happened to see him go under and dragged him from the hole. For years great heaps of *Ulva* – the French equivalent of Puget Sound's sea lettuce – had washed ashore and gathered in piles, liberally fertilized by runoff from pig farms. Sometimes the sun dried their upper surfaces so they became impermeable lids that trapped poisonous gas below. Petit's experience raised the profile of marine pollution in France and spurred a municipal clean-up that a few weeks later cost the life of a bulldozer driver who was overwhelmed by hydrogen sulphide. You know things are bad when beaches turn lethal. The French authorities have yet to tackle the farm pollution at source. Until they do, beaches will continue to accumulate stinking seaweed and the risks it brings.

These stories are shocking but still unusual. Many people remain puzzled by environmental horror stories because they don't match their daily experience. If things are going to hell, they say, why does my own life feel like it is getting better? According to the Millennium Ecosystem Assessment, a compendious review of the state of the world environment, they are right.[6] Indicators of human well-being are on the rise, even as habitats and species are lost around us. This has been called the environmentalist's paradox. It helps explain some people's forthright scepticism of climate change and why they have concluded not only that it doesn't exist, but that there is a massive conspiracy among scientists and greens who peddle this nonsense.

Well-being has certainly improved by several measures: childhood survival, GDP and levels of education have risen steadily the world over, with the exception of sub-Saharan Africa where they lag badly.

These measures don't capture all of the dimensions of well-being, but where aspects of life like happiness or gender equality have been measured they bolster the view that life is getting better for many of us. There is one measure, however, that might give even the most ardent climate sceptics pause, and that is the apparent increase in exposure to natural disasters. Catastrophic floods and landslides are made worse by loss of habitats like marshland and forest. The impacts of Hurricane Katrina in New Orleans in 2005 were worse than they would have been had the city not been built on drained marshes that had sunk below sea level. When the levees were breached, the city became a brackish-water lake. But, while our exposure to disasters has gone up, our ability to deal with them has too. We are richer and better prepared, so the risk of death from natural catastrophe has gone down globally.

There are several possible explanations for the environmentalist's paradox. The first is that food is overwhelmingly the key to human well-being, and food production is up. Production of fish, meat and cereals has outpaced human population growth globally, and health and life expectancy have benefited. (Growth has not been even and there are many places where rapid soil erosion or water scarcity has triggered famines.) A second explanation for the paradox is that our inventiveness has spared us the adverse consequences of our altered environment. Today, if production falters in one place due to drought or pestilence, we can bring food from somewhere else. Soil erosion and lost fertility can be countered by chemical fertilizers. Fish can be taken from distant seas to fill larders as local stocks decline.

The final possibility is more ominous: we have not yet felt the real cost of our activities but payback is on its way. Prudent financial managers will tell you to spend your profit and not draw down your capital. But for decades now we have been depleting natural resources at an unsustainable rate and literally eating into our assets. Times are still good, but we will feel the impact of our spendthrift ways when our savings run out. Our astonishing population growth is speeding the arrival of that day, although the payback will be staged as some resources will run out before others, and people in some places will have to pay sooner rather than later. The hit has come more quickly for a Filipino fisherman than for a supermarket customer in the US or

Europe, but one day it will happen to us too. We are like debtors living the high life until the moment our creditors force us into bankruptcy.

Already, there are credible estimates that suggest we are using something like one and a half planets' worth of natural resources, based on how much area our present consumption would occupy were extraction and energy use sustainable.[7] The only way this overshoot is possible is by living on our savings – in other words, by depleting stocks of natural resources that have built up over thousands or millions of years. Technological innovations and globalization have so far spared the wealthy, but, as environmental debts begin to be paid, we will all be forced to confront the costs of our lifestyles and those of previous generations. The 2006 Stern Review on the economic costs of climate change, commissioned by the UK government, estimated that there could be 200 million climate refugees by 2040.[8] We can see them already in the tide of refugees that flows to Europe today, on the run from drought, famine, flood and despair in places like Somalia, Sudan and Pakistan. One widely cited report estimated there were already 25 million environmental refugees in the 1990s, and the figure was expected to double by 2010.[9]

Human relations with the sea have shifted with the passage of time. For tens of thousands of years our ancestors frequented coasts and estuaries, where they hunted and gathered seafood. Close to the places they lived, the remains of shells and fish bones piled up over millennia tell a story of how they reduced the abundance of some species and altered the structure of ecosystems. But their influence was minuscule against the virgin canvas of a measureless ocean. This was the era of carefree use. Humanity neither cared nor needed to worry about what it took from or added to the sea. Ocean resources were limitless against the scale of human need.

From a human perspective, there was a 'golden age' of the sea, when people first began to harness the oceans for their own ends at a large scale, but before these uses had compromised their abundance and fecundity. In the Mediterranean, this period came in antiquity when commercial fisheries developed for tuna, marine aquaculture was invented and large-scale ports and harbours were built. In the last few hundred years, the steep growth of our influence and impact led people to realize for the first time that their activities could cause problems, like pollution or overfishing. At first those troubles were local and people's responses, if any, piecemeal. Often problems such as declining fish catches were ignored or overlooked as people could still obtain what they needed from further afield. But over time, despite a growing awareness, problems snowballed.

Will we overcome the difficulties we have created? We don't yet know how matters will play out. I can find reasons to worry, and reasons for optimism. There is the enormous scale and accelerating rate of change, the bewildering multiplicity of our impacts and our inexperience in dealing with global problems that are causes for concern. But there are good reasons for hope. People are ingenious and can rise to challenges; we have found some solutions and can devise others. Thus far, I have focused on the problem. In the remainder of this book I will concentrate on what we can do to improve the state of our seas and keep our ocean planet healthy. What is inescapable is that we must change. Continuing on the present course will lead to an environment that is increasingly hostile, both to aquatic life and to our own well-being.

I am an optimist. What gives me hope is the thought that we have never before had so much raw power with which to solve our problems. Of course it is our power over planetary processes that brought the difficulties in the first place. Nonetheless, it puts us in a different position from our less fortunate predecessors whose civilizations failed to survive their own transformations of the environment.[10] What we do share with past societies is our humanity. We possess qualities honed by the struggle for existence that predispose us to short-sighted and selfish behaviour. If we can overcome these evolutionary quirks, then a sustainable future may be possible. The Harvard scientist E. O. Wilson has summed up the challenge: 'We have Palaeolithic emotions, Middle Age institutions, and God-like technologies.' No wonder we struggle to shift from overuse to sustainable use of natural resources. If we are to survive, we will have to replace our present ways of organizing society, forged over thousands of years, with new means, forged in less than a century. The longer we delay action to reduce stress on sea life, the harder it will become to reverse the unwelcome changes we have unleashed. It is easier to fall into the pit of environmental damnation than to climb back out. The risk is that this backward slide will be tough to reverse. We treasure and love our seas and coasts. They are places to have fun, relax and go for inspiration. None of these pleasures can be guaranteed, however, unless we recognize and address the dangers now faced by marine life – and, through them, the dangers we face.

How will it feel for our children's children to live with changes we have imposed upon them? To get an idea of what the future has in store, look no further than Iceland, with its hardy people scratching a living on a scrap of barren volcanic rock.[11] It wasn't always this bare. When Scandinavian seafarers discovered Iceland, they found a country covered in soil, shrubs and trees. Overgrazing by their livestock destroyed the fragile vegetation and the soil was lost to the sea, leaving the grey desert we know today. Until a few years ago, nobody there realized their island had once been green. But now they can look over the monotonous rocky landscape and in their imaginations stroll through woodland and heath, and breathe the scent of rowan flowers in spring. If we fail to control climate change and find ways to live that are less demanding of our planet, I doubt that our descendants will so easily forgive us the problems we bequeath to them.

16

Farming the Sea

'The sea is inexhaustible, and there can never be a general and simultaneous depopulation. The ocean fisheries will always be copious and easy, and their yield will be greater as the ocean becomes more familiar and the methods employed more perfect.'

The French marine biologist Marcel Herubel wrote these lines in 1912.[1] Herubel was a gifted thinker. He was well aware that fisheries could be overexploited locally, but like many others of his time he could not imagine that humanity could ever dent the productivity of oceans on a global scale. Now, landings statistics collected by the UN's Food and Agriculture Organization show that we have attained the unimaginable. After a thousand years of virtually uninterrupted increase, the underlying trend of global wild fish catches has been down since 1988.[2]

The picture looks worse if we set fish landings against the explosive growth in human population. Wild fish available per person has fallen by more than a quarter since it peaked in 1970. If fish had been equally available to every citizen of the planet and all of the catch had been eaten directly – rather than fed to livestock – everyone could have eaten 190 grams of fish per week in 1970; today that has shrunk to 136 grams per week.[3]

Since at least the early twentieth century, governments have exhorted us to eat more fish to be healthy and slim. Some governments publish guidelines on how much we should eat. New Zealand and Greece think people should eat 500 grams or more per week, while Canada and Austria think 150 grams is fine. The US government recommends people eat 340 grams per week, while UK citizens are urged to eat 280 grams. Averaged across the world, we are told we should eat about 260 grams of fish and shellfish every week, an amount nearly double the availability from wild fish catches. There isn't enough wild fish to go round.

Ours is an ingenious species and we love technological fixes. If there are no longer enough wild fish in the sea to feed us, surely it makes sense to farm the oceans? Aquaculture, the production of fish and shellfish in captivity, is a no-brainer. Many now talk of a blue revolution in aquaculture to match the much-vaunted green revolution in agriculture. Farmed fish and shellfish already make up around 46 per cent of the fish sold for human consumption worldwide.[4] So is aquaculture the answer?

Aquaculture has a long history. The earliest representation of fish being kept in ponds was painted on the wall of an Egyptian tomb in the Nile Valley 4,000 to 4,500 years ago.[5] It shows a man seated beside an artificial tank with a central drainage channel, fishing for tilapia with a rod and two hooked lines. Floating plants shelter the fish, while in the background men pick fruit from trees that may have been irrigated by used pond water. Behind the angler, a woman receives the fish.

At about the same time (give or take 500 years), aquaculture is thought to have developed in China, where freshwater common carp were reared in ponds using fry collected from rivers and lakes. The first manual of fish culture was written in 475 BCE by the Chinese author Fan Li, who is better known today for his popular guide to the rules and pitfalls of doing business. The Chinese refined their methods over the centuries into a sophisticated system of polyculture, raising several other species of carp with different diets, mainly in ponds, but also in rice paddies and lakes. There is a possibly apocryphal story that this diversification was a direct response to an edict by the seventh-century Emperor Li Shimin, who declared that nobody was allowed to keep common carp because the fish shared his name. The different species used in polyculture systems eat algae, insects, food waste and even sewage. They benefit the rice by keeping the mud well aerated and the weeds and pests in check.

In Europe, the Romans, whose appetite for fresh fish is the stuff of legend, took aquaculture to the sea over 2,000 years ago.[6] Fish ponds were mainly built using a Roman invention, hydraulic concrete, which sets underwater. They have proved incredibly resistant to wave and tide – so much so that many remain intact and almost appear ready to receive fish today. These ponds were not intended to feed the masses. They were expensive to construct and costly to keep stocked. Lucullus, perhaps one of the greatest fish gourmands, reputedly spent more on his fish ponds than he did on his coastal villa.[7] He cut a tunnel through a hill to bring water from the sea for his fish.

Roman ponds were remarkably sophisticated. Some were built into dining rooms to form the centrepiece for banquets. Proximity to the fish they ate was important to Romans. In one dish, the fish were slaughtered at the table and guests would enjoy the colour changes of the dying animal. Some ponds had viewing platforms, while others had fishing towers from which an angler could pursue his lunch. Many ponds had ceramic pots built into their sides to provide hiding places for eels, which were highly prized at the time. (Some eels were kept as pets and adorned with gold jewellery inserted into body piercings.[8]) Coastal ponds were often built around natural coves or grottos and were connected to the open sea by channels closed with grates. Waves and tides renewed the water over the tops of the walls and through these grates. Like all previous attempts at aquaculture, Roman fish ponds were really just ranches. Fish were taken from the wild to be stored and grown in captivity, thus ensuring a ready supply no matter what the weather or season.

When Captain Cook first reached Hawaii in 1778 he found an extensive network of several hundred fish ponds.[9] Some were enormous, with rock walls over a kilometre long, up to two metres high and eleven thick at the base. In Cook's day, freshwater and marine ponds together supplied the people with an estimated 900 tonnes of fish per year. Hawaiian aquaculture is believed to have originated a thousand years ago and contained what might be called an entire ecosystem. Larval and juvenile fish entered through grates that opened to the sea, and grew in size within until they were trapped and periodically harvested.

Necessity has long been a wellspring of creativity. In the Middle Ages, between the eleventh and thirteenth centuries, the avalanche of soil washed off lands newly ploughed for agriculture in Europe triggered the crisis in freshwater fish supply which I described in Chapter 2. At the same time, sturgeon, salmon and whitefish found their migration routes newly blocked by mill dams. The fall in supply was set against an increase in demand from a larger and more urbanized population. Medieval Europeans had come to value fish more highly due to the spread of Christianity, which prohibited believers from eating the meat of quadrupeds on certain days. The result was the development of pond culture for carp sponsored by monasteries and the aristocracy.

European freshwater carp culture was more sophisticated than the Roman sea-fish ranches because it encompassed the entire life cycle, from egg to adult. However, it was not until the nineteenth century that efforts began in earnest to close the life cycle of marine fish in captivity. Again, necessity provided the impetus. Salmon runs in European rivers had for centuries been in private ownership, with fishing rights mainly reserved for the aristocracy. The Industrial Revolution caused many of those rivers to become so polluted and blocked by dams that salmon numbers plummeted. Fish hatcheries were established; salmon were stripped of their spawn in captivity and their offspring raised to fingerling size before being released into rivers to complete their life cycle in the wild. The growing sophistication of these efforts is attested by the fact that, from 1868 until the early twentieth century, salmon and brown trout eggs were taken by ship to New Zealand for release into rivers and lakes for the sport and sustenance of expatriate Brits.[10] Brown trout got going quickly, but the salmon were troublesome, eventually establishing in the watershed of only one river.

The late nineteenth-century revolution in sea fishing put wild fish stocks under intense pressure again and stimulated the first aquaculture experiments with plaice and turbot. As with salmon, these efforts concentrated on spawning fish in captivity to supplement wild populations with large-scale releases of young fish. Marcel Herubel captured the spirit of these pioneering efforts in 1912:

Piscifacture may legitimately claim to be of social value. It goes further than pisciculture, which limits itself to rearing fish; it endeavours to 'manufacture' fish. The first art is a nurse, the second a mother ... piscifacture will restore to the sea what it would have produced had man not multiplied his catches.[11]

For fairly obvious reasons, these well-meaning attempts ended in failure. Juvenile fish suffer extremely high mortality rates in the wild and few reached adulthood. Even the release of millions of young fish had little impact on the numbers that eventually found their way into fishing nets, especially since captive-reared juveniles tend to be less fit than their wild-spawned brethren. So the direction of aquaculture shifted.[12] A parallel development concentrated on raising wild spawned animals in captivity from egg to adult. A third strand saw the refinement of methods to raise shellfish such as oysters and mussels. Shellfish mariculture has ancient origins. In the Broughton Archipelago of British Columbia, First Nations people erected rock walls across bays to trap sand to the depths favoured by clams, which were an important winter food.[13] Nobody is sure when these walls were first built, but they may go back 5,000 years and show just one of the many ways these people transcended the simple hunter-gatherer stereotype to nurture nature to support their needs.

In less than a century aquaculture has evolved from a frustrating practice pursued by mavericks into a highly successful international business. From 1950 onwards, when the FAO began to collect statistics, aquaculture production has expanded from less than 1 million tonnes a year to more than 50 million tonnes, just over half of which is from freshwater fish like carp and tilapia. To that we can add another 16 million tonnes of seaweed. In the last decade or two, the growth of aquaculture has outpaced every other form of food production except organic (which is rising fast from a low base). Since 1950, it has increased at three times the rate of growth in world meat production, and even outpaced world population growth. So, when aquaculture is included, the total amount of fish available to eat per week has risen from 197 grams to 207 grams per person, better than the picture from wild fish catches alone but still a fifth less than the average of health recommendations. Unfortunately, about a third of wild catches are fed to livestock, pets and aquaculture, leaving just 160 grams per person per week to eat.

While nearly 400 different species of fish and shellfish are cultivated, a handful dominates production. On the marine side, the major
players are mussels, clams and prawns. Shellfish are cultured all over
the world, mainly on ropes strung across bays and mud flats in operations that can cover tens of square kilometres. Mussels and clams are
the mainstays of these farms, but even free-living molluscs like scallops are grown on ropes in bags.

Atlantic salmon is the species most Western consumers associate
with aquaculture, but in 2008 salmon farms produced 1.5 million
tonnes of fish, only about a twelfth of the volume of molluscs and
prawns sold that year. Salmon are joined in the top ten farmed fish by
a cast of lesser characters, including rainbow trout, seabream, seabass,
Japanese amberjack, barramundi and the unfortunately named bas-
tard halibut, whose forename is usually dropped by supermarkets! But
other highly lucrative species are also now farmed, such as bluefin
tuna, groupers and sturgeon.

In 2008, four out of five fish eaten in China were raised in farms.
Outside China, one in four fish eaten was farmed, up from one in
twenty in 1970. It seems that there is hardly a sea loch or fjord in the
developed world that does not support, or is not earmarked for, aquaculture. There are parts of Japan and China where you can stand on
the beach and see nothing but fish pens or seaweed strings all the way
to the horizon. In terms of food production, aquaculture has been the
roaring success of recent times. Putting that aside, the industry has
many problems. Some are acknowledged even by its most enthusiastic
advocates, but others have long lurked in the background and have
only recently begun to cause concern.

While filter-feeders such as shellfish can feed themselves, the awkward fact is that farmed fish must be fed, and many of the species
most prized by consumers in developed nations, like salmon, seabass
or tuna, eat flesh. People often think of fish farming in terms they are
familiar with from the cultivation of cattle or sheep. We raise animals
to convert things we can't eat, like grass, into things we can. But this
simple logic doesn't apply to aquaculture of predatory fish, which
happen to be the ones most Western consumers like best because they
are large and tasty with firm, succulent meat. While some freshwater
fish such as carp and tilapia eat a largely vegetarian diet, almost all of

the marine fish raised in farms are predators. So it is that we must catch wild fish to feed to captive ones, mainly so called 'forage fish' like anchovies, capelin, herring and menhaden that live in large schools. These animals would make perfectly good food for us if we ate them directly rather than passing them through salmon or seabass first. Here, then, is the first of many controversial aspects of fish farming: it can sometimes take several kilograms of wild caught fish to produce just one kilogram of farmed fish.[14]

Several of the species we farm are ferocious predators and come from very high in the food chain. Bluefin tuna are caught as juveniles in Australia and the Mediterranean and then grown to market size in pens, during which time they eat an eye-watering quantity of wild fish, up to twenty times the weight of flesh they produce. You get an idea of the problem by considering the idea of trophic levels. They are a shorthand way of assigning a species' place in the food chain. Think of them as steps on a ladder that ascends from plants and organic muck at the bottom, to sharp-toothed, beady-eyed carnivores at the top. Plants and detritus are level 1; herbivores and detritus feeders are level 2; meat eaters are level 3; and carnivores that eat other meat-eaters are level 4. Few food chains go longer than five links because something like 90 per cent of the energy passed upwards is lost at each step. It gets dissipated sustaining life and producing the next generations. Only about 10 per cent builds bodies. This energy loss explains why lions are rare compared to antelope and eagles less abundant than rabbits.

This is a simplified view of a food chain, since it assumes that species occupy just one trophic level. Most animals have a mixed diet, eating foods from more than one rung of the ladder. We can reflect this by giving them intermediate values. Most people eat a mix of meat and vegetables, so the average trophic level of humans is about 3.2. The trophic levels of Atlantic salmon and bluefin tuna are 4.4, which gives them a position on the food web higher than African lions, which come in at less than 4. If you are ever stuck for conversation at a party, you might drop the startling nugget that most domestic cats have a higher trophic level than lions. All that salmon and tuna that Tiddles eats places her well ahead of most big cats in the league of carnivory. Because predators sit much higher in food webs than sheep, goats or cows, the ratio of what you get back to what you put in is much less favourable.

In the last ten years, producers have worked hard to rebalance this equation. Across the industry as a whole, the ratio of wild fish going in to farmed fish coming out has fallen from roughly 1:1 to 0.63:1 (excluding species like mussels that feed themselves from the water).[15] Mostly their motivation is profit rather than concern for the environment, although there are refreshing exceptions. Fishmeal and oil from wild sources is expensive and prices have risen faster than inflation as demand has grown. About half of the fishmeal and nearly all the fish oil produced globally are now consumed by aquaculture (the rest goes to pigs, hens and even cows). Since wild fish catches have levelled off, aquaculture will likely swallow all the world's fishmeal by 2050, if not sooner. Those who can squeeze more fish out for every one put in gain a powerful kick to their bottom line. And, while there have been improvements in feed conversion ratios, for some species the figures remain bad. Because of the high fish oil content of salmon feed, pressed out of animals like anchoveta, salmon still average about 5kg of wild fish to produce 1kg of farmed.

There are several substitutes for wild fish flesh in aquaculture feeds. Some manufacturers blend scraps from processed fish into their feeds, but their oil content is low so it is only an answer for some farmers. Others grow marine worms for fish to eat. The most widely used substitutes are protein and oils from plants, such as soyabeans and rapeseed. The problem is there is a limit to how much plant matter a carnivore can take before its health suffers, and (of more immediate concern to consumers) its taste. Moreover, the much-vaunted health benefits of eating fish stem from their content of omega-3 oils. There have been many claims over the years about the wonders of fish oils, including that they reduce cholesterol, protect the heart, sharpen the mind, ease depression, lubricate the joints, improve vision, prevent Alzheimer's and on and on. The protective effects on the heart are well-accepted, although many other presumed benefits remain unproven. Omega-3 oils are only produced by aquatic plants, so feeding terrestrial plant substitutes to farmed fish will rob them of their health-food cachet. The German agro-chemicals giant BASF has an answer in the form of rapeseed plants that have been genetically modified to produce marine oils. These plants have been approved in the US, but many people remain wary of genetic manipulation so this alternative is unlikely to satisfy all.

In some places, the struggle to find enough fish to feed aquaculture and farming has taken a troubling turn. Bottom trawl fisheries can have fiendishly high rates of bycatch of unwanted species, most of which are thrown away dead. In India and Venezuela, a new use has been found for this bycatch.[16] Instead of chucking it over the side, it is dried and converted to fishmeal that then goes to chicken farms and aquaculture. At first this seems like a good idea, but it can keep trawling profitable long after the practice has driven the original, high-value target animals to economic extinction. The one last check on a fishery, which might spare an ocean habitat from being completely destroyed by overfishing, is its inability to turn a profit. If every scrap of life brought up in the trawl can be sold, then the trawlers will continue to drag their nets until there is nothing left at all.

Not all farmed marine fish are as predatory as salmon or tuna, and many shellfish eat plankton, plants or detritus. Nonetheless, finfish culture consumes more than one kilogram of wild fish for every kilogram produced, and across the board, aquaculture swallows two-thirds as much wild fish as it produces. Hardly a solution, then, to problems of overfishing. Why don't we just eat the wild fish directly? This is exactly what many people, celebrity chefs included, have tried to encourage. The main ingredients of fishmeal are animals like anchovies, sardines, sprats, herring and blue whiting, fish whose flesh simply drips healthy oil. The problem is that most are small and therefore fiddly, inconvenient and full of fine bones that wedge between your teeth or catch your throat on the way down. In a world of ready-meals and convenience foods in developed countries it is hard to persuade people to get intimate with heads, tails and slime when the alternatives are plump, bone-free fillets from farmed stock. Supermarkets love farmed fish too because they can guarantee supplies of even-size, uniform-quality fish fillets no matter what the season or weather. Farmed fish are definitely here to stay.

There is concern that fishing for 'forage' fish to sustain Western appetites for predatory fish robs the world's poor of their staple diet. These small fish made up over half of the fish protein eaten in a sample of thirty-six developing countries from Africa, Asia and other parts of the world.[17] In something like half of African nations, fish protein makes up a quarter to over a half of the total animal protein eaten. Industrial fishing by other countries in African waters under access agreements threatens fish supplies and has already driven up demand for bushmeat, to the detriment of wildlife on land.[18]

One alternative to these fish is krill, a finger-sized, bright-carmine shrimp that lives in polar seas. Estimates suggest that Antarctic waters harbour enough krill to sustain all of the world's aquaculture with little difficulty. Already, hundreds of thousands of tonnes of krill are caught annually to make into fishmeal.[19] Nothing is that simple, sadly. Fish, penguins, whales and seals depend on krill and industrial fishing could harm them if it dents krill stocks. And krill are already on the way down as a result of climate change. Krill survive over the winter by grazing algae attached to the underside of floating ice. If the ice melts, krill supplies will crash.

Aquaculture also suffers problems of husbandry. Farmed fish, like any animals kept at high density, are troubled by diseases and parasites. To combat infection and infestation, most farmers dose animals in ponds and pens with antibiotics, pesticides and fungicides. Since pens are mostly open to the sea, these chemicals spread pollution far beyond the bounds of the farms. Nor are these problems confined to captive fish. Take salmon farms, for example, which are mostly located in estuaries that lead directly to rivers inhabited by wild fish. In the wild, adult salmon rarely come into contact with juveniles because they either die after spawning or return to sea before the young hatch. Now, young salmon must run a gauntlet of salmon farms on their migration to the sea, bringing them into contact with the diseases and parasites afflicting captive adults. Sealice, for instance, feed on salmon skin and muscle. The lesions they create don't normally kill adult fish, but they can kill the young, who wouldn't normally have encountered them, at least not in such densities. In recent decades wild salmon runs in rivers that empty past farms have declined to a far greater degree than in those that don't. Estimates put the fraction of wild salmon smolts killed by sealice from farms at up to 95 per cent in one part of British Columbia.[20] In the Atlantic, salmon are ghosts of their former abundance in the wild, so such killing rates could wipe them out entirely. Since each river has its own salmon stock that has adapted to local conditions, every disappearance is a step on the road to extinction.

Farmed animals suffer from more aggressive illnesses than their wild brethren. These diseases have emerged as a result of the unnaturally high densities at which animals are kept, and their prevalence is encouraged by stressful conditions within pens and genetic uniformity of the stock. Infectious salmon anaemia seems to be a new strain of a virus closely allied to influenza that has adapted to exploit the high densities of susceptible fish in salmon farms.[21] Before the advent of pen culture, there had only been a single report of this virus in the wild. When an outbreak occurs it sweeps through captive fish and many countries now require all the fish in affected farms to be destroyed. Ironically, the industry itself has been the main vector for disease spread. Outbreaks in Scotland in the 1990s were closely linked to visits between farms by boats that carried salmon stock.

Infectious salmon anaemia can be carried by fish that have recovered from the worst effects of the illness. It can also be transferred from fish to fish by sealice, which raises serious concerns for the viability of wild fish stocks that have come into contact with infected farms. The disease has now spread to the US, Canada and Chile. It was first detected in Chile in 2007 and at the time of writing had halved Atlantic salmon production there.

In shrimps, another disease has jumped from pond to pond all the way across the Pacific. White spot syndrome virus emerged in China in the 1990s, then spread to Japan and Southeast Asia. Affected shrimps become lethargic and discoloured, but in truth there is hardly time to notice these symptoms since the virus can kill the entire stock of an affected farm within a few days. By the mid-1990s, the virus had nearly eliminated Chinese shrimp output, and, by the late 1990s, it had made it to North and South America. Amid all the death and destruction, white spot syndrome virus has at least had one positive effect. Before the disease emerged, shrimp farming often relied on the capture of wild juveniles to stock the pens. Now the industry has mainly moved to spawning its own shrimps from certified disease-free broodstock. Pens also used to exchange highly polluted water with the open sea, but there has been a shift towards greater separation between pens and the sea to minimize possible spread of infection.

Isolating farms from the open sea is a good idea for many reasons. When faced with disease, farmers have three choices: prevent, treat or risk losing everything. Options to prevent disease include vaccination, prophylaxis or disinfection. Vaccination is clean and clinical but only available for a few species and diseases, and often only in developed countries. Prophylaxis usually involves feeding animals antibiotics, while disinfection can mean dosing ponds with toxic compounds such as malachite green, which is known to cause cancer in people. There is growing disquiet about the use of antibiotics in aquaculture. Chile uses over 100 tonnes of quinolones every year, mostly on fish farms.[22] These antibiotics are important in human medicine but such heavy use risks the development of antibiotic resistance in microbes. Bacteria tested from sediment and water samples near fish and shrimp farms have begun to develop such resistance, and some can shrug off several antibiotics at once.

You might think that there is little chance of marine bugs causing trouble for us, but bacteria are able to swap genetic material, and genes for resistance have already made the leap from sea to land, and from bugs that affect animals to ones that infect people. A cholera outbreak in South America in the 1990s was of a strain that had picked up antibiotic resistance from contact with a bacterium that owed its enhanced resistance to the heavy use of drugs in Ecuadorian shrimp farms. The sometimes fatal gut bacterium *E. coli* has also acquired antibiotic resistance via aquaculture. For some human pathogens, exposure to antibiotics in aquaculture can be direct. Asian carp ponds are often praised as a model of sustainability because they raise fish that eat vegetables, dirt or insects, and they recycle human wastes. In Vietnam, some cities rely on vast aquaculture complexes to treat their sewage, which might make you pause the next time you spot some plump fillets of basa, a type of catfish, in the supermarket.[23] So human pathogens swirl around ponds where they can gain antibiotic resistance before being passed back to us when fish are harvested.

Many developed countries have recognized the dangers of antibiotics and have banned their use in aquaculture. In Norway, most salmon are vaccinated against problem infections. Things that happen in Asia or South America or Africa can seem comfortingly remote to many of us. But this is a globalized world and what happens in one place soon spreads to others, as outbreaks of bird and swine flu show. More than 90 per cent of world aquaculture production comes from developing countries, so until the problems are solved there little progress can be made. In Asia, where most of the world's farmed fish are grown, farms routinely use industrial quantities of antibiotics, pesticides, disinfectants and antifungals.

Toxic chemicals released to the sea are a worry, but they highlight another of the ways in which aquaculture does damage. Where pens exchange water with the open sea, as most do, the wastes from excess food and faeces simply washes away. If there are few farms and plenty of water to slosh past them, we might see little impact and easily live with the consequences. But farms are often put in sheltered bays, estuaries and fjords where water exchange is limited, and the growth of fish farming means that some places are packed with pens, each of which can contain tens or hundreds of thousands of animals. That adds up to a huge amount of waste. By 2000, Scotland's salmon farms released as much nitrogen – one of the main fertilizing nutrients that fuel plankton blooms – as contained in the untreated sewage of 3.2 million people, over 60 per cent of Scotland's population.[24] Output of phosphorus, the other major plant nutrient, was equivalent to the waste produced by 9.4 million people. In the early years of the twenty-first century, Chinese shrimp ponds released 43 billion tonnes of effluent into coastal seas, compared to just over 4 billion tonnes of industrial effluent and sewage from land-based sources.[25] Shrimp farms produce mind-boggling amounts of nutrient-rich sludge: in intensive aquaculture, tens to hundreds of tonnes per hectare for every shrimp crop.[26] These farms have doubtless played a critical role in the problems of oxygen depletion, harmful algal blooms and fish kills that have plagued regions like China's Bohai Sea. Blooms in China now affect an area bigger than the Mississippi dead zone. Toxic algae provide a strong motivation to find cleaner ways to farm fish, since they can wipe out stocks and profits within hours when they strike. In Philippine bays so crowded with cages and pens there scarcely seems any open water, since 1999 yearly mass kills of milkfish and tilapia have destroyed hundreds of thousands of tonnes of stock. Such problems not only threaten production, they pose real health concerns to consumers.

If the Bohai Sea is not the most polluted sea on the planet, it must come close. It lies downstream of the heartland of the Chinese economic miracle. This transformation of fortunes, like the past industrial revolutions of Europe and America, has been bought at the price of severe pollution. Industrial effluents mingle with sewage discharges from tens of millions of homes. Maps produced by the Chinese government show almost every part of the Bohai coast suffers from excessive levels of contaminants like lead, chromium, mercury and arsenic, alongside the fertilizing nutrients.[27] Off the city of Tianjin, water quality is so bad that most falls into the pollution category of 'offensive in colour, smell and taste'. This pollution threatens the burgeoning aquaculture industry. So filthy has the water become that there is widespread use of antifungal and antibacterial compounds to keep the animals alive, like nitrofuran, a known carcinogen banned throughout much of the Western world. So, farmed animals are contaminated in multiple ways – with DDT, flame retardants and mercury from the water, and with prophylactic chemicals applied by farmers. As China has become a major exporter of farmed fish and prawns, these problems are a concern for consumers far beyond the shores of the Bohai.

In previous chapters I have talked about how aquaculture has contributed to the spread of invasive species and diseases around the world. This 'biological pollution' is considered by some to be among the worst effects of fish farming, because it is usually irreversible by the time the problem is noticed. Atlantic salmon have already established themselves in several rivers in British Columbia. Huge Pacific oysters, three times the size of a man's hand, are now embedded in Northern European seas after they were introduced to replace native oysters. In the Gulf of Maine, imports of European oysters probably brought in the invasive seasquirts and finger-like *Codium*, a vibrant green seaweed that competes aggressively for space with native fauna. While no one knows how these or hundreds of other alien species incubating in new environments will turn out, some will definitely go rogue. But there is another form of release which is just as irreversible and scares some people even more: the use of genetically modified fish.

Like any form of farming, aquaculture has its high technology enthusiasts who can spot ways to make better animals. At the time of writing (2011), the first transgenic salmon was nearing approval in the US.[28] Called Aquadvantage™, this Atlantic salmon has been given a chinook salmon growth hormone gene, and another from ocean pout that prevents the growth hormone from being switched off in winter. So these fish pile on weight year-round to reach market size twice as fast as normal salmon. Protestors argue that escaped animals will interbreed with wild salmon and set free Frankenstein fish. But the company that makes them says it is highly unlikely, since 99 per cent of the fish have an extra copy of the salmon genome to ensure infertility and they will be grown in closed containers on land. Protestors do have a point though, since reassurances from the biotech industry have several times proved less robust than claimed. In any case, this salmon will have to clear regulatory hurdles outside the US to gain a foothold on the market since most salmon farming is in Canada, Chile and Europe. But, despite passionate resistance among environmentalists, it seems inevitable that in the long-run aquaculture will embrace genetically engineered fish.

The explosion of marine fish farms across the world has transformed coasts, estuaries and deltas in dozens of countries. China has pursued one of the most aggressive aquaculture development programmes of recent years. In 2003, marine aquaculture covered 15,000 square kilometres, about half of China's 32,000-kilometre coastline.[29] Most of these farms were carved out of mangrove forests, mud flats, saltmarshes and seagrass beds. Satellite photographs of the Bohai Sea coast, one of the most intensively farmed regions, show the toll taken. Straight-edged ponds incised in blue and turquoise pack the coast to depths of several kilometres inland, and crawl seaward across mud flats. The Philippines and Vietnam have lost three-quarters of their mangrove forests in the last few decades, half of them to aquaculture. Sadly, many of these ponds have been abandoned. When mangrove soils are exposed to the air, they become acidic. The acid leaches into pond water together with toxic quantities of aluminium, so the ponds cannot be used unless they are lined.

Mangroves and saltmarshes are self-repairing buffers that defend coasts against storm and flood. If they remain healthy, there is a good chance they could also ameliorate the worst effects of sea level rise by trapping sediment and building upwards. In many places, aquaculture has not only removed this benefit, it has caused the land to sink by sucking freshwater from below ground to create brackish-water ponds for shrimp and milkfish.

Ironically, the loss of these habitats starves fish farms of two things they need most: clean water and animals to stock their ponds. Coastal wetlands draw nutrients and pollution from the water that washes through them. They are also nurseries for a huge variety of fish and shellfish. Although disease has forced many shrimp farmers to switch to hatchery-raised fry, countless farms throughout the tropics still depend on wild sources to stock their ponds, despite its catastrophic environmental cost.

Tiger prawns are much favoured for their large size and rapid growth, but they represent a tiny fraction of wild shrimp fry. Most wild prawn fry are caught by nightlighting in shallow water under a moonless sky. Within minutes, netters are surrounded by a confusing buzz of hundreds or thousands of tiny animals, fish and fry, all drawn to the light like moths to a flame. For every individual tiger prawn caught in Malaysia and the Philippines, several hundred fry of other species are wasted. Half of the hundreds of millions of tiger prawns grown in Bangladesh are from the wild. The waste doesn't stop at other shrimp species either. In one study, scientists found that every tiger prawn fry collected cost the lives of hundreds of finfish fry, and over a thousand other animals that live on the plankton.[30]

In developing countries, the wetland grab has robbed the poor of the common lands from which they once eked a living by fishing and gleaning. Aquaculture provides jobs and income for some of the displaced. But the divide between rich and poor has grown, and the latter's quality of life suffers with every waterway blocked, every forest tree felled and every litre of pond sludge pumped into the sea. In several countries, for instance Bangladesh, Thailand and Honduras, some of those bold enough to protest about the injustices have been murdered to quash opposition.

So far, aquaculture has sheltered most of us from the effects of overfishing. Throughout the developed world, supermarket shelves creak under the weight of an ever-expanding range of farmed fish and shell-fish. There is something for everyone if you can afford to pay: sturgeon caviar, bluefin tuna sashimi and oysters on the half shell tempt high-end diners, while there are salmon and shrimp for the masses. And yet, the amazing expansion of fish farming since 1950 has come at a terrible cost to coasts, wetlands and shallow seas. Governments, hell bent on foreign exchange or job creation, have encouraged aqua-culture heedless of warnings. It is not hard to see that it can't go on like this.

The drawbacks of aquaculture are such that one might question whether it really is the solution to overfishing. Wouldn't it be better to protect fish in their natural habitat? If we were to manage wild fisher-ies well, we might be able to increase supplies from the open sea by a third to a half. (I will come back later to how we can do this.) But a 50 per cent increase falls far short of the needs of 9 billion hungry people expected by 2050. So if the world aspires to a healthy diet of fish, aquaculture will be essential. Like any kind of farming, there are better and worse ways of doing it. The present blue revolution will need to turn blue-green for aquaculture to become a net contributor to human well-being. What would it take to do this?

Better farming practice comes in many forms. In some countries, shrimps are grown in ponds at low enough densities that nature can feed them without the need for supplementary food. But such ponds take up more space than intensive farms, at a greater cost to wetlands and coastal ecosystems. In the Philippines, the value of mangrove forests is now beginning to be recognized, thanks to the tireless campaigning of scientist and activist Jurgenne Primavera. Based on decades of re-search, she says that shrimp ponds should not exceed a quarter of the area of mangrove forest if we are to preserve the ecological function of coasts. Her work has led efforts to restore forests and move ponds behind sheltering buffers of trees, and she was hailed a Hero of the Environment by *Time* Magazine in 2008.[31] (Shellfish and seaweeds can also be cultured in natural mangrove and saltmarsh channels.) At the other end of the spectrum, experimental farms in Belize have gone for high-technology, super-intensive methods. There shrimps are raised under covered raceways and fed on biofloc, clumps of microbes formed around starch grains and sprinkled into the water. This reduces the need for expensive feed and helps recycle nutrients from shrimp excreta, which reduces sludge production.

We may need to say goodbye to the succulent predators favoured by Western consumers and rich Asians alike. Marine aquaculture will have to learn instead from the ancient art of carp polyculture practised in the freshwater ponds and rice paddies of Asia. We need to find species that grow well together, so that one will clean up the waste produced by another. Some fish eat seaweed or detritus and can be raised more sustainably than those that crave fish flesh. Mullet grub around on the seabed for their food and might reduce pollution problems if farmed together with more predatory fish. Likewise, sea cucumbers are considered a delicacy throughout much of Southeast Asia and have been seriously overfished in the wild. These animals are a bit like vacuum cleaners with a hole at one end where detritus goes in and a bum at the other where something that looks nearly identical comes out. Years ago, before I was aware of seafood problems, I tasted sea cucumber soup. I have to admit I'm no connoisseur and it felt a bit like chewing on a rubber band. But the Chinese and Japanese love sea cucumbers and it could make sense to grow them beneath pens of fish to help recycle their waste.

The industry will need to work hard to raise standards and improve sustainability. I have met many fish farmers who are committed to doing just that. With their energy and enthusiasm aquaculture could indeed help feed the world. But there are challenges ahead. Growing shellfish has always been touted as one of the most environmentally friendly ways to produce seafood. Mussels, clams and scallops feed on plankton and other organic matter filtered from the water around them. They don't need to be fed wild caught fish and can improve water quality. But there is a catch. They depend on their carbonate shells, and life is going to get much tougher for them – and would-be aquaculturists – as the seas become more acidic. If you are fond of mussels and scallops, you may want to think hard about what we can do to reduce carbon dioxide emissions.

17

The Great Clean-up

There is an old adage that 'dilution is the solution to pollution'. Follow that logic and you would think that there is no better place to dispose of waste than in the sea. But even the sea can reach saturation. Over the last twenty years, estuaries, bays, gulfs and enclosed seas have been overwhelmed by the volume and variety of waste, both of the kinds you can see, like plastic, and the kinds you can't, like toxic chemicals and fertilizers.

In the past, nature could be relied on to cleanse our effluents as they passed through coastal marshes and mangroves. Once at sea, they were filtered by the gaping mouths of millions of suspension feeders that packed the seabed or swam in shoals so large they darkened the water for tens or hundreds of kilometres. But as the sea has been stripped of its consumers and its coasts cleared of the wetlands that once lined creeks and waterways, so its power to regenerate waste has waned. Dilution is no longer the solution. Today, the emphasis in pollution control must swing to reduction, prevention and, in some cases, removal.

For persistent pollutants like heavy metals and PCBs, reduction and prevention are essential. There has been some progress over the last two decades, but past generations have left us a toxic legacy that we must pay for, and every year we are adding more. In Europe and North America, bans have long been imposed on the use of the most toxic and lingering compounds like PCBs and organochlorine pesticides like DDT. These measures have begun to reduce the burden of pollutants in the Baltic Sea and the Gulf of Maine, but DDT and its toxic breakdown products are still with us, and these bans will have limited global impact until all of the developing world follows suit. Left to their own devices many of these chemicals bind to particles of silt and clay suspended in the water. When sediments sink to the bottom the toxins are buried, and in an ideal world that is where they would stay. Unfortunately, storms and tides mix sediments back into the water column and circulate the toxins back into the food web to haunt us again. And there is also a human force mixing mud and its toxic hitchhikers back into the water: bottom trawling and dredging.

It amazes me that more has not been made of the way that trawlers and dredgers kick up mud from the seabed. Trawlers fishing shallow water are visible from space because of the 'contrails' of mud that stream from their nets. The Gulf of Maine is one of the most heavily trawled seas on the planet. Oceanographers have known for years that a thick and permanent layer of suspended mud overlies much of the seabed, but they weren't sure why it was there. When worms that normally live within seabed sediments showed up in traps set twenty-five metres above the bottom, the mystery was solved.[1] Bottom trawlers churned mud, worms and pretty much everything else into the water above. Parts of the Gulf of Maine contain high levels of persistent pollutants washed downstream from the factories and heavy industry of New England. Here and elsewhere, trawlers are revisiting upon us the ghosts of chemical industries past.

It seems only recently to have dawned upon many developed nations that the oceans are a common pool shared by all, and that the pollutants released by one nation can swiftly slosh to the shores of another, even thousands of kilometres away. Plastics are easily seen, but pollutants like mercury or DDT, harder to detect and easily overlooked, have spread to every corner of the globe. Bizarrely, many developed countries still permit companies to manufacture and sell toxic compounds like DDT overseas. DDT itself is permitted for the control of mosquitoes that spread malaria, and 3–4,000 tonnes of it are sprayed inside people's houses in Africa and Asia every year. In India and North Korea, people still use DDT illegally for general agriculture. Malaria is terrible, but can't we come up with a less toxic alternative?

Like many others, this problem requires a global solution. The Stockholm Convention on Persistent Organic Pollutants came into force in 2004 and at the time of writing 176 of the world's 195 nations were party to it.[2] Currently it commits signatories to eliminate seventeen of the most toxic and accumulative chemicals, and limit production and use of several others, but we still have a long way to go. There are still too many countries where highly toxic pollutants are used and ultimately wash into the sea.

Much of the mercury emitted, although toxic, falls outside the umbrella of international conventions, or even national laws. Perhaps because of past successes in controlling mercury releases from industry, it is coal burning and waste incineration, not chemical use by manufacturers, that releases the bulk of mercury today (although precious-metal mining is a big contributor in some places). Mercury gives us another excellent reason, alongside greenhouse gas reduction, to move away from coal as a source of power. Given the rising mercury contamination of marine predators and the role of methyl mercury as an insidious poison, I don't think this problem has yet got the attention it deserves. Following close on mercury's heels, pharmaceuticals are emerging as another class of pollutants that needs to be taken more seriously. Again, the only way to clean up pharmaceuticals is to prevent them getting to the sea in the first place. Sophisticated scrubbing technologies are being trialled on sewage-treatment plants in the European Union, but they are expensive and energy intensive. The search is on for cheaper and more environmentally friendly alternatives.[3]

Despite my concerns, I am encouraged by recent efforts to deal with pollution. Clean air and water laws are continually introduced and upgraded, and these directly benefit marine life and water quality. The tide even seems to be turning against plastics. By one estimate, nearly 6 million plastic bits and pieces are discharged into the sea every day; if we include microplastic particles like the ones we find in facial scrubs, that number rises to billions. In a fine example of the way in which an individual can make a difference, a UK wildlife camerawoman named Rebecca Hoskins became concerned about the impact of plastic bags on marine turtles who mistake them for jellyfish. She persuaded her home village of Modbury in England to do away with plastic bags, beginning a fightback against the tyranny of plastic waste that has since led to a countrywide reduction in plastic-bag use by UK supermarkets. In the US, first California, then other States and cities all the way to Washington DC, started charging customers five cents per plastic bag – the impact has been quick and impressive. The back-lash against plastic is also reaching the developing world. In 2005, the Indian state of Maharashtra reached breaking-point when plastic refuse choked the streets, creeks and storm drains, and the authorities banned plastic bags. South Africa similarly banned the thinnest plastic bags in 2003; they had become so common blowing around the countryside that they had been dubbed the country's 'national flower'.

For several decades now, environmental groups the world over have mounted beach clean-ups, often to great effect. The rubbish piles heaped up at the end of the day are horrific testaments to how messy we are. Recently, efforts have moved offshore to tackle rubbish at sea and on the seabed. In Europe a project called 'fishing for litter' aims to get the fishing industry to bag up the junk they catch in their nets and bring it ashore for disposal, reversing their previous practice of shovelling it back over the side with the bycatch. Were this approach to be applied more widely, it could make a big difference to the amount of trash lying at fishable depths. The European Commission announced in 2011 that it would trial a scheme to pay fishers to catch plastic in the Mediterranean, Europe's filthiest sea.[4] The Mediterranean has earned this epithet because it is surrounded by populous countries visited by millions of beachgoers and is almost completely enclosed, so junk has accumulated. The State of Hawaii also has a programme to collect lost and discarded fishing gear, which is then burned to produce power for the islands.

What has shocked the world into taking plastic more seriously are the great ocean garbage patches circling endlessly on voyages without destinations. Unfortunately, these are not of much interest to fishermen. They are short on critical nutrients and have low levels of plankton productivity, so there aren't many fish. Unless we do something, these places will continue to suck in rubbish like oceanic black holes. The Great Pacific Garbage Patch is said to be twice the size of Texas: enormous at nearly 700,000 square kilometres, but not beyond the bounds of control. In a way, the ocean gyres do us a favour by concentrating floating garbage into places where we could gather it up without bankrupting ourselves. Twenty boats equipped with hundred-metre-wide skimmers could sweep the gyre in the North Pacific in five years if they worked round the clock. Such boats don't yet exist but more conventional clean-ups have been attempted on smaller scales.

This would only work for stuff that floats, and only for fragments above a certain minimum size. To save the albatrosses, we would have to fish up everything larger than a bottle cap, which would be a technical challenge. Is it worth going to so much trouble for a bunch of silly birds? The answer is yes. Even if you can't care less about the albatross, self-interest might compel you to do something. Plastics crumble into ever-smaller fragments over time, picking up toxins and passing them up the food chain, and potentially ending up on our plate as we enjoy a tuna sandwich or sushi lunch. They take hundreds, in some cases thousands of years to degrade, so unless we clean up the ocean garbage patches, they will continue to accumulate and break down until we stop making our plastic so durable. Fishing for litter will make little difference in the long-run, however, unless it is married with efforts to stop littering in the first place.

The vast majority of stuff picked up by beach clean-ups is local. At a remote beach in Brazil, collectors found bleach and detergent bottles, hypodermic syringes and a host of other things washed downstream from towns and villages by a nearby river. In Majorca, beach trash is dominated by cigarette butts, plastic water and soda bottles, and tampon applicators. Litter peaks in summer, when the island is busiest. A former student who went to work in Asia wrote to me of her horror at the huge quantities of plastics and other rubbish that choked the banks of rivers in Nepal. She asked a villager if it was ever taken away. He said, 'Yes, yes ... when the monsoon rains come, it is all taken away.' She met this detritus again when she sailed across the Bay of Bengal, where junk carried from the Himalayas piles up in the ocean. Local efforts to tackle garbage would make a big difference, and in some places skimmers have been strung across rivers to catch things before they head to sea. But skimmers are really a last resort, and create problems of their own, such as blocking the passage of boats.

It would be far more effective to tackle pollution at source. In the case of plastics, the three 'r's apply: reduce, re-use, recycle. To them we can add recover and redesign.[5] By 'recover' we mean energy recovery using waste plastic to generate power, as in Hawaii. But burning produces other kinds of pollution, so it is better to recycle where possible.

Anyone who has wrestled three layers of plastic from their box lunch knows manufacturers could do much more to reduce their use of plastic and design recyclability into their products. In a promising development, some new plastics are made from biodegradable materials like corn starch. More problematically, others are designed to crumble rather than biodegrade. They break down when exposed to ultraviolet light from the sun. That isn't much good in a landfill or out at sea, where floating plastic bags are quickly covered in algae that block much of the light and prevent breakdown.[6] Setting this difficulty aside, making a bag that shreds into small pieces will only thicken the broth of microplastic particles circulating in the sea.

Preventive measures will ultimately do more to reduce the problem of plastics than laws that are impossible to enforce on the open sea. Despite the fact that it is illegal to dump plastics from ships, the amount of rubbish picked up from British beaches in clean-ups sponsored by the Marine Conservation Society increased 88 per cent between 1994 and 2010, much of it jetsam from boats, and nearly two-thirds of it plastic.[7] In South Australia, a ban on dumping from fishing vessels failed to reduce wildlife entanglement.[8] The number of sealions from Kangaroo Island wrapped in bait box packing straps increased six-fold and more than twice as many fur seals were trapped. The results of a similar scheme in South Georgia in the Southern Atlantic were more encouraging, as incidence of seal entanglement halved.

Just before the Beijing Olympics, a flotilla of small boats set forth every day into a green sea near Qingdao. For weeks their crews forked great heaps of seaweed into the boats, piling it up like marine haystacks. Qingdao had become so polluted by nutrients from sewage, agriculture and industrial runoff that thick masses of seaweed obstructed the passage of boats. The Olympic sailors would have recorded some pretty feeble times without their efforts.

It is relatively easy to regulate sewage or factory effluents that can be traced back to their sources, and much harder to regulate nutrients washed to the sea from farms and city streets, or blown offshore from ploughed fields. Nonetheless, success is possible, as an unintended experiment with the Black Sea proved.[9] By the early 1980s, fertilizers bled from the plains of the southern Soviet Union had turned the

Black Sea into a green soup with few fish and an alarming number of jellyfish, but the collapse of Communist rule at the end of the decade triggered a slump in fertilizer use, as state-controlled farms ran out of money to buy agrochemicals. Since the early 1990s, water quality has significantly improved, with fewer and less intense plankton blooms and less oxygen depletion in coastal waters. Better water quality seems to have made life harder for the invasive comb jellyfish *Mnemiopsis*. Its population has collapsed and fisheries for small open water fish have recovered. The Black Sea experience offers hope for other enclosed coastal seas that now suffer from pollution-induced dead zones, jellyfish explosions and harmful algal blooms.

Nutrient pollution has choked the Baltic Sea, surrounded as it is by fourteen industrialized countries with farms that rely on agrochemicals, heavy industries that produce copious discharges and a combined population of 85 million whose fertile sewage effluent ends up downstream. As I explained in Chapter 9, this modest inlet of the North Atlantic is shallow, with an awkward circulation that leaves puddles of dense, salty water to stagnate near the bottom. Over the years there have been many plans to restore the Baltic to health but none has yet worked. Indeed, as living standards improve in Eastern European countries their downstream impact on the Baltic grows worse. In desperation, perhaps, engineers have begun to ask whether it is possible to fix the Baltic by brute force – or technological brilliance as they might prefer to call it.[10]

In recent years, dead zones have swallowed up to 60,000 square kilometres of the Baltic Sea. Because of its difficult circulation, thick layers of undigested organic matter accumulate in its basins in water that has been drained of all oxygen. If we could just stir some oxygen from the surface back into these lifeless waters, life could return and take care of the heavy burden of waste. That might happen if we could bulldoze Denmark out of the way to improve circulation, but the Danish wouldn't approve. So, instead, how can we mix oxygen from surface waters to the seabed? Engineers have calculated that it would take between 2 and 6 million tonnes of oxygen a year to resuscitate the dead zones. It is hard to imagine a figure like this, but easier if you think of it in more concrete terms: 6 million tonnes of oxygen would fill 55,000 railway carriages!

One scheme envisages a sea of wind turbines that would drive pumps to flush surface water to the bottom. Another idea is to install vast arrays of vertical pipes. Waves washing over them would push aerated water down with the help of gravity to exit near the seabed. Another idea is to throw in some alum or powdered limestone to lock up the phosphorus so that plankton blooms will run out of fuel, but the volume is daunting, and the addition would contravene the London Dumping Convention. There is now so much phosphorus in Baltic sediments from fertilizer and sewage runoff that there have even been suggestions it should be mined. Some have proposed we should intensively fish sprats, which eat zooplankton, to spare the zooplankton that graze on phytoplankton blooms. I am sure this would be popular with the fishing industry, but it hardly seems a persuasive solution as it would create as many problems as it would solve. In fact, close scrutiny shows none of these ideas to be realistic or viable right now, so it is back to Plan A: reduce nutrient pollution flowing into the Baltic. Countries around the Baltic have committed to a reduction of 50 per cent, but it will take strenuous effort to meet this target.

Bottom trawling multiplies the problems. Organic matter and nutrients can get buried on the bottom so they are no longer available to power blooms of plankton or poisonous algae. Trawlers churn sediments to release those nutrients back into the water so that plankton blooms on and on. A small trawler fitted with two 7.5m-wide nets and a chain that cuts 2cm into the seabed can raise approximately 2,000 tonnes of sediment per hour of trawling, of which about 200 tonnes will remain in suspension for days.[11]

Why do we do it? Why do we foul and defile our own living space? To take one recent example, in the 1990s China embarked upon a programme of economic expansion based on coal-fired power stations in full knowledge that the country would soon choke on its own prosperity. Perhaps the Chinese didn't anticipate just how fast that would happen or with what suffocating vehemence the pollution clouds would envelop them. Throughout the world, rivers and beaches are clogged with the jetsam of civilization; air and water filled with the noxious hallmarks of progress. I think our minds are still stranded in the past. For most of history, the number of people on the planet has been small and the unoccupied space seemed almost limitless. We used natural materials that when discarded soon crumbled back into water or soil to feed new life. Today we are proliferating faster than the adaptive capacity of our biological natures. Before the twentieth century, nobody lived long enough to witness a doubling of the world's population. It is a mark of just how much the rate of growth of human societies has accelerated, that some people alive today have witnessed not just one doubling, but two.[12] Adapt we must, because the tide of progress is not about to recede any time soon.

Although cleansing the oceans is a task bigger than any other clean-up attempted in history, pollution control is perhaps among the easiest of the oceans' problems for which to gain political will. People place health concerns high on their political to-do lists, especially in democracies. Action will only follow public awareness and demand, but the latter is on the increase. However far we get, and there is a long way to go, pollution control and prevention will always remain a work in progress. New pollutants are constantly introduced as by-products of our inventiveness and industry. We must be ever watchful.

18

Can We Cool Our Warming World?

Wouldn't it be wonderful if the world had a thermostat that could be reset a couple of degrees? How many of our problems would disappear! In one sense, it does, for greenhouse gases such as carbon dioxide and methane act as a control. Add more to the atmosphere and you dial up the heat. Remove them, and we could regain our cool. Technologists and governments, while searching for ways to reduce emissions, are also on the hunt for engineered fixes to our climate problem.

The planet has ways of dealing with high levels of atmospheric carbon dioxide. The greenhouse worlds of the past did eventually cool through reactions that removed carbon dioxide and methane from the atmosphere. The carbonic acid formed when carbon dioxide dissolves in the oceans is neutralized in a reaction that converts carbonate to bicarbonate. This reduces the amount of dissolved carbonate and so expands the layer of deep water that is undersaturated with carbonate towards the surface. This exposes billions of tonnes of carbonate sediments on the seabed to corrosive water, which can dissolve them. As the carbonate dissolves, it buffers the effects of acidification, so the oceans keep taking carbon dioxide from the atmosphere. Above water, the weathering of silicate rocks also takes up carbon dioxide.

The awkward problem for us is that these feedbacks take tens or hundreds of thousands of years. In this respect, the effects of fossil-fuel burning could outlast pollution from nuclear energy. (After 10,000 years, mixed nuclear waste becomes about 3,000 times less radioactive, whereas the effects of a massive release of carbon on world temperatures will have declined far less, to between a half and a quarter of the peak.[1]) Conditions in the sea took 100,000 years to return to normal after the Palaeocene-Eocene spike in world temperatures and ocean acidity 55 million years ago, and probably a million years after the Permian-Triassic greenhouse episode 251 million years ago. We can't wait that long. This means we must prevent the release of further carbon dioxide and find ways to take some back.

There is an emerging global consensus that we should restrict the rise in temperatures from greenhouse gases to 2°C at most, which means we cannot go above about 450 parts per million (ppm) of carbon dioxide in the atmosphere. Beyond that the risk of dangerous climatic upheaval is deemed too high. But even that target leaves us exposed to a small chance that we will pass an unknown tipping point and suddenly find ourselves on a headlong course towards catastrophe. To date the world has concentrated on ways to reduce emissions. These include greater efficiency in energy use and tapping more 'green energy' from renewable sources. The oceans already play their role in the fight against climate change, as the biggest and most reliable sink for carbon dioxide, but maybe could do more. They cover more than two-thirds of the planet and are, for the most part, uninhabited by people, so they offer considerable opportunities for engineered solutions.

Hydropower and wind have long provided us with energy, but there are severe constraints to their expansion on land. Most major rivers have been dammed, at least in the developed world, and new megadam projects are highly controversial because of the harm they cause to people and the environment. Wind farms are sprouting around us but are often bitterly opposed by those who feel they disfigure the landscape. The restless sea, out there and, once offshore, not in anyone's back yard, offers a less obtrusive world of opportunity. Because the sea is pretty much flat and covers 71 per cent of the planet, nearly 90 per cent of the world's wind energy is offshore.[2]

Already wind farms dot the horizon around many countries and huge blocks of shallow sea are being licensed for development. In one extraordinary case of fractured thinking, the UK Crown Estate, which owns the seabed, licensed Britain's entire share of the Dogger Bank, a shallow hill in the middle of the North Sea, for wind-farm development. The Dogger Bank also happens to be one of the richest fishing grounds, but it seems nobody thought to ask for fishermen's views on the plan, nor conservationists, who have now declared the place a

Special Area of Conservation. Wind farms have yet to be built there, and it looks like there will be tense discussions when that stage is reached. Completed and proposed wind farms will soon grow to cover more than 20 per cent of the UK's territorial waters, which extend up to twelve nautical miles offshore. They are expected to supply about 25 gigawatts of electricity by 2020 (a gigawatt is 1 billion watts), roughly 7 per cent of projected demand. If all of the world's accessible offshore wind energy were tapped, it could supply around 5,000 terawatts (a terawatt is 1,000 billion watts), which is equivalent to about one-third of the world's current energy use.

Plans are forging ahead for wind farms in US waters, with the first likely to be a 130-turbine farm in Nantucket Sound. As in many countries, the US path to offshore wind power has not been without controversy. Critics contend that wind farms are bad for wildlife, such as birds that fly into the turbines. However, the evidence from Holland, where offshore wind farms have been in use since 2006, points the other way.[3] Turbine footings created new habitat for creatures like mussels, anemones and hydroids, and shelter for schools of fish. Since the Dutch farm was put off limits to fishing, it may in time come to act as a protected area of sorts. While some birds avoided the farm, such as common scoters and gannets, others were attracted, like the great cormorant. Harbour porpoises were heard calling more often within the wind farm than outside, perhaps because they could find more to eat there. The potential for offshore wind energy has also been recognized in China, with the first farm completed in 2009. Developers there have plans for many more, given the great length of the Chinese coast. This move into clean energy is welcome for a country that has relied so heavily on coal up to now.

Waves and tides afford other possibilities for power generation whose potential has barely been touched. Two notable exceptions are a hydropower barrage across the Rance estuary in France, which has been in use since 1966, and a tidal turbine at the mouth of Strangford Lough in Northern Ireland, which has two sixteen-metre-diameter blades. Tidal dams have existed since at least the Middle Ages, when they were used to power coastal corn mills. But our capacity to take advantage of tidal power has grown beyond the wildest dreams of the millers of long ago. Today plans have been drafted for colossal schemes like the Severn Barrage in south-west England. At its most ambitious, this scheme would bridge eighteen kilometres of the Severn Estuary which has, at fifteen metres, the second-greatest tidal range in the world. It could supply over 10 per cent of the UK's electricity needs in 2020, but comes with a large environmental price tag. Like dams further upstream, barrages change water levels. Tidal range would be reduced and the area of rich saltmarsh and mudflat habitats vital to hundreds of thousands of migratory birds and commercial fish would fall. So far, the environmental costs of the Severn Barrage have been perceived as too high, and the project was shelved (again) in 2010 while eight new nuclear power stations were given the go-ahead. But as the costs of climate change become clearer, resistance to tidal barrages may wash away. The pace already seems to be picking up. There has been a small tidal generator in Canada's Bay of Fundy since 1984, and in 2006 a demonstration project was installed at Race Rocks in Vancouver Island. South Korea has several tidal power stations and ambitious plans for more, while China has installed one scheme, and the first Indian station is planned for 2013.

Wave power schemes are more palatable to conservationists. To stand at the edge of a cliff, stiff to the wind, and watch the waves hurl themselves ashore, and feel them rumble through your body is to know the power of the sea. The immensity of the energy is palpable. But many engineering challenges must be overcome before waves will be viable competitors to wind and tide. Many ways have been devised to harness their power. In one, waves rise and fall in an air-filled tube to push air through a turbine. Another involves flexible floating sausage strings whose bending pressurizes hydraulic fluid to drive motors that feed a generator. There are several prototypes off the European coast, one named Pelamis after the seasnake (a better name than 'sausage'). A third method relies on waves breaking over a wall to fill a reservoir which then drains through a turbine. None of these methods is commercially attractive as yet.

Generators might also tap the power of strong currents to turn underwater rotors, following the principle of tidal barrages. The mighty Gulf Stream as it pours through the narrows between Florida and the Bahamas might be an ideal location, or the Kuroshio Current as it races past Japan. Such generators would have to be moored in deep water, probably thirty metres or more below the surface to avoid causing a shipping hazard. To date, none has been built.

Progress towards restraining emissions to keep them below the goal of 450 ppm of atmospheric carbon dioxide has been desperately slow. As of 2011, emissions were rising in line with the worst-case scenario of the IPCC. With every failed negotiation the likelihood rises that we will overshoot this target. An increasingly vocal group of scientists has begun to think the unthinkable. If we can alter the planetary climate by filling the atmosphere with carbon dioxide, perhaps we can find another way to dial down the heat. Some are now looking for ways of removing carbon dioxide from the atmosphere, or finding methods to cool the planet. Such 'technofixes' remain highly controversial, and there is a real risk that tinkering with the planetary thermostat could do more harm than good. Ken Caldeira, the American scientist who was among the first to point out the risks of ocean acidification, put it well in his testimony to a British Parliamentary Committee in 2008:

> Only fools find joy in the prospect of climate engineering. It's foolish to think that the risk of significant climate damage can be denied or wished away. Perhaps we can depend on the transcendent human capacity for self-sacrifice when faced with unprecedented shared, long-term risk, and therefore can depend on future reductions in greenhouse gas emissions. But just in case, we'd better have a plan.[4]

Carbon dioxide levels may already be too high for the most sensitive habitats on Earth and the most delicately balanced parts of the world's climate system. I keep coming back to coral reefs. They have suffered badly from global change already, and their very survival is now in doubt because of the one–two punch of warming and acidification. In the summer of 2009 I spent a day with a distinguished group of experts to talk about what we can do to save coral reefs. Late on a sunny London afternoon, David Attenborough and the Australian scientist Charlie Veron presented our conclusions. They told the waiting press that coral reefs as we know and love them are probably doomed unless we can reduce carbon dioxide levels to less than 350 ppm – about 10 per cent less than where we are now and

100 ppm less than target levels negotiated for unsuccessfully in Copenhagen in 2010. If these magnificent habitats are to survive, we will have to find a way to suck some of the carbon dioxide we have already released out of the atmosphere.[5] Our only hope is carbon sequestration.

There is no shortage of ideas of how to dial back our carbon footprint, some madcap, others serious. Every meeting I go to these days seems to attract a collection of entrepreneurs touting their schemes for climate geo-engineering. Glossy brochures are handed out, breathless talks given and promotional videos screened. There is a smell of money in the air, and these people are in hot pursuit of cash from carbon sequestration. At the moment they pursue money from companies who want to offset their carbon footprint, but government cash may follow in future if binding emissions targets are negotiated. It is hard to fault their enthusiasm, but most of these schemes seem to have been poorly thought through. Engineering fantasy often seems to trump common sense and suppress the desire to test whether the machinery on offer can possibly live up to the hype, or whether it is even sensible to use it in the first place.

One company plans to break down the two-layered structure of our tropical oceans where warm, low-nutrient water floats on top of cooler, nutrient-rich water below. Their idea is to speed up carbon dioxide removal by getting nutrients back to the surface, where there is more light, to stimulate plankton growth. Their scheme would deploy vertical pumps powered by waves to suck water from a few hundred metres down and release it near the surface. To make a difference, you would have to seed the oceans with millions of these pumps (remember, they see piles of money ahead). As the company rightly points out, global warming will increase the stability of these ocean layers by heating the surface more. Their plan, they explain, could help to feed the world, since more productive oceans produce more fish. The problem is most of the carbon taken up under this scheme would simply be released again when the plankton and fish in the surface waters died. In most places, only a tiny fraction of the extra production would be taken out of circulation for the long term. The effect would probably be piffling. Producing electricity from their wave generators might be a more reliable means of reducing carbon emissions. The company believes that their pumps could be used to stir up dead zones, or cool coral reefs threatened by bleaching. Since hurricanes form in places with high sea temperatures and are sustained by drawing heat from the sea, they have also suggested that their upwelling pumps could reduce the frequency and intensity of tropical cyclones. Their entrepreneurial fire burns bright.

Other means of making the oceans more fertile and productive have been proposed. Far from coasts, enormous swathes of the sea are not short of the nutrients that usually limit plankton growth – nitrogen and phosphorus – but they have very low productivity nonetheless. Their problem is a lack of the micronutrients that are needed in trace quantities to create the building blocks of life. Iron is one essential micronutrient, whose value to life was established billions of years ago in iron-rich water as metabolic pathways evolved. Mostly the sea gets its iron from river runoff or windblown dust, like the red dirt that blasts from the Sahara across the North Atlantic. Places like the south-eastern Pacific and mid-Indian Ocean are short of other nutrients essential for life, but about a quarter of the oceans are limited by iron alone. The whole of the Southern Ocean around Antarctica is chock full of nitrogen and phosphorus, but far from any source of iron.[6] Likewise, the tropical eastern Pacific has plenty of nutrients but little iron. Perhaps, some have speculated, plankton production could be given a push if iron were added.

One of the first to notice the link between plankton growth, iron and climate was the US oceanographer John Martin, from Moss Landing Marine Labs in California.[7] In the late 1980s he noticed an inverse relationship between the amount of iron in deep sea sediment layers and the amount of carbon dioxide in the atmosphere at the time those sediments settled to the seabed. Periods of low atmospheric carbon dioxide corresponded to ice ages, while higher levels were present in the mild periods between glacials. Martin thought that iron was the key to low carbon dioxide levels, because it boosted plankton growth which pulled carbon from the atmosphere. The bodies of dead plankton and fish carried some of that carbon to the deep sea. In an experiment in the tropical eastern Pacific in 1993, Martin proved that a soluble iron compound mixed into the sea could trigger a plankton bloom. In the flush of enthusiasm that follows success in science,

Martin made a quip that became famous. 'Give me half a tanker of iron, and I'll give you an ice age.' Luckily not!

Entrepreneurs and other freelance enthusiasts have since taken a leap of the imagination and proposed that ocean fertilization on a massive scale could rid the atmosphere of excess carbon dioxide. A niggling flaw in their plans is that nobody is sure where carbon dioxide absorbed by plankton will end up. Indeed, many scientists now believe that the great majority of extra carbon dioxide absorbed will simply be re-released into surface waters as the plants and animals decompose. Surface waters are pretty good at recycling nutrients through microbial action before they sink to depths beyond the reach of mixing by winter storms. How deep that is varies from place to place and depends on the violence of winter weather and vigour of local currents, but in most places it is somewhere between 200 and 1,000 metres. Until carbon gets beyond this mixing zone, it hasn't really been taken out of the atmospheric system.

As of now, there have been twelve attempts to fertilize the sea with iron. Most caused a pleasing spike in phytoplankton growth, but offered little evidence of carbon export to the deep sea. Early on, several companies set up shop and began to pull in venture capital money on the grounds that they could sell 'carbon offsets' from artificial ocean fertilization. One company, Planktos, even got around to doing its own fertilization off Hawaii, dumping a paint-based iron compound into the sea along a fifty-kilometre track from the rock star Neil Young's antique yacht. While Young might have been pleased to have offset his massive carbon footprint, the findings of this 'experiment' were never made public.

The idea of dumping iron to offset our carbon consumption is highly questionable, since fertilization contravenes the terms of the London Dumping Convention. In recent years two UN Conventions (on biodiversity and marine pollution) have come out against ocean fertilization. Planktos folded in 2008 and its founder bitterly attacked environmentalists for what he saw as a hate campaign. There are still others in the game, such as Climos and the Ocean Nourishment Corporation (who propose to fertilize with urea as a nitrogen source in areas not limited by iron), but enthusiasm for this approach is fading. Projections suggest that even the most optimistic scenarios could only increase ocean uptake of carbon dioxide by about a ninth of present emissions. It turns out that John Martin's half tanker of iron was way off the mark. To trigger an ice age you would have to continuously fertilize the entire Southern Ocean, all 20 million square kilometres of it. If you stopped that fertilization, things would swiftly bounce back to where they were.

Another possible downside of fertilization is the effect on deep water oxygen. As I have already explained, when dead plankton sink beneath the mixed surface layer of the sea they fall into a world where oxygen is scarce. Rotting plankton will use up precious oxygen and could enlarge the area of the ocean where there is too little to sustain anything but the simplest life forms.

A different approach to tackling our climate problem is to extract carbon dioxide from the atmosphere and pump it into the deep sea. Below about 3,000 metres, the high pressure and low temperatures turn carbon dioxide into a liquid denser than seawater. The idea is to inject liquid carbon dioxide through long pipes into ocean trenches or pool it on the seabed. While these ultra-deep trenches would conveniently hold lakes of carbon dioxide, each sustains its own distinctive community of life, with many species found nowhere else, and it would likely cause extinctions. Experiments with small puddles of carbon dioxide placed on the deep sea bed off the California coast suggest that the approach is feasible, but again this solution has fallen foul of environmentalists. Video footage of puddles show fish swimming nearby apparently unharmed, but capping large areas of ocean sediments with liquid carbon dioxide would certainly kill most of the animals that live there. There are also concerns that carbon dioxide stored in this state would find its way back to the atmosphere. Calculations suggest that only about 6 per cent would mix back into the water over 200 years (though up to 20 per cent more would return later), but the technique would still buy us precious time to reduce emissions. However, it would enhance the problem of acidification in the deep sea.

A more realistic and practical solution to carbon dioxide disposal can perhaps be found in active or spent oil wells. One Norwegian oil company has extracted carbon dioxide from natural gas and pumped it back into wells below the North Sea since 1996.[8] The method was introduced to reduce emissions taxes and transport expenses and to increase yields from the well, rather than to reduce emissions of a greenhouse gas. This concept, known as carbon capture and storage, has become a cornerstone of the much vaunted 'clean coal' technology recently proposed for the next generation of power stations. There is little chance that we will abandon fossil fuels in time to avoid serious climate change so we urgently need to find ways to generate power at less cost to the natural environment. There are several ways to capture carbon dioxide at the point of emission. The gas can then be pressurized, piped as a liquid and forced back underground. Active or spent oil and gas fields seem to offer a means to lock away carbon dioxide for thousands to millions of years. Water from salty aquifers deep beneath the Earth's surface could also be displaced by carbon dioxide (the excess water would leak back into the sea). Such carbon capture and storage is now a central plank of international negotiations to avoid climate change. It could save up to 20 per cent in emissions.[9] Unfortunately, current underinvestment in the technologies required means we will probably overshoot emissions targets and so risk dangerous climate change.

There are still other ways in which the oceans can help us engineer a cooler climate which remain the stuff of daydreams. One idea is to enhance the weathering of silicate rocks by exposing ground-up rock to hot carbon dioxide and water, and either storing the reaction products as solid rock, or dispersing it throughout the oceans. Silicates make up most of the rock in the Earth's crust and include sand, clays, quartz, olivine and diatomaceous earth, a substance made of the ancient remains of planktonic diatoms whose shells were built of silicon. Of these, olivine is especially suitable. Ocean dispersal of the carbonate reaction product has the advantage that more carbon dioxide is used up and it could help reduce acidification because of the bicarbonate produced when it dissolves. However, the huge expense and disruptive impacts of mining the silicate rocks is a potential killer, regardless of any scruples about possible harmful effects on marine life.

Another scheme foresees a future in which floating platforms out at sea blast columns of salty vapour into the atmosphere to trigger the formation of denser and brighter low-level clouds.[10] The extra cloud cover would increase the Earth's albedo, or reflectivity, and help bounce some solar heat back into space. As well as keeping temperatures down, these clouds would help prevent and might even reverse polar ice melt. The idea sparked controversy almost as soon as it was first proposed. Some contended that if clouds were seeded in the Atlantic, it might dry out the Amazon, while others suggested that rainfall would go up, provoking floods and landslides. The problem is that many of these ideas for climate re-engineering inhabit the margins of our understanding. Predicting their effects is almost impossible and there are serious risks of unexpected consequences.

A lower cost and perhaps less intrusive solution may be to place microscopic particles into the stratosphere, high in the Earth's atmosphere, where they would form an aerosol and bounce some solar radiation back into space.[11] Huge volcanic eruptions do this sporadically and reduce temperatures fast. Mount Pinatubo in the Philippines erupted in 1991 and released a vast cloud of ash high into the atmosphere. It caused the biggest injection of sulphur dioxide into the atmosphere of any twentieth-century eruption and depressed global temperatures by about 0.5°C for the following two years. In engineered cooling schemes, sulphur dioxide or hydrogen sulphide gases would be released which would then react to create sulphate aerosols of the right particle size. Others say we should release sulphuric acid itself at high altitude for a better effect. Some have suggested that the world's commercial airline fleet could be used to seed sulphuric acid far above the land. While this would wrap the earth in a reflective blanket, it would do nothing to reduce either the build-up of greenhouse gases or ocean acidification. The same is true of cloud brightening. Sun-blocking strategies may be cheap and quick, but they are far from perfect.

High-altitude sulphate particles could also slow the repair of the ozone hole, which has gradually been mending since we reduced emissions of ozone-destroying CFC gases in the late 1980s. It would also lead to a whitening of the skies, so say goodbye to those gorgeous cloudless mornings with heavens of brilliant azure. On the other hand, things will liven up in the evenings as the particles scatter light to produce sunsets of rich vermilion and carmine.

To make any kind of difference to carbon dioxide levels in the atmosphere, climate-control schemes would have to radically transform the oceans. They would introduce a whole slew of new human impacts on the sea and so are highly controversial. And yet, as Charlie Veron, the Australian coral expert, put it, 'when your house is on fire, you don't worry about wetting the wallpaper to put it out'. As evidence of climate change mounts, our appetite for geo-engineering fixes might grow.

One problem with all geo-engineering schemes is that their effects are somewhat unpredictable. The Earth's climate system is complex and we struggle to reproduce the nuances of its behaviour in simulations. Doubtless, not everyone would benefit from climate geo-engineering. Filling the atmosphere with sulphate aerosols to reflect sunshine back into space would reduce rainfall, since solar energy drives evaporation and cloud formation. Some places experienced a severe drought after Mount Pinatubo erupted, so the ethics of deliberately simulating the effects of an eruption are questionable. The fact that there will be some losers in geo-engineering need not disqualify the use of such technology outright. We vaccinate children against disease knowing that a small minority will be harmed by the vaccines, because the great majority will benefit. However, the distribution of benefits has to be overwhelmingly skewed towards the positive to justify a risky intervention. A sensible guiding principle is the basic creed of medical ethics: 'First do no harm.'

Most geo-engineering schemes have so far foundered on the shoals of scepticism and fear. The costs may be enormous and the perceived benefits too uncertain to warrant full-scale investment. But some technologies have been tried and tested and even make commercial sense, such as stuffing carbon dioxide back into the oil and gas fields from which it was released. If this process were allied to an efficient system for extracting carbon dioxide from the air it could begin to push the climate-change genie back into the bottle from which we have called it. Middle Eastern countries often fret about what to do once their oil runs out. With all that sunshine, empty desert and those spent oil fields they have the energy, space and carbon dioxide storage capacity to shift from being the purveyors of climate change to planetary saviours.

Before we embark on risky experiments that could backfire and are likely to be difficult to reverse, it is worth remembering that there is another method of carbon dioxide removal from the atmosphere that doesn't involve any risk of making things worse, and can be achieved using technologies tested over millions of years. Healthy, natural ecosystems are carbon sinks that draw carbon dioxide from the atmosphere and sequester it in sediments, peat and carbonate rock.[12] It has been estimated that every year healthy saltmarshes, mangroves and seagrass beds collectively remove carbon dioxide equivalent to half the emissions of the world's transport network (which totalled about 13.5 per cent of global emissions in 2000).[13] Despite covering only 1/200th of the area of the world's terrestrial vegetation, these habitats remove a comparable amount of carbon from the atmosphere. This makes them some of the most intensive carbon sinks on the planet.[14]

And yet, just when we need them most, these habitats are being lost at an alarming rate of 2–7 per cent a year. If we were to halt our losses today, we might have the same impact as the 10 per cent reduction in emissions that would be required to stabilize warming at 2°C or less. If we embarked upon large-scale habitat restoration efforts, like those undertaken for Vietnam's majestic mangrove swamps following the widespread use of defoliants during the war, they could contribute more. So if you can do just one thing, protect the saltmarshes and mangrove swamps! We limit our options if we think only in terms of global emissions. I hope you will by now agree with me that a compelling case can be made for protecting, nurturing and rebuilding ocean life to assist our safe passage through the climate crisis ahead and to safeguard the planet for our children's children.

19

A New Deal for the Oceans

I had lunch not long ago with a high-ranking official from the European Union's fisheries directorate. We chatted amiably about politics and fish and why the Common Fisheries Policy was so feckless. Over coffee he became reflective. As a boy, he said, he remembered paddling the tide line of a beach near his home in Spain: 'Every so often I would step on a flatfish and feel it wriggle away in a puff of sand and indignation. Now when I go there with my children, we never see them anymore.' It's a familiar story, and one I hear more and more often everywhere I go. Unwittingly, he had described a once-common method of fishing that dates back to prehistoric times. Our early ancestors would paddle the tide line, spear in hand, ready to pounce when they disturbed hidden fish. There are still places with enough flatfish to stumble across one here and there, but, as life dwindles around us, it is depressing to consider how easily forgotten such simple pleasures are.

Throughout this book I have touched on the many different ways in which humans have changed the oceans since our arrival on the scene. Like it or not, we are stressing and killing marine life, making it harder for all manner of species to sustain and reproduce themselves. The struggle for existence has become a lot tougher.

It baffles me that conservation bodies don't pursue more energetically the recovery and rebuilding of nature. Instead, they often wage battles to hold the line, with some wins but frequent losses. Each skirmish for this or that species is fought with passion and vigour, but while we are winning individual battles, overall we are losing the war. We need to defend nature if we want to protect our own long-term interests. Nature conservation is too often perceived as a luxury, a view that has become embedded in attitudes and policies. Many people think of the oceans as a remote and incidental feature of our world. Their importance is felt in a physical sense, but people don't realize how much we all owe to *life* in the sea. Human well-being and economic prosperity cannot be unhitched from nature. Climate change is exposing the folly of our neglect for the ecological underpinnings of life. All over the world, humanity is in retreat. Once productive landscapes have lost their soils, fertile valleys have become too salty to grow crops, lakes and rivers are drying up, groundwater has been drained far beyond the reach of roots, and now the sea is rising again to reclaim some of the world's most fertile and populous lands. In the oceans, rich fishing grounds have been emptied and life choked from estuaries, bays and vast areas of the seas. Not since the last ice age has there been less living space for people. Our scant regard for nature means that we have shrunk the ability of the planet to support us just when, with 7 billion mouths to feed, we most need to expand it. Contrary to what some might imagine, enlarging the planetary life-support system doesn't mean pressing nature into an ever-smaller area.

The solution to our present difficulties can only come from adopting strategies that rebuild nature's vitality and fecundity. Refurbishing the natural world is not an antidote to all the ills facing the sea, but it is as close to a cure-all as we are going to find. This doesn't mean that we all have to become vegan or give up holidays abroad. Nor does it mean we have to turn off our computers and return to the technologies of yesteryear. We just have to become smarter about how we think about and use nature.

What we need in times of great planetary stress is a vibrant biosphere. The biosphere – that part of the planet that supports life – is sustained by four elements: energy, raw materials, the variety of life and its abundance. Let's take energy and raw materials for granted for a moment and consider life's variety and abundance. Both are fundamental to the sustenance of a habitable planet. Together they constitute the engine of the biosphere. Life's variety makes up its component parts, but the speed at which the engine runs depends on its abundance. In many places, wildlife has been so depleted that ecological processes now only just tick over. Keeping a few water-filtering oysters around will not maintain good water quality in estuaries; for that we need dense beds of filter-feeding invertebrates and aquatic vegetation.

To date, wildlife conservation has concentrated on sustaining variety and has neglected abundance. The world's wonders would be less today without the efforts of thousands of people who, since the nineteenth century, have expended incredible energy to protect rare species and save wildlife on the brink of oblivion. Many creatures are now so rare that they depend on intensive care for survival. In some places, entire nature reserves have been established to protect a single species – even a single individual. We must never give up such work, but now that nature is in retreat all around us we have to shift our focus and give greater attention to abundance. Conservation, of course, is all about increasing numbers; the more common a species is, the less its extinction risk. But we have to set our sights higher, because extinction prevention and life as a museum will not keep the oceans healthy and sustain humanity through the difficult times ahead.

There are two bywords for resilience in the natural world: diversity and abundance. More enlightened economists refer to them as key elements of natural capital. Diversity begets flexibility and adaptability. Greater diversity enhances capacity to respond to change in positive ways; to keep the machine of life running if you like. Abundance dictates how fast the machine runs. If humanity is to survive into the future we must come to appreciate that we cannot prosper without revving the engine of life back into a higher gear so that life's processes run swift and sure: the uptake and locking away of carbon; water filtration and detoxification; the construction of biological structures like reefs, marshes and forests that resist sea level rise and protect coasts; indeed the production of oxygen itself. If life were a multinational company, many of its subsidiaries would have gone under by now as a result of lost productivity. The whole business would be at risk of failure. When banks or countries fail, we are quick to step in to prop them up, recognizing that we cannot afford a systemic breakdown. We are reaching that point with nature. It is time that we apply a little shock therapy to our environment. So allow me to propose a New Deal for the oceans, a bold and ambitious plan to rebuild and sustain the abundance of life at sea. This New Deal should know no boundaries and should apply to every ecosystem and region on this planet.

Where should we start? People love technological solutions. If nature is going bust, then surely we can build our way out of trouble. We can throw up seawalls to hold back the ocean, farm fish in ponds, mix up nutrients artificially in barren areas to boost productivity, even plant plastic seagrass (God forbid!) to recreate nursery habitats for fish and prawns.[1] To many, the answer to improving fisheries is simple: build an artificial reef and they will come. Better still, fill the sea with artificial reefs. Scuba divers have long sought out wrecks where glittering shoals of tiny fish attract animals with bigger mouths. A wreck adds a third dimension to a flat seabed and has the added benefit of discouraging trawlers and dredgers from risking their gear nearby. Anglers love wrecks too. There is always the chance of enticing some grizzled giant from among their rusting pipes and panels. But a nagging question remains. Do they simply concentrate existing fish, or do they add living space and so increase the amount of fish a habitat can support?

This debate has bounced around the world of fisheries for decades without any conclusive answer. Perhaps that is because, like so many questions, there is no definitive answer. Sometimes they do; sometimes they don't. But even if artificial reefs did provide some uplift in fish stocks, is that a good enough reason to fill the sea with junk? Many of the most ardent advocates of artificial reefs seem to have large quantities of bulky waste to dispose of – tyres, power station ash, railway carriages, ships, oil rigs. I once saw a newspaper photograph of three garbage trucks being pushed off the side of a boat in Thailand. The caption read, without any hint of irony, that this was a scheme promoted by Queen Sirikit to 'rescue the sea from years of overfishing and pollution'. Ever since I have carried an image of these trucks chugging around the bottom of the Gulf of Thailand collecting rubbish!

The Gulf of Mexico must be one of the most junk-filled seas in the world. All of the states bordering the gulf have artificial reef programmes and appear to be dedicated to getting rid of trash under the guise of environmental protection. Here is what the State of Alabama Department of Conservation and Natural Resources has to say:

In 1993, the U. S. military in addressing the need to de-militarize obsolete battle tanks realized that immersion in seawater was an acceptable
method. The idea of using these obsolete materials to create artificial
reefs was born ... As plans developed the full extent of the impact that
this program will have on the reef fisheries and associated economy of
Alabama became obvious. The conservative estimate for the life span
of the tanks is 50 years as artificial reefs. The potential economic impact
of these tanks as artificial reefs during this time is millions of dollars.
Even this conservative estimate far outweighs any other method of
removing these tanks from military service. It is an outstanding and
creative way to convert swords to plowshares.[2]

I love the language, 'immersion in seawater was an acceptable method'
of disposal. It sounds so much better than 'offshore dumping'. Southern
US states now use the sea as the repository of choice for all manner of
waste, including old highway bridges, oil and gas platforms, mine spoil,
school buses, you name it.[3] The states of Alabama and Florida have gone
further than others in that they permit individuals and companies to
dump in designated areas. Only the people involved know the precise
locations where they have sunk their junk so they will benefit most
from possible improvements in fishing. Mostly these new 'reefs' consist
of old vehicles or bits of concrete. I don't want to suggest that creating
artificial reefs has no merit. They keep bottom trawlers out, and there
is some evidence that recreational catches of red snapper are higher
near the reefs than in areas of open seabed. But I have to say the idea
makes me deeply uncomfortable. There are better ways to recover
productivity than to dump our discarded trucks and bridges in the sea.

In Japan, artificial reef creation has been married with aquaculture
in 'sea ranching' whereby areas of coast are used as semi-enclosed fish
ranches within which habitats are 'enhanced' with concrete reef structures. Fish are fed at certain places and times, and in some cases music
is piped into the sea so they come to associate the music with feedings.
I can't help but think that, rather than being a sensible way to enhance
fish production, this is just another way in which Japan offers subsidies to its politically well-connected concrete industry. The majority
of fishery subsidies in Japan go to the cement industry, to fund construction of harbours and wharves.

The natural world is far better at generating the services ecosystems provide than we are at engineering them. Unlike real reefs, Florida's and Alabama's underwater scrapyards disintegrate within five or ten years of disposal to litter the bottom with old car parts and crumbling panels. Seawalls cost millions of dollars per kilometre to build and have to be regularly maintained at great expense, whereas saltmarshes, mangroves and coral reefs protect the coast far better and look after themselves. Fish ponds produce one or a few varieties of fish, usually with large subsidies from wild nature in the form of feed, clean water, land and waste disposal. The mangroves, saltmarshes and tidal flats they replace are nurseries to dozens of commercially important species that grow and disperse to sustain fisheries more widely. In New Zealand, tests with plastic seagrass showed that it attracted a considerable variety and abundance of fish. Many estuaries in New Zealand have lost nearly all their natural seagrass since the 1960s. But real seagrass is far better at the job, with the added benefits of oxygenating the water, filtering wastes, binding sediment, capturing carbon and providing food for a host of animals from turtles to snails. Natural habitats are solar powered and will continue to deliver their multiple benefits as long as the sun shines and we take care not to harm them. So why are we finding plastic and concrete solutions rather than preserving and protecting our natural resources?

There are many ecological services that we cannot artificially recreate or which would simply be too expensive or impractical to re-engineer. It would be impossible to create a high-capacity filtration system that worked as well as the carpet of invertebrates that once covered much of the seabed. Stirring oxygen into dead zones would be an enormous undertaking. And no man-made reef will ever come close to the richness of a healthy estuary. Restoration efforts are most worthwhile when they work with nature. In America's Delaware Bay, intensive efforts to re-establish living oyster reefs are beginning to pay off as transplanted oysters are seeding the bay with their own offspring. And restored beds provide the foundation for rich communities of other invertebrates and fish that the natural beds of old did. In the Philippines, replanted mangroves help protect coasts from erosion while also creating nurseries for wild fish and shrimp that later on fill the nets and baskets of fishers. In Britain, farmland long ago reclaimed from the sea is being re-flooded to encourage saltmarsh growth to replace mud-flats that are disappearing under sea level rise.

Although I spend much of my life behind a desk, wild nature inspires and renews me. I carry the memory of it with me wherever I go, and I return to the well whenever I can. A few years ago I was lucky enough to witness the elemental struggle between two great creatures of the ocean in the Galapagos Islands. I was with a small group in a dinghy when some twenty metres away we spotted a hunting orca plunging through the face of a wave as it cornered a green turtle against a hidden reef. White foam streamed over its dark flank lending it a frightful beauty. Another orca crested beside it, then lashed its tail as the king of predators snatched the turtle and flung it across the surface of the sea. The orcas seemed to play with the turtle after that, rising and diving, arching their fins against a backdrop of volcanic slopes, black on black. It was a timeless moment: the millennial struggle for survival played out over and over in these waters. Storm petrels flitted above in excited flocks, while far overhead frigate birds circled and waited, hoping for a share of the spoils. Nature offers us things that no artifice could ever create. If we give it a chance, it can heal many of its wounds.

I am not foolish enough to think that we can wind back the clock to some primordial condition when our species first walked out of Africa. Change is inevitable. Species rise and fall; ecosystems shift and warp with time. But if our biosphere is to weather the storms ahead, then we must strengthen the twin pillars of life's variety and abundance.

How can we deliver this New Deal for the oceans? Marine reserves are a critical part of the solution. Placing areas off limits to exploitation and excessive use has proven time and again to be a powerful tool to protect and enhance life at sea. But if reserves are to make a real difference, there must be lots of them in extensive networks that are well enforced. They cannot simply safeguard a smattering of beauty spots. We should think of these interconnected safe havens as the new foundations underpinning everything we do in the sea. Fortunately, there is a growing appetite for this kind of protection in the world at large.

The mountains of Arran rise up from Scotland's Firth of Clyde in a ragged ridge of brown and gold. The island is known as the 'sleeping warrior' for its profile from the mainland. Its warrior spirit must infuse some of Arran's 5,000 inhabitants for it was here that the first battle against the overfishing and degradation of the Firth of Clyde was fought, and eventually won.

In 1995 a small group of islanders got together to campaign for the protection of a short stretch of coast in Lamlash Bay. They were shocked by how, within a few decades, the bay had been nearly emptied of fish and its fragile seabed habitats destroyed by scallop dredging. For years they were rebuffed by fisheries managers, government officials and politicians. Even Scotland's own nature conservation agency wouldn't support them because they felt that the bay was not special enough to warrant protection. The islanders refused to give up. Over the years, their campaign group, the Community of Arran Seabed Trust, grew to include half of Arran's population.[4] Votes count and political support for the protected area followed. Through dogged persistence and a conviction that the destruction of the Firth of Clyde should not be ignored, the people of Arran achieved their goal. In 2008 Lamlash Bay became Scotland's first marine protected area to be closed to all fishing. A small community had taken the future of the sea into their own hands.

The Arran islanders' faith in the regenerative powers of the sea quickly proved well-founded. Within two years of protection, habitats inside the reserve began to take on a more lush and healthy appearance, and juvenile scallops had become more numerous under protection than in nearby fishing grounds.[5] Other benefits, like the recovery of big fish, will take longer to accrue, but reserves elsewhere show they will come if protection is maintained.

Across the world, other communities have trodden parallel paths to Arran's. In the Philippines, a few villages troubled by the growing scarcity of fish established small protected areas in the 1970s to help rehabilitate reefs and fisheries. Those areas were studied by researchers from Silliman University under the direction of Angel Alcala, an affable and determined man who later became a government minister, and Garry Russ, a scientist at Australia's James Cook University. Over time, fish and corals rebounded in the reserves. A change of leadership in one village led to the Sumilon reserve being reopened to fishing in 1984. Within a year stocks fell by half, and the connection between protection and benefit was firmly established. By way of contrast, support for the reserve at Apo Island remained strong and fish stocks there kept on going up, year after year.[6] Twenty-five years of sustained protection has since increased Apo's stocks eleven times over.

Word spread about these islands of protection in what are otherwise some of the most intensely overfished seas on the planet (remember pa-aling). With the encouragement of Alcala and others keen to revivify Philippine seas, groups of fishers who lived near reserves travelled the country to share their experiences. Other villages copied their approach. The Philippines now has hundreds of community-managed protected areas. Even though the fishers involved were desperately poor, they realized that, if they didn't make some short-term sacrifices, they would have no fish to eat in future.

In Fiji, local chiefs have for centuries jealously guarded patches of the coral reef exploited by their communities, always keeping some areas off limits to fishing (called *tabu*, which is where our word 'taboo' comes from). In recent years, their efforts have been harnessed into the creation of a national network of community-managed reserves. Fiji has committed itself to the protection of at least 30 per cent of its seas by 2020 and, with nearly 200 protected areas already, it looks likely to reach this target long before then. Big fish are still easy to come by in Fiji, and the islands have some of the most exciting shark dives in the world. Examples of effective protection are multiplying across the world. The Hol Chan reserve in Belize, Cabo Pulmo in Mexico, Ras Mohammed in Egypt, Exuma Cays Land and Sea Park in the Bahamas, Tsitsikamma in South Africa, the list grows longer month by month, all thronging with life and testament to the spectacular effects of protection. Time and again, when people realize what they have lost, and understand that careful stewardship of their resources will bring some of their riches back, this kindles enthusiasm for action.

Marine protected areas represent an indispensable building block of our New Deal for the oceans. In the last century and a half, we have expanded our exploitation of the oceans from a narrow strip around the coasts to the deepest and most remote seas. In the early nineteenth century, the untapped sea exceeded the areas that were fished by at least a hundred to one. Today, that position is reversed. This expansion, like our earlier spread to every corner on land, has left far less space for wild nature. The difference between land and sea is that we depend on wild nature to supply the majority of our seafood. To ensure continuity of this food before it is too late, we must set up parks at sea, and lots of them. If well protected, they could harness nature's resilience to produce a resurgence of life. As Apo Island proved, exploited species can increase five, ten, even twenty-fold within a decade of protection, and sometimes in just a few years. After eleven years of protection in Spain's Cabo de Palos reserve, dusky grouper, a popular Mediterranean eating fish, had leapt in abundance over forty-fold, while in Florida's Dry Tortugas Ecological Reserve, mutton snapper rebounded by three times in only four years.[7]

Watching reserves refill with life over a span of years, as I have done many times in the Caribbean, is an uplifting experience. Caves, ledges and overhangs repopulate with animals that had formerly been scarce, while the blue waters above thrill again with the flash and strike of predators and prey. After an absence of ten years, I recently revisited a small marine reserve at Molasses Reef in the Florida Keys National Marine Sanctuary, which has been protected from fishing since the 1970s. I was delighted to find many more fish than on my first visit, including several kinds that are usually absent from exploited reefs. Metre-long blue parrotfish grazed the bottom in amiable groups, their pale flanks a pleasing contrast to the indigo shades of the midnight parrotfish mingling with them. I even spotted a thick-set Nassau grouper between two coral heads; once the mainstay of Caribbean fish dinners, it is now missing from much of its former range.

Refuges like these cannot solve all ills. They will not address the root causes of global change, nor reduce pollution, but they can alleviate some of the effects of these problems. This is how. Marine reserves – places off limits to all exploitation – have proven to be spectacularly good breeding grounds. This is a simple consequence of two things: increased abundance and extended lifespan of the animals they contain. Early on in this book I explained that overfishing had caused a collapse in fish and shellfish stocks because it culled the big, old, fat females that were the engines of egg production. A big fish can produce hundreds of times more young than a small fish, and many marine creatures keep on getting bigger as they grow older and remain fertile most of their lives. So having craggy and ancient giants around is a brilliant way to ensure masses of offspring. If you think of a marine reserve as a fountain of microscopic eggs and larvae pouring into the sea you won't be far off. As populations recover and individual inhabitants grow larger, the flow increases from a trickle to a stream to a gush. Long-established reserves offer up countless millions of eggs, larvae, seeds, spores, planulae or whatever other means their inhabitants use to reproduce. Like dandelion seeds caught in the wind, currents carry offspring away from their parents for distances of metres to hundreds of kilometres, depending on the species. So the new life is spread far beyond the limits of the protected habitat.

Stresses like climate change and pollution reduce survival rates and make reproduction harder; by increasing the flow of offspring, protected reserves can mitigate these effects. Bigger populations that reproduce vigorously will be much better able to battle through the battery of stresses that we throw at them, tipping the balance between life and death back in their favour.

Protected areas do things that fishery managers cannot achieve without them. The Leigh Marine Reserve in New Zealand is just five kilometres across and yet it produces more young snapper, a burly grey fish much loved by fishermen, than a hundred kilometres of fished coast. For some animals in long-established reserves the multiple is even greater, and protected areas out-produce fishing grounds a hundred to one.

Marine reserves can boost the resilience of fish populations and fisheries in another way too. One of the effects of fishing, if you recall, is that it increases the variability of fish stocks, which induces swings in the fortunes of fishers. This happens when fishing drives down populations and shortens lifespans so far fewer offspring are produced. Reserves reverse this effect by securing a steadier and richer flow of offspring.[8] If you fish for a living, that difference matters. When the stream of new life flows fast, there will always be something to catch. When it is weak and intermittent, catches are unreliable.

Reserves can also bring back habitats swept away by trawls or dredges, and reverse habitat collapses induced by plagues of grazers like sea urchins let loose by the removal of their predators. If you could fast forward through the years after a reserve was created, you would see kelp forests rise from the seabed, seagrasses thicken and spread and sediments disappear under crusts of invertebrates, like oysters and sponges, that elbow their way above the bottom. The gradual rebuilding of these underwater metropolises increases the capacity of the sea to sustain life. There are more ways to make a living and more places to hide.

There probably aren't too many people who, if given the choice, would want the fish on their plate to come at the price of extinction for other species. But that is the unseen cost of many fisheries today. Most people are unaware that some of the species that show up on the fishmonger's slab simply cannot sustain productive fisheries in the long run. They grow and reproduce too slowly. Most sharks and the bigger skates and rays fall into this category. So does almost everything caught more than 500 metres down – deep-sea beasts like Chilean seabass, orange roughy or roundnose grenadier. They are caught because they are there, and when they are gone they disappear from markets. If you are old enough, you will remember the sudden appearance of orange roughy in supermarkets in the 1980s and 1990s. Their firm white fillets were similar enough to cod and haddock to find a ready market. Just as swiftly they vanished early this century as their fisheries collapsed, the seamounts from which they were pulled left empty. There are good reasons why we farm animals that are highly productive and feed low in the food web, like chickens and cows, rather than bears or pumas. But it is the bears and pumas of the sea that we have grown used to eating.

Many exploited species became rare so long ago that we can no longer remember a time when they were abundant: angel sharks and giant skates in Northern Europe, bluefin tuna in the North Sea (even Iceland), halibut in the North Atlantic, goliath groupers in most of the Caribbean, rosefish in the Gulf of Maine. Celebrity chefs and environmental organizations sometimes confuse variety with sustainability. They exhort us to eat less cod, for instance, to ease pressure on an overfished species, while urging alternatives like ling or wolffish. Here a little historical perspective may help us judge the wisdom of these substitutions. Both ling and wolffish are impressive beasts. Fully mature they can grow longer than a man is tall. The wolffish has a face like a seasoned rugby player and teeth to match, their gappy grins and broken canines testament to a lifetime of hard knocks gained crunching clams and urchins. We are not the first generation to appreciate wolffish steaks: they were much admired in the nineteenth and early twentieth centuries. Trawl fishing records from the UK show that since 1889 catches of wolffish per unit of fishing power have fallen by 96 per cent.[9] Ling catches have fallen by the same amount. My student Ruth Thurstan once gave a presentation in which she showed a photograph of fish landings made at the UK port of Grimsby in the 1890s. Two-metre ling were piled on the dock like logs. A fisherman in the audience rebuked her for confusing tuna with the trawl catches she had mentioned. But he was the one who was confused. He had never seen ling so large or abundant, so he mistook them for tuna. Neither ling nor wolffish is a sustainable alternative to traditional favourites. Today they are rarities that were once common. So if you want to find a substitute for cod, a better option would be something that is still common, like haddock, whiting or pollock.

Marine reserves offer a way out of the dilemma of whether to fish hard to benefit from high-production, high-turnover species like scallops, prawns and plaice, or to fish lightly to prevent the disappearance of the big and vulnerable. At the moment, species like the 'no-longer-common skate' hang on in scraps of rough ground beyond the reach of trawl and dredge. But these de facto refuges are shrinking under the onslaught of new technologies and fishers chasing better catches as traditional grounds fail. In the west of Scotland, scuba divers report scallop dredgers scraping the bottom right up to the edge of submerged reefs, guided by sophisticated underwater sensing equipment. They lift tonnes of sediment that settles over the reefs and smothers life on the rocks, so their impact extends even into places of apparent safety. Reserves provide refuges where fragile and vulnerable animals and plants can thrive without hindering fishing operations. Intensive fishing and susceptible wildlife can coexist, but not in the same places. In this respect the sea is no different from the land. Sensitive animals and plants that cannot cope in landscapes dominated by industrial agriculture find safety in protected woodlands, heaths and meadows. We just haven't got our heads around the simple fact that we need to set aside parks in the sea. We can't easily see what is going on down there, but many modern fishing technologies are equivalent to clear-cutting and monocrop farming rolled into one.

With so much in their favour, why is it that marine protected areas are so often opposed by the group that stands to gain most from them – the fishing industry? In 2010 the UK embarked upon a grand project to create a national network of marine protected areas to redress over a century of accumulated damage to marine life, and to fulfil its commitment made in 2002 at the World Summit on Sustainable Development. Like other coastal nations of the world, the UK promised to create a national network of marine protected areas by 2012. Sadly, many within the fishing industry have chosen to oppose this. Lurid headlines and exaggerated claims have filled the fishing press; 'NO TAKE ZONES ARE THREAT TO FISHING' from *Fishing News* is typical.[10] In reality, the biggest threat to fishing is fishing itself. Overexploitation is putting people out of business, not protected areas. All the evidence suggests that protection keeps fishers at work. In developing nations, marine reserves have been proven to boost the food security of people who live nearby.[11] In the Mediterranean, artisanal fishers, the very lifeblood of coastal culture and community, make a better living in the waters near marine reserves than those in places where there are none.[12]

Three-quarters of the US haddock catch is caught within five kilometres of the boundaries of large protected areas on Georges Bank, east of Cape Cod, and in the Gulf of Maine.[13] Lobsters have bounced back to seven times their previous abundance in a tiny scrap of protected area that was declared in 2003 next to England's Lundy Island, and the reserve is spilling young lobsters into surrounding fishing grounds.[14] Every year the examples multiply and yet some people seem to stick stubbornly to views that science has proven to be unfounded.[15] I hear the same arguments against protected areas that I heard fifteen years ago, about how they presage the collapse of the fishing industry and will eviscerate thriving ports. But ports are closing because a lack of protection means that every year there are fewer fish. Here is just one example local to me: the port of Whitby on the Yorkshire coast was a bustling centre of fishing in the mid-twentieth century and long before that. Old photographs show the harbour crammed with boats and the dockside stacked high with boxes full of fish. In 2010 there were ten boats left, all for sale.

One old chestnut that pops up again and again is how marine reserves will not provide any benefit to mobile fish like cod, which are the mainstays of cool water fisheries in Europe, North America and Japan. This view is based on a misunderstanding of how they live their lives. In recent years, cod have been fitted with tags that track their movements. Many of them are remarkably sedentary and for months on end hardly move at all. Some stay put pretty much all year round. There is a telling example of what protection can do for cod in the convoluted entrance to the Baltic Sea.[16]

Between Copenhagen in Denmark and Malmö in Sweden there is a short and narrow strait called the Öresund. It varies from four to forty-five kilometres wide and is the main route for ships in and out of the Baltic. Because of its importance to shipping, Öresund has been closed to mobile fishing gears like trawls and seine nets since 1932. Catches in the Öresund are made mainly by fixed nets. To the north, a region ten times its size, called the Kattegat, is fished by bottom trawls. The fortunes of their respective fisheries in the last thirty-five years could not be more different. In the late 1970s, fishers trawled 15–20,000 tonnes of cod from the Kattegat while gillnetters in Öresund took about 2,000. In 2008, the Kattegat yielded 450 tonnes of cod, compared to 2,140 tonnes from the Öresund. Research catches made in Öresund that year showed cod there to be fifteen to forty times more abundant than in Kattegat and much bigger too. Some of the cod pulled from these waters today rival the behemoths of old. Similar size differences were found in several other kinds of fish, including lemon sole, haddock, plaice and whiting. The Öresund and its fisheries have remained in great shape because they are protected from the most destructive kinds of fishing. Despite the fact that it covers just 2,000 square kilometres and the coasts that border the strait are home to nearly 4 million people, marine life flourishes.

Fisheries benefit from complexity. As it transpired, not all cod are the migrants we took them for, in endless transit over thousands of kilometres between nursery, feeding and breeding areas. Populations like those in the Öresund, or the Firth of Clyde for that matter, are locally adapted and more sedentary. Indiscriminate fishing has laid waste to this population diversity and in the process it has undermined the capacity of populations, and the fisheries that depend on them, to continue to thrive as conditions change. This adaptability becomes ever more valuable as the world around us changes.

Alaska's Bristol Bay salmon fishery is a perfect example of how population variability can work to stabilize fisheries.[17] Bristol Bay is 400 kilometres by nearly 300 wide and opens into the eastern Bering Sea. It is fed by dozens of salmon rivers and streams, each of which supports a population of sockeye salmon adapted to local conditions. Salmon stocks vary in size and productivity as the climate shifts. Under any given set of conditions some fare poorly while others thrive. Fifty years of salmon fishing records show ups and downs but collectively salmon production has remained strong, much stronger and less variable than it would have been if the bay had only supported a single sockeye population.

This has been dubbed the 'portfolio effect'. Just as investors spread their risk by keeping a variety of stocks, having a variety of differently adapted fish populations imparts stability to ecosystems and fisheries. George Rose, who works out of Newfoundland's Memorial University and saw Canada's cod collapse at first hand, knows more about cod than almost anyone alive. Even as fishermen scooped up the last of the great cod shoals in the early 1990s, he found several populations local to different parts of the coast that appeared unaffected. He describes nature as like an orchestra where not all the instruments play loudly at once. 'You have the horns playing for a bit, then the strings come in.'[18]

The same effect is apparent across species. Every animal and plant makes a living in a slightly different way. Each one responds differently when conditions change; each reacts to stress from human activities in a dissimilar way. Something that one species might barely notice, or easily overcome, could fell another. Where there is a broad portfolio of species to draw upon, ecosystems are much more likely to continue to function as the seas change. Simplified ecosystems of the kind that now predominate in many places have little of this resilience. Their variety has been winnowed to a handful of species such as prawns and scallops. We will not be able to restore life to these places by tinkering with mesh sizes or fishing gears alone. Marine reserves are essential to defend and promote diversity, and thereby help both wildlife and the fishing industry weather turbulent times ahead.

One thing that is seldom appreciated about marine reserves is that strong protection is needed for strong benefits. Early on I described the phenomenon of 'fishing down the food web' as top predators are depleted and fisheries turn to smaller animals lower down the food web. Effects of exploitation on marine habitats parallel this trend, with the progressive loss of structure-forming animals and plants; complex habitats crumble into simple ones. To be truly effective, marine protected areas must push ecosystems back up the slope of collapse. How far they can go – in other words, how much recovery is possible – depends in large part on how strongly they are protected. A little protection, such as a small reduction in fishing pressure, will nudge things back slightly. Modest protection, such as a halt to scallop dredging, while continuing the use of trawls targeting fish, will be met by modest benefits. Animals of middling size will probably do well with middling protection like this, but the big beasts of old will not return. Only strong protection – the removal of all exploitation and the alleviation of other sources of harm – will drive ecosystems back towards their most prolific and productive. Because protected areas generate controversy and often inspire highly vocal opposition, politicians often seek compromise not in whether to create them in the first place, but in how much protection they will be given. The result is that the seas are full of protected areas that aren't really protected at all. Many of Europe's Special Areas of Conservation are like this (they should really be called Suspicious Absence of Conservation). In a way 'paper parks' are worse than no protection, because they give the illusion of conservation without the reality. America, whose politicians are no less weak-willed, suffers the same problem. Some of America's National Marine Sanctuaries still offer little real protection from fishing, although times are changing there for the better.

Marine protected areas will only contribute meaningfully to raising the variety and abundance of life if they spread and multiply. There is little value in protecting a few beauty spots to a high level if we continue business as usual in the rest of the seas. The UN's Convention on Biological Diversity set a target in 1997 of 10 per cent of the oceans to be protected by 2012. At the time of writing – the tail end of 2011 – we had only made it to 1.6 per cent, so the target date has been pushed back to 2020. Is this ambitious enough? An honest scientist will say no.

Looking at the question of how much to protect from a wide range of angles gives remarkably concordant answers. Studies have looked at how much coverage is needed to sustain or maximize fisheries production, to protect populations of all species in more than one place (to avoid losses from random catastrophes), to prevent the loss of genetic diversity, to ensure protected areas are large and numerous enough and sufficiently interconnected to sustain themselves. Across the board the answer is that we need to protect several tens of per cent of the sea to achieve these goals, not single figures. The middle of the range of answers is about 35 per cent,[19] so the 10 per cent target proposed by the UN is too low. In my view we need to protect a third of the oceans from direct exploitation and harm, and manage the rest much better than we do today. Hitting the UN target means a six-fold increase over the next decade. I accept that it is unrealistic at this stage to expect more. But this target only represents a waypoint on the path to a more sustainable future, not the endpoint.

20

Life Refurbished

I often come across people who think that we can't afford to cut back fishing when every day there are more mouths to feed. But simple maths tells you that restocking our seas makes economic sense. Think of it this way: if you have a million fish in the sea and can catch 20 per cent of them every year without depleting the stock, that stock would give you 200,000 fish a year. Now imagine that you nurtured your fish and gave them a chance to grow so that you had 5 million. Your 20 per cent would come to 1 million a year. The interest rate on your capital is the same, but the yield is much bigger. And with fish more abundant they would be easier to catch, so you would need fewer boats and each would cost less to run.

Wishful thinking? Not really. In 1889, there were ten to fifteen times as many large bottom-living fish like cod, haddock and halibut in the seas around the UK as there are today. As I explained, a late-nineteenth-century fleet, made up mostly of open-decked sailing boats, landed twice as many bottom-dwelling fish as we do today, with seventeen times less effort. Less fishing, more fish, bigger catches, food for more people. A World Bank report, aptly titled *The Sunken Billions*,[1] highlighted the madness of overfishing when it calculated that major fish stocks of the world would produce 40 per cent more if we fished them less. It sounds paradoxical – fish less to catch more – but that is the simple message. In Europe, the benefits of fishing less could be even higher, perhaps 60 per cent, since stocks there are in such a poor state.[2]

There are people out there who insist that all is well with the seas and that those who say otherwise are just bleeding-heart greenies or scientists who have drunk too deep from the cup of environmentalism.

Foremost among them is Ray Hilborn, a well-respected fisheries scientist from the University of Washington in Seattle. The flavour of his argument can be garnered from the title of a *New York Times* op-ed he wrote in 2011: 'LET US EAT FISH'. In it he claims that 'On average, fish stocks worldwide appear to be stable, and in the United States they are rebuilding, in many cases at a rapid rate.' Hilborn has been on a one-man mission to face down 'doom-mongers', as he calls them, tramping the world in search of headlines. He finds a ready audience – who wouldn't rather hear that we're doing just fine? But Hilborn's reading of the state of fisheries is highly selective, and he seems immune to any amount of historical evidence of decline. Sure, some fisheries are doing well in some parts of the world (and after recent reforms, many are on the increase again in America after years of mismanagement).[3] On a narrow reading this is good news, but these productive fisheries often come at the expense of a host of other species, as is the case of scallop dredge fisheries, which thrive in the open, sandy, predator-free bottoms that are left when other fisheries have carted off the rest. As Canada's cod shows, pulling down fishing rates (Hilborn's main metric of success) doesn't always lead to recovery. Hilborn seems to place more faith in his mathematical models than he does on evidence, like someone who insists on going out in the rain without an umbrella because the forecast said it would be sunny. I admire his optimism, but his relentlessly upbeat reading of the state of ocean life is Panglossian.

Overfishing has been called one of the biggest soluble environmental problems in the world. We know what must be done and it takes no more than a sentence to say it. We have to fish less, waste less, use less destructive methods to catch what we take, and provide safe havens where fish can reach their full reproductive potential, habitats can flourish and vulnerable species can be protected. So why hasn't it happened?

Every year Europe's political leaders set ludicrous fishing quotas which over the last twenty-five years have averaged a third higher than the safe catch levels recommended to them by their own scientists.[4] The folly of this approach is obvious. A forester who cuts down more trees than he plants will soon run out of wood, while a farmer who sends ten more cows to market every year than he produces will

run out of livestock. There is a simple omission in the complex political calculus of quota-setting: you can't cheat nature out of more than
it produces. Europe is running out of fish in large part because politicians have chosen to ignore what scientists are telling them they need
to do to get the most from the sea. In serving the short-term interests
of the fishing industry or their own narrow electoral goals, our political leaders are condemning the industry to death in the medium to
long term.

The same broken reasoning and flaw in collective decision-making
has poisoned the management of fisheries on the high seas. Away from
the coasts in international waters, catches are regulated by Regional
Fisheries Management Organizations. These bodies consist of representatives from all of the nations with an interest in fishing the area in
question. Instead of following scientific advice, these political appoint-
ees often pursue their perceived national interest at the expense of
common sense. The car crash of bluefin tuna management is a perfect
example of all that is wrong with the system. Bluefin tuna are hugely
valuable (in part because they are now rare), so nations are reluctant
to stop fishing them even when scientists are virtually unanimous in
saying that this is what must happen for the stock to recover. Countries are so busy competing with one another for a slice of the pie that
they seem blind to the fact that every year the pie is smaller. Prince
Charles summed up the folly of this behaviour beautifully: 'It seems
one thing to destroy a species out of ignorance; but it is totally another
to destroy it in the full knowledge of what we are doing.'[5]

The perversity of political decision-making is mirrored within the
industry itself. Both are manifestations of 'the tragedy of the commons'. Fishermen will try to take as much as they can to maximize
their own gains, even though they would all be better off in the long
run if they exercised restraint. According to economists, people are
more willing to look after what they own, and so the idea of 'catch
shares' was born.

In New Zealand, Iceland and some of North America's fisheries, the
fishing industry has been given shares in the take. For example, the
New England groundfishery for haddock, flounder and other bottom-
living fish was converted to catch share management in 2010, while
the Canadian halibut fishery has been managed this way for over

twenty years. Own 10 per cent of the shares, and you will be able to take 10 per cent of a given year's catch. Catch shares have the advantage of ending the race to fish. Prior to catch shares, the seasonal opening of Canada's halibut fishery had been shortened year after year to the point when the entire catch had to be landed in just six days. Catch shares have become very popular with some environmental organizations, which see them as a remedy for all fishing's ills, and with many of the philanthropists who support them. But I think they are oversold. I worry that giving away something that all of us own – the resources of the sea – to a few will set us up for problems in the future. I am not alone. One of the system's most prominent critics is Daniel Bromley, a professor of Applied Economics at the University of Wisconsin. In a stinging polemic against catch shares, Bromley abandoned the neutral language of academia:

> The universal policy response to this [fishery management] failure seems to consist of nothing more imaginative than the free gifting to the commercial fishing sector of permanent endowments of income and wealth under the utopian claims associated with [catch shares] ... This ... can only compound the tragedies of past malfeasance by the dangerous endorsement of this bundle of confusions, contrivances, and deceits.[6]

We seem to have blind faith that new private owners will look after fisheries properly and see to their health and preservation. Their incentive is that their shares will increase in value as fish stocks recover to more healthy levels. But these catch shares often go beyond this and can be bought and sold, with the result that small family businesses have sold out to larger firms and ownership of the fisheries has become vested in fewer and larger hands. Within one year of the New England catch share scheme coming into use, the Gloucester fleet had lost 20 per cent of its boats. Before catch shares 80 per cent of the fishery value was landed by 68 per cent of the boats, afterwards it was landed by 20 per cent. There is little evidence that the new owners of catch shares feel any responsibility for long-term stewardship of the seas – or indeed that catch shares offer up any wider environmental benefit.[7] Having given away public property, it will cost society dear to get it back if we change our minds. Imagine how you would feel if your government gave away all the national forests to industry, and then twenty years later used your taxes to buy chunks back to turn them into nature reserves or return them to the public amenities they once were.

Seth Macinko, a colleague of Bromley's from the University of Rhode Island, pointing to how politicians usually ignore the scientific advice they are given, says, 'Catch shares are seen as a solution to the problem of fisheries management, but we haven't tried management yet!'[8] In my view, some of the downsides to catch shares could be solved if they were leased to the industry for set periods and the leases were sold to the highest bidders, much as governments do with other public franchises such as running rail networks. Even then, on their own, catch shares cannot solve the problems of overfishing. They don't protect habitats from destructive fishing practices, they don't stop other wildlife from being taken as bycatch, and they won't save species like groupers or sharks that disappear from places intensively fished for more resilient animals.[9] We need enlightened and effective public management of fisheries for that, and a willingness to establish protected areas to safeguard the fragile and vulnerable.

People often ask me, 'What can I do to help?' One place to start is to avoid eating fish that are overexploited in the wild or taken using methods that harm other wildlife. It is easier said than done. In the winter of 2011, I was invited to dinner at one of London's top seafood restaurants in the heart of Mayfair by some wealthy philanthropists to talk about ending overfishing. My fellow guests included government ministers, the heads of conservation groups, journalists, media tycoons and business leaders. Needless to say, our hosts had gone to a great deal of trouble to produce a menu of tantalizing delicacy and exemplary sustainability. Then came embarrassment served with the very first course. The menu said 'hand-dived Loch Fyne scallops' but the three half shells on my plate spoke of an altogether rougher journey from sea to plate. Their edges were worn and chipped from tumbling in the belly of a scallop dredge. They had nothing like the porcelain fragility of the frilled shell edge shown to me by a Scottish scallop diver who offered up a handful of freshly caught scallops. My hosts had been duped. Despite their best efforts, even a high-class restaurant with experts in seafood sourcing couldn't do the right thing. So what hope does a confused customer have contemplating the fish counter at the supermarket, trying to figure out what to get for dinner?

You may have heard of the Marine Stewardship Council, an organization dedicated to helping you choose seafood from sustainable fisheries. It has a blue-and-white logo in the shape of a fish with a tick mark for a tail that implores you to take this packet home rather than some other dubious fish whose background cannot be trusted. Rupert Howes, the chief executive of the organization, certainly hopes you know the logo, but you would be in good company if you did not recognize it. Three-quarters of all shoppers quizzed in 2010 didn't know what it was. In gaining visibility for the Marine Stewardship Council, Rupert faces a dilemma: the label will make little difference to seafood-buying habits if people cannot find any certified fish, but fisheries have to pass tough standards before the label is awarded. They are judged against three criteria: is the population healthy; does the fishery harm other wildlife; and is it well managed? From what I have described in this book, you won't be surprised to hear that not many fisheries achieve these standards. So some fisheries are being certified on a promise to get better, which makes environmentalists uneasy. As of 2011, the Marine Stewardship Council's label was carried by more than a hundred fisheries that together made up 12 per cent of the world's reported catch. The organization now has visibility and the fishing industry is falling over itself to gain accreditation after big customers like Walmart, McDonald's and the German Kaufland chain pledged only to sell certified fish (McDonald's has already fulfilled its commitment). Holland has promised to sell only fish that carry the logo 'as soon as possible'. If others follow suit, it could make a big difference.

There is a price to this success. An increasing number of fisheries are being certified that cannot possibly be regarded as doing little harm to other wildlife or habitats. Ask any sponge that has been shredded by a scallop dredge, or juvenile skate carried away by a prawn trawl. Where fisheries for scallops or prawns seem reasonably healthy and well managed for the target species, it seems that the certifiers have simply shrugged their shoulders about wider environmental damage.

The report for a Scottish prawn fishery acknowledged that prawns could not be caught without damaging bycatch species and the seabed, but concluded that, since the seabed had been damaged like this for some time, there was no harm in carrying on. Here is an example of the fish bycatch from four days of prawn trawling off the west of Scotland: to land 28,300 prawns, 12,500 fish from 38 species were caught and all were thrown away, the vast majority of them dead.[10] The destruction of seabed life went unrecorded.

The trouble is that some of the most robust fisheries today are for species that have high population growth rates and can live in heavily degraded environments. The peculiar fact is that it is possible to have 'sustainable overfishing' in which resilient species manage to hang in there after others have long departed. Scallops and prawns, for instance, have benefited from the removal of most of their predators by overfishing. Cod, angel sharks, giant skates and a bevy of others once crunched their way through legions of scallops and prawns. What remains are fisheries of last resort. They are all we have left when everything else is gone. Scallops and prawns are the last commercially viable fisheries in Scotland's Firth of Clyde (and even then, by 2011 the prawns had become so small from overexploitation the fishermen called them 'beetles'). I was left breathless with indignation when a leading environmental organization teamed up with the prawn trawl fishery there to help them get Marine Stewardship Council certification. I asked the chief executive of this organization whether he would support the destruction of rainforests if it could be shown that the palm oil that replaced it was produced sustainably. In his enthusiasm to find industry-led solutions to overfishing, he had forgotten that his core business was to be nature's champion.

Does that mean the Marine Stewardship Council logo cannot be trusted? While its reputation is dented, I wouldn't go that far. It is a pretty solid guide to fisheries that are well managed and where the target animal is still reasonably abundant. But that is as far as it goes. Until fisheries of last resort and those that use destructive methods like dredging stop being certified, you cannot be sure of the wider harm that is being done. There is a simple way out of this bind – the MSC needs to embrace the idea of 'mitigation'. In my opinion there is only one circumstance in which fisheries that use trawls or dredges should be certified, and that is if managers establish extensive protected areas alongside fishing areas so that fragile and less-resilient species can find refuge. I recently had an awkward conversation with Rupert Howes and Canadian fisheries scientist Daniel Pauly, who shares my concerns about the MSC. I put this suggestion to Rupert, but he said the MSC was not in the business of recovering overfished species. Good grief! Given the depleted state of so many of the world's fisheries, it should be. The MSC needs to think beyond sustainability in a narrow fisheries sense because it isn't good enough. In light of the many pressures the sea faces, we need to manage the sea for resilience, which means 'sustainability +'. What that entails is the promotion of increased abundance of life across the board, not the survival of the dregs left behind after we have plundered the rest. This shift in approach would multiply the MSC's ability to roll back the harm done by fishing many times over. Then we could eat dredge-caught scallops with a clearer conscience.

So, to get back to the stumped customer standing by the fish counter, here is my advice: try to avoid prawns or scallops and other bottom feeders fished up by dredgers and trawlers, such as plaice, cod and hake. If you want to eat such fish, try to find them line caught: hand-lines and trolls (hook and line towed from a boat) have less bycatch than longlines. Eat low in the food web, so favour smaller fish like anchovies, herring and sardines over big predators like Chilean seabass, swordfish and large tunas (you will be doing yourself a favour as these predators also concentrate more toxins). If you can't give up tuna, choose pole and line caught animals which have virtually zero bycatch. 'Dolphin friendly' versions alone may not be very dolphin friendly, since tuna are often caught with purse seines, walls of net that surround and stress dolphins and snare sharks, turtles and other wildlife. Farm-raised fish and prawns often come at a high environmental cost in destroyed habitat and wild fish turned into feed. Vegetarian fish like tilapia and carp are better than predators like salmon and seabass. Organic is better too, since your fish will have been dosed with fewer chemicals.[11]

Can the oceans continue to feed us? Will marine plankton survive their acid future well enough to support the rest of life at sea? Can we keep dead zones from expanding further? The exponential growth in human population, together with our ever-increasing aspirations for material comfort, is taking us into the unknown. The population explosion over the last half century means that rates of change ahead will be greater than anything we have experienced up to now. The only thing we can know for certain is that problems will get worse, perhaps much worse, before they get better. Greenhouse gases already emitted commit us to unavoidable climate change that could push some species and ecosystems to the limit of their endurance, and perhaps beyond them in the case of coral reefs. If we are to avoid the more disastrous of foreseeable futures, we must use every means at our disposal to lower stress and boost the abundance and variety of life in the sea. It could just buy nature enough time for us to stabilize our own population, switch to energy sources other than fossil fuels, and find ways of living within the limits of a finite planet.

For that, we need therapies complementary to protected areas and lowered fishing intensities. Stress, as any sufferer will testify, can affect every aspect of life. Stressed people are more tired, suffer more illness and are less able to cope with everyday life. Chronic stress can eventually lead to psychological or physical breakdown. Multiple stresses in the sea have analogous effects, triggering lower fertility, higher rates of injury, disease and parasite infection. Stressed ecosystems are more susceptible to invasive species. Because ecosystems are a product of the interactions among the animals and plants from which they are constituted, the influence of impacts reverberates. If our underwater worlds are to cope with the vicissitudes of climate change, we will have to do everything we can to minimize other stresses. Research in Australia suggests that coral reefs could cope with a couple of degrees more global warming before throwing in the towel if the water that bathes them is clean, clear of sediment and free of nutrient pollution.[12] We can't perform miracles overnight, but some countries have understood that it behoves them to do everything humanly possible to protect their natural resources. In 2010, in recognition of the compounding effects of multiple stresses, the Thai government closed eighteen popular dive sites when coral reefs bleached due to seawater warming.[13]

Many stresses affecting marine life have local origins, so local remedies can be effective. Coral reefs illustrate graphically what is at stake, and what the options are. The Stern Review of 2006 made it clear that dealing with the problems climate change may bring could cost a great deal more than tackling greenhouse-gas emissions at source.[14] You can build seawalls to hold back the ocean, but they are massively expensive, they don't work as well as natural habitats, and their costs put them beyond the reach of most of the countries that depend on coral reefs for their survival.

I recently came across a study of historic sea levels in the Maldives in which the authors found that sea levels had been 50–60cm higher than now on a couple of occasions in the past few thousand years.[15] They concluded that 'there seems no longer to be any reasons to condemn the Maldives to become flooded in the near future'. This is encouraging perhaps, but before Maldivians take too much comfort from the news they need to keep in mind that today's coral reefs face a fusillade of other stresses that didn't exist in the past. The Maldives has led on the world stage in negotiations to curb greenhouse-gas emissions. Their voice will have to be multiplied to be heard – we need a coalition of 'wet-footed' nations from across the world that by dint of geography are on the front lines of global change. These negotiations have proven extraordinarily difficult and will take time to conclude, perhaps decades. But time is in short supply for coral countries. Perhaps the only thing that could buy them enough time to survive the transition to a world of reduced emissions will be to raise their standards of reef protection to the very highest levels. It may not work, but it is the only thing that might.

Many people I talk to are pessimistic about the future, feeling that human nature cannot be changed: we will just carry on depleting and despoiling. It is easy to feel overwhelmed, but it would be wrong to give up and do nothing. Mahatma Gandhi once said, 'Be the change you want to see in the world.' If we all shoulder some responsibility for improving the oceans, at whatever level of influence we can muster, then change will come. Small effects multiplied millions of times make a great force.

The future is an unknown and sometimes frightening place. When I was a child the Cold War was at its height and I grew up fearing that nuclear annihilation was just around the corner. In preparation for a post-apocalyptic future, I devoured books on survival and self-sufficiency in the wild. Young people growing up today fear a slow-motion apocalypse. The creeping gears of climate change and an ever-greater burden of a growing population mean that future worlds will be different in ways that defy accurate prediction. Today, whole societies must hone their survival skills if they are to prosper. We have to work out how to live within our means and there isn't much time to do it.

Yet we are adaptable and resilient. Our success is testament enough. In the industrial age, so far, human resilience and adaptability have been directed towards maintaining the use or extraction of our natural capital in the face of dwindling resources. In the case of fishing this has included increased fishing intensity and power, the invention of new gears and methods, and periodic switches to less-favoured species. What we need now is to redirect our energies towards maintaining and rebuilding our resources. Prepare for change, and you have less to fear. It is our good fortune that there are so many ways to rebuild the variety and abundance of life. We have the means; do we have the will? If we are to refurbish our planet, we need to reinvent our relationship with the natural world.

From the perspective of society as a whole, the best policies are those that maximize long-term benefits and minimize costs. But politicians are human. They tend to serve the interests of those who most ardently court their favour, and weight short-term interests far more highly than long-term considerations. I constantly encounter politicians who view nature conservation as an extravagance, which affluent societies can afford to indulge in a little, but no more. Perhaps this is why we so often see policies that pay lip service to nature rather than recognize it as indispensable to human welfare. Their logic is that economies compete with nature for space and nature should yield to economic interest. We have to reverse that logic. More space for nature means more space for us. Our interests overlap far more than most people recognize.

21

Saving the Giants of the Sea

After a lecture in Glasgow one evening, I was collared by a grouchy Scot. Fixing me with an unfriendly stare, he spat out, 'Whit are we goin tae do aboot all them grey seals then? They're eatin moor fush than we catch!' My talk had been punctuated with old pictures of monster catches and bleak graphs of fish landings over the last century, whose lines plunged like jagged alpine slopes towards zero. This man had picked out what for me is one of the few bright spots in a sea of decline: the fact that grey seal numbers around the UK have been on the increase since we stopped hunting them in 1970. So too have common seals. Seabirds like gannets and guillemots also bounced back after we stopped eating them in the late nineteenth century.

What this proves is that conservation works. If we don't kill animals they live longer and become more common. In the eastern Pacific, grey whales have made a comeback from near extinction in the 1930s to a population of 20,000 today. Blue whales are slowly fighting back in Antarctica, the Atlantic and Pacific. Several species of fur seals have struggled away from the precipice of extinction in the last century, and Kemp's ridley turtles in the Gulf of Mexico have multiplied out of the risk zone. Florida's goliath groupers are mounting a comeback after fishing for them was banned in 1990. There is much joy to be had from these examples, for which the credit is widely shared.

Many ocean giants are quintessential world travellers, so protecting them is complex. In the aftermath of the Japanese tsunami of March 2011, Fukushima's nuclear disaster brought some of those connections to public attention in America. Albacore tuna winter in the seas off Japan before they head for California 8,000 kilometres away, where they support valuable fisheries. Northern bluefin tuna make the

288

same journey but more sporadically. Suddenly, Fukushima didn't seem so far away. Humpback whales migrate thousands of miles from chilly polar waters in both southern and northern hemispheres to warm-water breeding grounds in places like the Caribbean and the Mozambique Channel between Africa and Madagascar. Grey whales enliven the seas all the way along the west coast of North America as they shuttle between breeding lagoons in Mexico and summer feeding grounds in the Bering Sea. White sharks and whale sharks swim thousands of miles between feeding and breeding grounds. Some albatross circle Antarctica many times in their lives. Protected areas are often dismissed as a waste of time for such nomads. The truth, as you probably already know from migrant birds, is that, put in the right places, they can do a lot to help even the most peripatetic species.

Hanifaru, a coral ring on the eastern edge of Baa Atoll in the Maldives, is home to one of the great spectacles of nature. A ledge of coral traps and concentrates plankton as water pours in and out with the tide. When conditions of tide and plankton are right, dozens or even hundreds of manta rays glide in to feed. On a recent dive I was lucky enough to meet them.

Nothing prepares you for the first encounter with a manta ray. As I swam over the lagoon, the white sand bottom dropped away and the water darkened. Dense swarms of copepods coursed around me, causing my skin to tingle as they zinged against my body. I had never seen plankton this thick before. To a manta, it was an irresistible soup. Ahead of me, a large crescent shape loomed from the murk, followed by another, then another. Six mantas in all, head to tail, their broad oval mouths gaping wide. A pale flash of belly caught my eye and, turning, I saw another ray, mid-cartwheel, curling back on itself to make the most of a rich patch of food. Others appeared as I watched, mesmerized by their effortless grace. Soon twenty rays circled below, around and above me in a swirling silent vortex of planktonic gastronomy. Monolithic and impassive, they seemed oblivious to my presence, and yet not a fin brushed me though I was in their midst. Two giants headed directly for one another but at the last moment veered upward, belly to belly, before turning backwards. Another seemed certain to strike me on the downstroke of a wingbeat but delicately withheld the flap until it had passed by. The dance, for that is what it felt like, continued for an hour until the last of the plankton had washed over the reef sill on the falling tide.

Baa Atoll's mantas are lucky because they have a champion in the person of Guy Stevens. Guy is in his thirties, tall and lanky with sun-bleached hair that is usually wet from his latest dive. He spent years getting to know these mantas and soon found that every one had a unique pattern of black spots on its white belly. Since his discovery, he has recorded over the course of 10,000 observations more than 2,000 different rays, half of which visit Hanifaru.

Manta rays are the largest of all stingrays, although they have lost their stings and have quit the seabed for life in open water. They are among the largest fish in the sea, with a wingspan of up to seven metres and a weight of over two tonnes. For hundreds, perhaps thousands of years they have inspired awe and fear in those who have seen or imagined them. Few nineteenth-century books on ocean life were complete without a vivid description of some hapless pearl diver falling victim to a manta ray. But these were fanciful tales as manta rays had more to fear from people than the other way around. Many mantas basking at the surface were speared and hooked for sport by adventurers. Great amusement was had in the early twentieth century by hunters chasing mantas off the Carolinas in America. Sometimes the rays out-witted their pursuers, towing their boats with effortless strength until the lines broke.

Today, fishermen have the upper hand. Commercial fisheries have developed for mantas in Indonesia, the Philippines, Sri Lanka and Mexico, where thousands are landed every year. Elsewhere mantas are threatened by entanglement in longlines, capture in purse seine nets set to surround shoals of fish, and boat strikes. Guy Stevens' efforts to protect mantas led to Hanifaru Bay being declared a marine reserve. Although Maldivians have never hunted mantas, Hanifaru was their favourite place to catch whale sharks (now also protected) which were valued for their livers and fins.

Marine reserves can be useful for highly mobile species by protecting places where they aggregate, breed and raise their young, and in bottlenecks along their migration routes. Caribbean lemon sharks spend their first two or three years of life in coastal mangrove nurseries – without them the sharks would decline. The Mexican lagoons used by grey whales were nearly the species' downfall when whalers discovered them in the late nineteenth century, but protection in the 1970s helped boost numbers. In many places, salmon runs were virtually wiped out when nets were strung across river mouths as they returned from the sea to spawn. Close regulation of fishing on these migration highways has helped Alaska's salmon to a vibrant resurgence. Turtles haul themselves onto the same nesting beaches time and again over long lives. Judicious protection of these places can help save these animals from harm. But marine reserves offer only part of the protection that the giants of the sea need.

I recently came across a study that described efforts to reduce the bycatch from a longline fishery for mahi-mahi in Costa Rica that made my blood run cold. If you pay any attention to the sustainability of the fish you eat, you may know that mahi-mahi (also called dolphin fish or dorado) is considered a good choice. Mahi-mahi are supple-bodied gold-and-emerald fish spangled with blue that sport amid shoals of tiny fish wherever the sea is warm. They grow fast and reproduce early so they are resilient to fishing. Retailers like to label their products with enticing words like 'line caught in the clear waters of the Pacific', conjuring images of a gallant figure pulling on a rod to swing a single fish into the boat – no bycatch, clear conscience. The reality is usually very different. With the exception of pole-and-line-caught tuna, line-caught mostly means longlines. They can be tens of kilometres long and carry thousands of hooks. The problem is they catch far more than the target species.

The longline fishery featured in this study was at Playa del Coco and supplied mahi-mahi to America as well as more locally.[1] The collateral damage from the capture of just 211 mahi-mahi, which took 54 longlines with 43,000 hooks, was atrocious: 468 olive ridley turtles, 20 green turtles, 408 pelagic stingrays, 47 devil rays (close relatives of the manta), 413 silky sharks, 24 thresher sharks, 13 smooth hammerhead sharks, 6 crocodile sharks, 4 oceanic whitetip sharks, 68 Pacific sailfish, 34 striped marlin, 32 yellowfin tuna, 22 blue marlin, 11 wahoo, 8 swordfish and 4 ocean sunfish. To capture enough mahi-mahi to provide one lunch for five average-sized office blocks caused carnage in Costa Rican seas. What is even more disturbing is that this longline fishery had already switched to so-called 'turtle friendly' circle hooks! How friendly is that, hooking nearly 500 turtles for 200 mahi-mahi? Circle hook points are bent inwards rather than upwards as in the standard J-hook, so turtles are less likely to be hooked and fewer are gashed if they swallow the hook. But they often die nonetheless, as they tangle in the lines and drown.

Turtles aside, what about the devil rays, the sharks and the sunfish, the largest bony fish in the world about which we know almost nothing? The Playa del Coco fishery isn't just bad, it's absurd! It clearly isn't possible for a longline fishery of this kind to be sustainable according to any reasonable meaning of the word. Yet this is just one example of the reckless slaughter that pervades the open sea. How many images do we have to endure of sharks having their fins hacked off, their still-living bodies thrown back into the ocean before we do something about it? Next time you sink your teeth into a delicious mahi-mahi sandwich, spare a thought for the ghosts of all those others slaughtered to catch that fish. The ocean's big animals need protection beyond the limits of protected areas. Otherwise it won't be many years before this Costa Rican fishery and others like it close shop as there will be nothing left to take.

While I'm on the subject of good seafood choices, let's return to those 'dolphin friendly' signs I mentioned that most cans of tuna sport these days: you should know they are a little misleading. It all began in the eastern Pacific where pods of spotted and spinner dolphins associate with schools of yellowfin tuna. Fishermen found it easier to spot dolphins than tuna so they began to set their nets around dolphin pods, in the process drowning or mangling in their winches tens to hundreds of thousands of dolphins every year. The US banned the import of tuna caught using this method in 1990, so the fishermen changed their ways. Many still use dolphins to find tuna but now they haul their nets in a way that lets them escape, assisted by swimmers who jump into the water to herd them out of the back of the net. Today 'only' about 5,000 dolphins are killed directly by the fishery every year. Of greater concern is the fact that populations of the two species of dolphins that have been targeted are not recovering. Their reproductive rates are down, which can probably be attributed to stress from monthly near-death experiences at the hands of tuna fishers.[2] Calves may also be separated from their mothers in the confusion of fishing operations. Setting purse seine nets around tuna schools is a highly effective way of catching fish, but like long-lining it inflicts severe collateral damage.[3]

In the past when migrant animals left their seasonal haunts we could only guess where they went and what they did. For much of the twentieth century, our guesses were little better than the long-held belief, common in previous centuries, that swallows overwintered on the bottom of ponds. Now sophisticated electronic tags allow us to travel with them, often in real time. The sea is giving up its secrets and we can use this revolutionary new understanding of the oceans' biggest creatures to craft better ways to protect and manage them.

The most ambitious tagging programme, 'Tagging of Pacific Predators', is run out of Stanford, UC Santa Cruz and the Pacific Fisheries Environmental Lab in California. Since 2000, this team has attached electronic tags to over 2,000 animals, including seabirds, whales, sharks, turtles, sealions, even squid.[4] Tagged leatherback turtles in the eastern Pacific follow familiar paths from their nesting beaches to seasonal feeding grounds. These routes could easily form core areas for seasonal protection from the longline fisheries that have decimated them. Despite the name, the project also tags creatures in other parts of the world. They recently combined data from tagged bluefin tuna in the Gulf of Mexico with information about oceanographic conditions they experienced over the course of their migration to build a picture of preferred bluefin habitat. The findings show how fishermen who target other ocean-going predators like swordfish or yellowfin tuna could avoid places where bluefin are most likely to be present. The study showed that bluefin particularly like a region of the northwestern Gulf of Mexico that was severely affected by the Deepwater Horizon oil spill. Since bluefin go to the gulf to breed, this is worrying news.

The 400 northern right whales that migrate along the east coast of the US and Canada transit some of the world's busiest shipping lanes, and about one a year is killed by a boat strike. Satellite tags and arrays of underwater hydrophones now show areas where whales concentrate. In Canada, hydrophones listen out for these whales and warn ships to slow down and be more watchful when whales are close.

They need to be careful, because calculations suggest that for the northern right whale the loss of even two females a year could tip the balance towards extinction.[5]

Shipping is a major problem even in protected areas, as most allow the right of 'innocent passage' to boats that is enshrined in the UN Convention on the Law of the Sea. Stellwagen Bank National Marine Sanctuary lies just above the hook of Cape Cod off the coast of New England. It was created mainly to protect the humpback whales that visit every summer to gorge on thick shoals of tiny fish and krill that flourish where cool and warm currents meet. I visited Stellwagen on a charter boat in 2001 to see the place for myself. The sea was steel grey with a low, rolling swell but otherwise conditions were good and we soon spotted columns of spray from blowing whales. There were seven humpbacks, and as we approached I was thrilled to discover they were bubble-net feeding. In this technique the whales blow a curtain of bubbles around a shoal of fish to ball them into a tight mass before rising through the middle to engulf them. All at once a group of mouths burst through the surface in cavernous liquid yawns, fish and water streaming from their barnacle-knotted brows. These whales would long ago have learned to tolerate whale-watching boats, but much bigger and more dangerous vessels cut through their world, as Stellwagen straddles the shipping lane in and out of Boston Harbour.

More than 3,500 ships go through the Stellwagen Sanctuary every year, one every two and a half hours. To lessen the risk to whales, the shipping channels were narrowed and moved to avoid areas of highest activity, especially of northern right whales. The move is thought to have reduced the risk of ship strikes by 80 per cent. To the north, there has been a similar reorganization of shipping routes in and out of Canada. While welcome, this does little to reduce shipping noise, which occludes whale calls and breaks down social groups. When big ships pass through Stellwagen, the Sanctuary is so filled with noise that there is little space left for whales to converse.

I have been shocked by how rapidly populations of the seas' largest and most majestic animals have declined in my lifetime. Leatherback and loggerhead turtles, most species of albatross, and almost all species of shark are down by 75–90 per cent or more since the 1960s.

46. Aquaculture ponds for shrimp and fish cram the coast in China's southern Bohai Sea as this view from space shows. Much of the Bohai coast looks like this and natural habitats are hard to find in this crowded and highly polluted sea.

47. Coastal clean-ups have become regular events in many parts of the world, this one in a community near São Paulo, Brazil. Despite long-standing bans on dumping plastics at sea, the amount of rubbish picked up is still on the increase in most places, some of it dumped illegally, some blown to sea from the land and some washed offshore by rivers.

48. *and inset* Lamlash Bay on the Isle of Arran in Scotland became a marine reserve with full protection from fishing in 2008. For over a decade the local community fought prejudice, indifference and obstruction by government officials and industry in order to gain protection for their waters. Surveys in 2010 and 2011 showed that life in the bay is already on the rebound.

49. Cabo Pulmo Marine National Park in Mexico's Baja California underwent a spectacular recovery of marine life following its protection from fishing in 1999. By 2009 the weight of predatory fish in the park – like many of those in this photograph – had increased eleven times over.

50. It may take many years for protection from fishing to rebuild depleted populations of big, old fish like this black seabass in California's La Jolla Marine Reserve. Such animals are usually the first affected by overfishing and the last to recover.

51. European governments agreed in 2003 to create an international network of marine protected areas by 2010 through the OSPAR Commission. This is the result, showing protected areas in red and fishing closures in international waters outlined in yellow (NEAFC: North East Atlantic Fisheries Commission). The network is groundbreaking because protection has been extended beyond the 200 nautical mile limits of national waters to include six protected areas in the international waters of the high seas.

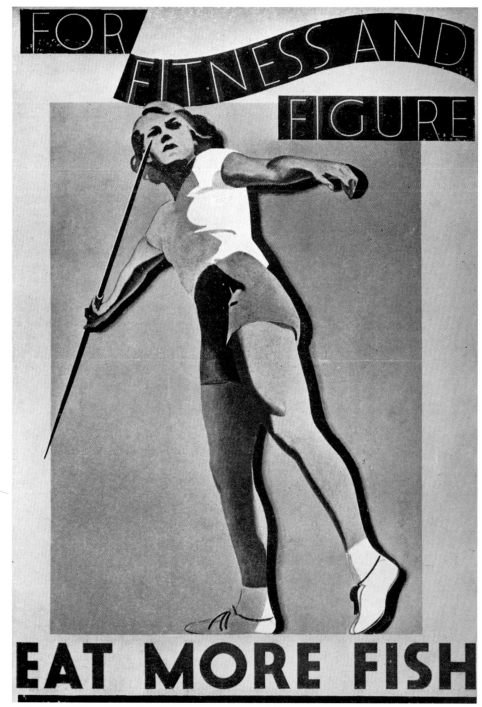

FOR FITNESS AND FIGURE

EAT MORE FISH

52. For decades, governments and fishing industries have exhorted people to eat more fish for good health, as in this 1930s poster. Recently, the benefits of fish oils for brain function and circulation have become better understood, leading to a renewal of such calls. However, supplies of wild seafood are in decline and these health benefits are also offset to varying degrees by increasing seafood contamination by pollutants and toxins from harmful algal blooms.

53. Leatherback turtle tangled in a longline set to catch fish. Leatherback numbers have plummeted in the Pacific Ocean in the last twenty years because of bycatch, pushing this animal to the brink of extinction there. Satellite tags show how the protection of key migration corridors in the Eastern Pacific might help stem their slide towards oblivion.

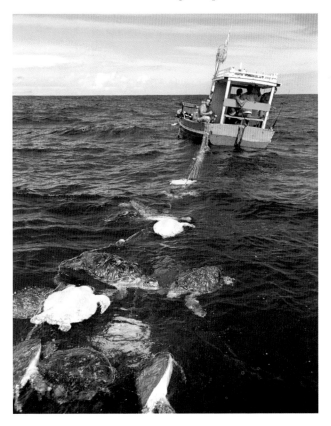

54. For centuries, green turtles were hunted for their eggs and meat. These days they are protected by law in most countries but nonetheless still often fall victim to other fisheries as unwanted bycatch. This abandoned fishing net with seventeen dead green turtles was discovered drifting off the coast of Bahia, Brazil, days after a storm.

55. Hanifaru reef in the Maldives regularly hosts remarkable numbers of huge manta rays that congregate to feed on thick concentrations of plankton. Each of the animals in this frame measures several metres from wingtip to wingtip. In 2011 Hanifaru reef was declared a World Heritage Site in recognition of its significance to manta rays and whale sharks, and has been made a marine protected area by the Maldivian government.

56. Shark populations across the globe have fallen into a steep spiral of decline in the last two decades driven by demand for their fins, the signature ingredient of a prestige food in Chinese cuisine: shark-fin soup. Economic growth in China has created a soaring demand for the fins, many of which are stripped from the animals on deck before their still-living bodies are dumped over the side to suffer a prolonged death. This animal was found among many others in a 'shark graveyard' on a reef in Sulawesi, Indonesia. Protecting sharks will require measures to curb demand for shark-fin soup alongside restraints on overfishing.

57. Although protecting the giants of the sea requires decades of dedicated work, there are encouraging success stories. In Brazil, the number of olive ridley turtle nests has increased by fifteen times in northern São Paulo since conservation efforts were launched in 1980 by a national environmental organization, Projeto Tamar.

58. In the mid-nineteenth century, when stocks of easily captured whale species had been all but exhausted by whalers, a large population of bowhead whales was discovered in the northern Bering Sea. They quickly came under attack and within a decade most had been killed. The population recovered slowly through the late twentieth century after protection was granted. Stone and steel harpoon points discovered in whales killed recently by Inupiat hunters led to the surprise finding that these animals can live to 170 or even 200 years old. We can only wonder at the upheavals they must have experienced in their ocean world within the space of a single lifetime.

Perhaps more enraging is how slow the world's response has been, even when the causes are obvious and reversible. The collapse of Atlantic bluefin tuna has exposed an institutionalized inability to instigate measures that will allow the species to recover. Here the will of a conservation-minded majority of nations is subordinated to the will of a handful bent on business as usual, regardless of the consequences. International forums are usually constituted on an egalitarian basis, where every nation has an equal vote and decisions are made by consensus. If unanimity cannot be achieved, the measure fails. This set up gives excessive power to minorities, and allows essential measures to manage the environment to be hijacked by self-interested groups.

A perfect example of this took place at the UN General Assembly in 2006. On the table was a proposal to introduce a global moratorium on deep-sea bottom trawling. This fishing method is highly destructive to vulnerable seabed habitats and its abolition had gained widespread support. The measure came within a whisker of being passed, but was vetoed at the last minute by the Icelandic delegation. A nation of 300,000 people stymied the introduction of protection critical to the survival of deep-sea life. I am a democrat, but that upset me. At some point we need to draw the line so that small minorities cannot torpedo the fate of the planet. The wildlife and environment of our children's generation will be under even greater pressures than they are today. Although it won't be easy or popular, reform is essential.

Bodies such as the United Nations often have a consensus decision-making process that gives equal weight to the views of all participants, be they large nations or small, strong or weak. In many of the bodies I have seen at work, such as Regional Fisheries Management Organizations, the European Union and the United Nations itself, important conservation measures would have gained approval if decisions had been made on a majority vote rather than needing full consensus. We would have an international moratorium on deep-sea bottom trawling and Atlantic bluefin tuna would be on the road to recovery if we shifted to majority voting.

Of course, majorities could always be gained by coercion or bribery, so decisions that weigh present greed or obstinacy over long-term responsibility could still be taken – it isn't a cure-all. Here is an example. Horrified at the scale and speed of the decline in bluefin tuna and frustrated by the deadlock in the International Commission for the Conservation of Atlantic Tunas, a coalition of conservation-minded nations and environmental organizations proposed at a meeting in Qatar in 2010 that Atlantic bluefin tuna be added to the list of the Convention on International Trade in Endangered Species (CITES for short). CITES regulates international trade in things like elephant ivory in an effort to ensure endangered species are properly protected. Until 2010 it had never dealt with fish of high commercial importance, as there was a tacit agreement that these species were best left to fishery managers, which is what pro-fishing nations argued in Qatar. It seems strange that other delegations swallowed this idea, since the only reason for bluefin to be proposed was the serial ineptitude of fishery managers. Japan, which consumes more bluefin than any other nation, lobbied hard to keep it off the list. In a move that left conservationists and diplomats steaming with indignation, their delegation served bluefin tuna at a banquet and gave away earrings and necklaces made of red coral, another threatened species considered for the list and later thrown out.

Imagine that a country served up roasted Amur leopard, or delicate fripperies of Miyako kingfisher with a Philip Island hibiscus garnish at a conservation conference. It was not a proud moment for Japan, and by these selfish acts they called into question the entire worth of CITES. There comes a point at which the world must rise above national hubris and unite to save the common heritage of humanity for the benefit of our own and future generations. Many felt that CITES was the place to do that; now they are not so sure. The scientific case for protecting bluefin tuna could not have been stronger, but it was voted down. Although majority decision-making would shift the odds in favour of conservation, it does not automatically lead to better environmental decisions.

I don't have a perfect answer; both consensus and majority voting have drawbacks. The latter will be anathema to some and unsettling to smaller and poorer countries. What is clear is that we must reinvent the concept of social responsibility for a crowded planet. As the world has filled to capacity, the freedom to exercise wilful self-indulgence or flagrant waste or to produce pollutants that spill into the living space of others has shrunk. The world can no longer afford to accept decision-making that puts selfish self-interest and short-termism ahead of consideration for fellow inhabitants of the planet, human or other, living or yet to come. At the global level, we need a Charter for Nature – a Charter for Life – akin to the UN Universal Declaration of Human Rights. Critics will no doubt be swift to condemn this as untenable in a world where human needs press ever harder. But a charter for wildlife is really a charter for human life and well-being. As Gaylord Nelson, a US senator and the founder of Earth Day, once said, 'The economy is a wholly owned subsidiary of the environment, not the other way around.' He meant that we depend on the environment for everything. Governments rarely consider the supporting roles of marine life when they devise management strategies for the sea. That omission is beginning to cost us dearly as oceans are stripped of life, breaking down their capacity to provide things we take for granted, such as protection from storms and clean, healthy water. We see this in the rise of toxic red tides off the coast of the southern United States, fuelled by polluted runoff that poisons fish and people alike. We see it in the spreading dead zones and in the heaps of decayed seaweed that clog fishing nets in China and the Baltic, and in the plagues of jellyfish that close resorts in the Mediterranean.

The laws governing the high seas are weak. Countries must first sign up to the UN Convention on the Law of the Sea to be bound by its regulations (the US still hasn't signed). They also have to sign up to become members of Regional Fisheries Management Organizations to be tied by their rules. Nations that remain outside these agreements are not breaking any laws if they catch fish or pollute or infringe any other legal measures set out by these bodies because they are not party to them. It seems mad and it is. It is as if a robber were free to break into houses and take whatever he wanted because he refused to acknowledge the law against theft. This legal loophole has spawned a trade in so-called 'flags of convenience', wherein countries like Belize, Panama and Liberia sell their flags to ships who wish to remain outside the law. Often this is to avoid restrictions on fishing, but it is also to avoid international labour laws so that crews can be made to work under horrendous, exploitative conditions. You can buy a flag online in minutes for a few thousand dollars.

It is hard to combat activities that are not overtly illegal. One approach which is enjoying some success is for Regional Fisheries Management Organizations to 'harass' the boats of non-signatory countries out of their areas by denying them access to ports. The high seas are enormous, so this can make it uneconomic for boats to operate beyond the law. However, even if conservation measures on the high seas were made legally binding on all, governance would still be left in the hands of organizations that place exploitation far above conservation.

Are there any good models of international collaboration to inspire us? The beginnings of a nature-first approach to management can be seen in the Convention for the Conservation of Antarctic Living Marine Resources. This agreement has been signed by thirty-one countries to date. The overarching aim of the Convention is to ensure that marine life remains healthy in the Southern Ocean. Only as much fishing as is compatible with this aim is allowed, and populations of penguins, albatross and seals, among others, are monitored closely to ensure there are no adverse effects. This body has been pretty successful in kicking out longline boats that were slaughtering albatross by the thousands while fishing for Patagonian toothfish. Victory in one place can mean disaster for another, however, as fleets of illegal or unregulated boats simply move elsewhere. What we need is governance like Antarctica's to be rolled out across all of the high seas and for it to apply to everyone, whether or not they have signed up to international conventions and treaties. The world is too small and crowded to let fleets flagged by irresponsible countries sow destruction wherever they wish.

The UN Convention on the Law of the Sea was conceived and written in the 1960s and 1970s, came into effect in the 1980s, and is badly outdated now. Article 62 gives a flavour of the motivations of those who drafted it:

> The coastal State shall determine its capacity to harvest the living resources of the exclusive economic zone. Where the coastal State does not have the capacity to harvest the entire allowable catch, it shall, through agreements or other arrangements . . . give other States access to the surplus of the allowable catch.

In other words, the default position is that all of the seas must be exploited. In fact, the Law of the Sea still has no provision to create protected areas despite wide recognition that they are needed, even within the United Nations. A couple of years ago I asked a legal expert close to the UN when we might expect a change. He rolled his eyes and groaned as if to say 'not in my lifetime'.

We don't have much time to reform our management of the oceans. Big creatures go down hard and fast and come back slowly, if at all. Most have few young and numbers can take decades to over a century to recover. Whales often give birth to a single calf every other year, and sometimes less. Manta rays in the Maldives appear to pup only once every five years.[6] With the megafauna, there are no quick fixes and populations can get stranded at very low abundance where they simply can't give birth fast enough to compensate for losses. But recoveries do happen. In 2010, for the first time in over a hundred years, possibly much longer, a southern right whale gave birth in the Derwent Estuary near Hobart in Tasmania. In the early nineteenth century, boats had to stay close to shore to get from one part of the bay to another because there were so many whales it was too risky to sail through their midst. By the 1830s there were nine whaling stations near Hobart, but by the 1840s their work was done. They had killed almost every right whale that used the bay. It has taken a very long time for them to come back.

Some regional bodies are moving forward to create high seas protected areas. I was lucky enough recently to work with OSPAR, a regional body that oversees environmental protection of the North East Atlantic. In 2003, ministers of the fifteen OSPAR countries agreed that by 2010 they would fill their seas with marine protected areas that would create a network of wildlife havens. By 2008 it was clear that there was a massive high seas void in the emerging network so they asked my team at the University of York to find places that warranted protection and to build a case for each one.[7] The outcome could not have been better. In 2010 OSPAR declared six places in the middle of the North Atlantic as marine protected areas; together they cover 285,000 square kilometres of high seas. For comparison, this is bigger than New Zealand or the UK and about four-fifths the size of Germany. When I heard the news I was at a conference. There can be few moments of such excitement in a scientific career. I felt like pulling my shirt off and running around the room in a footballer's goal celebration, but there were rather a lot of people there so I decided against it! Our science was the easy bit though. It took extraordinary political skill and tenacity by many people to see this through. The politics was complex and often frustrating, as we saw firsthand at a succession of international meetings. We had to revisit some sites several times to alter boundaries, or bolster the case made. We also had to drop two sites due to objections from some member states including, to my chagrin, the UK. There were last-minute complications when Iceland and Portugal made claims to the United Nations for huge areas of seabed to become part of their national waters. But in the end we won through.[8] Ours may be the first high seas protected area network, but another is being built in the Southern Ocean, and island nations of the Pacific have declared no-fishing zones for tuna in pockets of international waters between their exclusive economic zones. The world turns slowly, but at last a future is being built for life at sea.[9]

The nineteenth-century biologist Ernst Haeckel spent much of his time looking at plankton through a microscope. His exquisite illustrations were celebrated and inspired the organic forms of Art Nouveau.[10] There is immense beauty to be found in a thimble of seawater, but tiny animals and plants produce different emotions from the ocean's giants. I once went diving in a kelp forest off California's Channel Islands. On the boat ride from the mainland, we came upon two blue whales which captivated us for half an hour as they fed. As one sounded it threw its tail skyward and rivulets cascaded from a fluke as wide as a two-way road. I felt awed and humbled to be so near the largest animal ever to have lived on this planet. Nothing about my dive in the kelp forest could match that experience. On the boat ride back, 400 Pacific white-sided dolphins erupted from the water around us. The sea filled with leaping and twisting bodies ablaze with the warm glow of the setting sun. The megafauna worked their magic again.

The great beasts of the sea have long inspired wonder and fear. They exert a hold that few other creatures can match. More than this though, they have come to be symbolic of the state of our seas. If whales, dolphins, sharks and turtles disappear they will leave the sea incomplete; we lose something more important than mere flesh and blood with their passing. Slowly we have come to appreciate that megafauna are not merely embellishments: they are vital to the natural rhythms of life. Preventing these species from going extinct is not enough. Our aim should be to rebuild their numbers.

What my grumpy Scottish questioner had failed to grasp is that grey seals had only become a problem because of our failure to reverse declines in the fish that both seals and people enjoy. In his view, we should shoot the seals – a logic shared by the Scottish Government and disingenuous Japanese whaling officials who argue that their hunt, banned by most other countries, is necessary to spare more fish for people to eat. In the past there were fish aplenty for wildlife and us. That is the point we need to strive for once again. Culling the oceans' largest inhabitants because they eat fish treats the symptoms, not the disease.

22

Preparing for the Worst

We are on the cusp of one of the great reorganizations of planetary life. Five times our world has been plunged into turmoil that ended with the extinction of huge swathes of life. Asteroid impact and geologic upheavals that belched forth colossal quantities of lava have been implicated in several of these mass extinctions. Many scientists, myself included, now believe we are at the beginning of a sixth mass extinction, for the extinction rates of plants and animals now run at 100–1,000 times the background rate seen in the calm intervals between previous geologic extinction spasms.[1] Since we are the cause this time, it is still possible to avert catastrophe.

This is a remarkable century in which to be alive. According to the best estimates, the human population will reach a peak in the next ninety years at somewhere between 9 and 11 billion people.[2] As fertility rates fall, a run of population growth that has been virtually continuous for thousands of years will come to an end. Grave concern tempers this good news. There are important assumptions built into this prediction which we cannot take for granted. The final population figure is incredibly important to our ability to survive the apex of human population. It depends critically on quickly lowering fertility in developing countries. Demographers are concerned that people simply expect to watch human numbers rise to the peak and then gently fall afterwards. But lowering fertility fast enough to stabilize the world population at 9 billion will require us to ensure people in developing countries have rapid and secure access to family planning. The world's religions must rise to this challenge too. Just as the old ways of using resources are no longer tenable, so must religious doctrines, forged thousands of years ago, be reinterpreted for this century of global crisis.

Regardless of the specific level at which the human population peaks, demand for resources will be intense and there is no assurance that the world can feed a population this large. Enormous pressure will be heaped upon the natural world as we scramble for ways to live sustainably at this population level. While the curves of predicted population growth all level off reassuringly between 2050 and 2100, the curve of population multiplied by consumption does not; it just carries on skywards in an exponentially steepening rise.

All those new people added to the planet will not want to scratch a living for a dollar a day. They will aspire to better material standards of living, just like the billion plus people who live in abject poverty today. One of the development goals established by the United Nations at the beginning of this millennium was to end poverty and hunger. If, as I said earlier, the world is already using resources equivalent to the sustainable production of one and a half planets, how much more will a world need with 30–60 per cent more people, the poorest of whom will want a bigger share than is claimed by today's poor? This is what scares the pants off me.

If we just carry on with business as usual, humanity has a bleak and uncertain future. The direst predictions see us on a planet suffering from runaway global warming. Drought and inundation by rising seas will rob the world of its most fertile lands, while wildfire, flood and tempest will cause upheaval elsewhere. The oceans and lands will haemorrhage species as the sixth great extinction accelerates, leaving us with less and less from which to meet our needs. Civil disorder will rise and conflicts among nations will escalate as people struggle to survive. Refugees will flood the luckier nations as they flee war and famine and rising seas, and might overwhelm them. Standards of living will plummet in both developing and developed nations.

There is a less dismal future ahead if we quickly wean society off fossil fuels and apply every scrap of human ingenuity to find ways to live within the limits imposed by nature and to bring population growth down fast. There are many people who hope for the latter, but they will have to fight their case hard to win against the entrenched opposition and apathetic insouciance of others who fear the inconvenience such a move might impose on their lives or businesses. It is a natural reaction not to want to change long-established habits. As I have shown in this book, our relationship with the seas has developed over more than a hundred thousand years. It is incredibly difficult to shrug off mindsets that have such deep roots we scarcely question them: the mantra that there are plenty more fish in the sea, or the belief that a little collateral damage in catching them won't hurt, or that the oceans are a bottomless receptacle for wastes and can disperse and neutralize all of the toxins we release.

Every generation likes to think of itself as different from its predecessors. For much of the last thousand years this has been true as each generation has brought its creative fire to find new solutions to old problems. Over much of the course of this history we have harnessed nature to our own ends to secure comfort and prosperity. But now that relationship has begun to falter: we have surpassed nature's capacity to sustain our wants.

This change has overtaken us rapidly so it isn't surprising we are finding it hard to adjust. My grandparents never worried about their world running out of fuel, or space to grow crops, or fertile soil, or fish. It is us who must grapple with a planetary account that is in the red. We must rebuild our inventory or humanity faces bankruptcy. At the heart of the problem is a difficult dilemma: how do developing countries meet their aspirations from a dwindling pool of resources? It is not right for developed nations to dictate that they remain poor, or that they pay a higher price to gain access to energy and infrastructure, which is one of the main reasons climate-change negotiations have been fraught and, to date, unsuccessful. On balance we are most likely to tread a path somewhere between the best and worst extremes but there is one certainty: things will get worse before they get better. We must prepare for the difficult times ahead.

In the deep past, there were comparable changes in ocean acidity to those we are likely to face in the next hundred years which led to widespread loss of life. Some scientists, such as the Australian coral biologist Charlie Veron, see alarming parallels between those crises and the present. With increased ocean temperatures and acidity, he says, we will not just lose one ecosystem: we will set in motion a domino effect in which we will see one ocean ecosystem fail after another. It could be a replay of what happened 55 million years ago during the runaway global warming of the Palaeocene-Eocene Thermal Maximum, when carbon dioxide levels, temperature and ocean acidity spiked ferociously. The planet has recovered from previous extinction crises, but not without devastating upheavals and loss of life. As species fell others rose from the ashes of their empires, and ecosystems were remade from a few of the old parts and many new. But it took millions of years.

The geologic record gives us more than enough reasons to terrify us into action. But we cannot stop the chemical transformation of our atmosphere in its tracks any more than we can halt human population growth. Both have enough latent capacity to maintain the momentum of planetary change far into the twenty-second century. The seas are likely to become more hostile to life.

Even if we get serious about cutting greenhouse-gas emissions, with more and more mouths to feed the assault on the oceans could simply carry on. More fertilizer and sewage input would increase the frequency of harmful algal blooms, intensify oxygen depletion, create more dead zones and set the stage for the jellyfish ascendancy. The spread of aquaculture will eat away at natural habitats and aggravate problems of nutrient enrichment. More intense agriculture on degrading soils will flush extra mud into coastal waters, which might help build coastal wetlands like saltmarsh and mangrove, but would destroy sensitive habitats constructed by invertebrates like corals. Sea level rises to which we are already committed by past emissions will lead to more seawalls and other defences in a process of coastal hardening that will squeeze out productive habitats like mudflat and marsh. With the disappearance of these vital nurseries, wild fisheries will suffer, and there will be fewer feeding grounds for migratory birds. And, if we remain wedded to all the comforts that modern technology can give us and remain as wasteful as we are today, the oceans will continue to accumulate toxic contaminants and the soup of plastic particles will thicken.

It doesn't have to be like this, but we are sleepwalking in that direction. We can change the course of the future by what we do today. There is an old adage, much loved of self-help books, that says 'today is the first day of the rest of your life'. If we change course by a few degrees now, it will take us to a very different place in fifty years' time from where we are headed. Our children will reap the benefits. The great thing about the future is that it has yet to come so there is everything to play for. We don't have to accept our fate. In the few pages left I will explore what we can do to maximize the chances that our transition to sustainable lifestyles can be accomplished without impoverishing nature and destroying ourselves. I have to assume that we will work to control human population growth, switch to clean energy, reduce emissions and lower pollution of all varieties, because without these shifts there is no question the outcome will be bad; the only uncertainty is when will the end finally come?

Aldo Leopold, one of the fathers of America's environmental movement, said 'to keep every cog and wheel is the first rule of intelligent tinkering'. It is a phrase often quoted and for good reason. Even if there are some seemingly redundant parts in the mechanism of nature, we don't know which ones they are. Nature's variety is one of its great assets when it comes to coping with change, and therefore it is one of ours. The portfolio effect I spoke of in fisheries applies much more widely. If investors hold a diverse portfolio of assets, they can be pretty sure that it will continue to yield dividends no matter what the future brings. At any particular time some investments will do well while others languish. A species that is rare today and seems unimportant may be common tomorrow. Under different conditions, it could become the lynchpin of some vital ecological process. We have little or no capacity to predict which species that might be from the thousands around us.

In 2001 European environment ministers set themselves an ambitious goal: to halt biodiversity loss by 2010. To many people, and probably most of the politicians who set the target, this simply means preventing any more species from going extinct. Although that is important, halting biodiversity loss implies much more. It means stopping further loss of natural habitats, turning around population declines, and ending the piece-by-piece dismantling of food webs. Europe failed to achieve its goal, of course. Halting the collapse of biodiversity without tackling the drivers of loss is impossible. I would no more have expected a bon viveur with a cellar full of fine wine to give up drinking. Stopping biodiversity loss will take far more than bold words. At the global level, the European target was watered down to 'achieve a significant reduction in the current rate of bio-diversity loss'. We failed even to achieve that.[3] While those targets were welcome, they must be met with a complete change in our attitude to nature. Otherwise we will never achieve them.

Some people are preparing for the worst by stockpiling frozen stores of seeds or genetic material of plants, corals and other animals. These gene banks are a little too science fiction for me. They could work for plants, and maybe corals, but the approach has great limita-tions for most animals. Captive-bred animals struggle to cope when released into the wild, even after being raised by nurturing parents. Imagine how difficult it would be to resurrect a species from a tissue sample. There are dreamers out there who yearn to stitch mammoths back together from fragments of DNA deep frozen in the Siberian tundra, but these frozen libraries will do nothing to keep our planet-ary environment running. I think there are better ways to create arks in which species might tough it out until humanity gets to grips with the climatic convulsion we have set loose.

One approach is to identify the places where habitats and their spe-cies are most likely to survive, and then give them powerful protection. During past ice ages, the deterioration in climate forced many species from large parts of their ranges. In some places, conditions remained more clement. They formed refuges where animals and plants hung on. When the ice melted, the survivors spread forth to reclaim lands they had lost. Some people are searching for a modern equivalent of these refuges, especially for embattled habitats like tropical coral reefs. Rod Salm is one such hunter. He works for The Nature Conser-vancy and heads up its coral programme. For the last decade Salm has been on a quest to find places where coral reefs are most likely to wea-ther the warming and increased ocean acidity ahead. Some of the most likely places, it turns out, are ones that have few human inhabit-ants. George W. Bush protected several of them in the Pacific Ocean in an uncharacteristic frenzy of environmentalism at the tail end of his presidency. Rose Atoll, Palmyra and Kingman Reefs, the Northwest Hawaiian Islands, all these places support more vibrant and intact reefs than almost any that fringe inhabited lands.

The UK created the largest marine reserve in the world in 2010 when it placed the Chagos Archipelago off limits to all fishing. This British Territory rises like a vision of paradise from the middle of the Indian Ocean, with its white sand beaches, swaying palms and vivid turquoise lagoons. It seems an ideal refuge from which reefs around the Indian Ocean rim could be replenished with life in some more agreeable future. Aside from a US military base – actually because of it – the Chagos are uninhabited.[4] Its reefs were badly hit by the 1998 global mass coral bleaching event, which killed nine-tenths of its coral. But they have since bounced back better than any others in the Indian Ocean.[5] The near absence of any other stresses such as pollution, construction or significant exploitation enabled the corals to cope. The Chagos and places like it offer evidence that local efforts to reduce the cocktail of human stresses can make a difference.

One organization that championed the protection of the Chagos was the Alaska-based Pew Global Ocean Legacy programme. Their aim is to see some of the world's most intact and iconic marine ecosystems given the highest level of protection. The goal is bold, not least because the places on their target list are vast and protection at this scale has never been attempted before. But they have notched up several successes (with the help of George W. Bush and the UK) and have set their sights on Australia's Coral Sea and New Zealand's Kermadec Islands.[6]

Blue Marine Foundation, a UK-based environmental group, secured the financing for the first five years of protection of the Chagos. Blue plans to mobilize private capital to protect extraordinary and important places at sea wherever they may be. Possibilities include South Georgia and Gough Island in the South Atlantic. Gough is a windswept speck of heath and rock that is inhabited only by albatross, penguin and a handful of other hardy wildlife (and some monster mice that have tripled in size and taken to a diet of seabird chicks since whalers dropped them off in the nineteenth century). It was discovered by the Portuguese in 1505 and was claimed by the UK in 1938. Gough came with over 125,000 square kilometres of ocean around the island. Imagine what value it would have to wild nature if it were protected. Gough has been described as one of the most important seabird colonies on Earth, and its waters are home to numerous whales, dolphins, fish and invertebrates.[7] Though our world is crowded, there are still many opportunities to safeguard wild places and create a lasting legacy for future generations.

Other organizations are concentrating on protecting hotspots where a rich and distinctive variety of plants and other creatures can be found, which they believe will be of disproportionately high value in extinction prevention. Conservation International, a powerful US-based group, has identified hotspots on land, such as the Atlantic forests of Brazil, California (for its flowers), Madagascar and the wonderfully named Succulent Karoo of Southern Africa. They are pouring money into these places to save species before it is too late, with considerable success. Similar places exist at sea, although they have only been mapped for a few habitats and species groups, such as coral reefs and open ocean predators, and a systematic catalogue of ocean life is still years away.[8] Other organizations, such as the World Wildlife Fund, are targeting more diffuse areas for conservation known as 'priority' eco-regions. Their thinking is that, since life operates at large scales, protection efforts should encompass the same. Such eco-regions include the Gulf of Alaska, the North Sea, the South Kuroshio Current off Japan and the Adriatic Sea.

Protecting remote, intact areas like the Chagos Archipelago will give us a sporting chance of saving life's variety as the globe changes. I'm all for it. But nature at arm's length will do little to satisfy our aesthetic or spiritual needs. Places usually remain intact because they are hard to get to. It is lovely to know wild polar bears stalk seals on distant ice floes. It comforts me to think that there are isolated coral reefs that remain virtually as they were 200 years ago, like Kingman Reef in the mid-Pacific where sharks patrol in almost unthinkable numbers and thick shoals of fish like burly emperors and sweetlips wait for nightfall to fan out and dig for prey. But we need places nearby too.

Nature close at hand is almost priceless; look at any city park if you doubt the truth. New York's Central Park opens like a green oasis hemmed in by cliffs of stone, concrete and glass, a place of tranquillity and beauty away from the dust and din of traffic-snarled streets. A song thrush that sings from your own rooftop is lovelier by far than the unheard call of a loon on some distant northern lake. Nature nearby is what matters when it comes to benefits like clean and healthy water and air. I saw a study by a worthy group of scientists not long ago that suggested we could enhance the protection of bio-diversity if we decommissioned the least cost-effective protected areas, sold off the land and bought up a far greater area of cheaper, more effective land instead.[9] It sounds sensible, and at face value the figures add up. There is just one drawback – the places deemed most likely to offer poor value for money were parks close to cities, where land val-ues were sky high. To me those are the places where nature is highly valuable, where it is most accessible for us to enjoy. To decommission them in favour of more distant patches of habitat would amount to withdrawal of the lifeblood of human experience. I once flew into São Paolo in Brazil. On the descent we passed a near-endless sprawl of high-rises, apartment buildings and shanties. The buildings seethed up and over the sides of the low hills like mould. Hardly a breath of open space or green leaf broke a man-made vista that stretched to the horizon and beyond. It was one of the most depressing things I have seen.

In the race to save species, conservationists usually home in on the best remaining fragments of habitat. This is understandable in a world where money is tight. Such areas offer a way to maximize species pro-tected per dollar spent. But, in doing so, conservation neglects places less rich and remarkable. Sometimes this is because they are more degraded, but often it is because environment and evolution have not bestowed on them the variety gifted to other places. But society has a stake in low diversity regions too. In fact, these places are likely to be much more susceptible to loss of important ecosystem services than the highly diverse sites seen as conservation priorities. Caribbean coral reefs possess a handful of coral species that do the heavy lifting of reef construction; West Pacific reefs have dozens. Disease epidemics in the Caribbean have knocked out two of the main reef builders and are now hacking away at the third. As a result, growth of reefs in the Caribbean at best has juddered to a halt, and in most places is in reverse as erosion takes hold. It is easy to see the results underwater.

Where bright coral turrets garlanded in multi-hued fish once rose, all too often there is a low, hummocky panorama of green weed pock-marked with sponge and seafan. The results are visible above water too, as resort beaches wash away and condos fall into the ocean. Local conservation efforts matter, even in places that would add little to the catalogue of species protected.

We tend to be sceptical of government promises and targets. But I have been impressed by how great a motivator targets for marine conservation have been. Delegates at the 2002 World Summit on Sustainable Development in South Africa agreed to set aside more marine protected areas and to restore fish stocks to more productive and safer levels by 2015.[10] The latter target will not be met in full; indeed it cannot be, because some fish populations grow too slowly to rebuild in the three years left. Like Europe's plan to halt biodiversity loss, I doubt whether the people who agreed to the targets thought they could ever be achieved by the deadlines set. What these goals have done is to set the hares running, and in their wake there has been an unprecedented burst of effort to protect the oceans. Marine protected areas are proliferating. George W. Bush made a huge splash with his Pacific protected areas, and single-handedly added over 30 per cent to the total area under protection in the world oceans. That makes him one of the world's greatest marine conservationists! That expression might stick in some people's throats given his rather less glorious environmental record on land, but, whatever your feelings about President Bush, he left a wonderful legacy in the sea.

Other leaders have emerged who are shaping the global agenda. In 2005, the president of the tiny Pacific state of Palau, Tommy Remengesau, launched the 'Micronesia Challenge' in which he urged other states in the sprawling Micronesian archipelago to protect 30 per cent of their marine environment and 20 per cent of lands by 2020. *Time* Magazine lauded him as an Environmental Hero in 2007. But Remengesau was a flawed leader, embroiled in corruption and with rather thin environmental credentials, since he built his own house without a permit in the midst of a mangrove forest. Nonetheless, the Micronesia Challenge has gained the support of many of the island nations of the region, and has inspired countries further afield, such as Indonesia and the Caribbean island of Grenada. Island nations have the most to lose from mismanagement of the sea, and their example is one that nations better endowed with land should all follow.

There has been much head scratching over how big and how close together protected areas need to be to function optimally.[11] Based on how far animals and plants move and disperse, the consensus is that they should typically be about ten to twenty kilometres across and no more than forty to eighty apart. At its most risk prone – ten-kilometre reserves eighty kilometres apart – such a network would cover one-ninth of the sea. At its most risk averse – twenty-kilometre reserves forty kilometres apart – it would cover a third. In network building, we would do well to heed the investors' caveat that past performance is not always an indicator of future performance. In warmer seas, eggs and larvae will develop faster, which means they will spend less time swimming or drifting in the open sea before settling on a place to live and won't travel as far as they do now.[12] That means we would be wise to establish networks at the higher end of the protection range. Some nations and states have already started to act on these principles.

In 1999 California started to build a network of coastal protected areas that, when complete, will cover some 20 per cent of its seas (like all US States, California has control over waters up to three nautical miles out to sea, the rest is managed by the Federal Government). The UK network I mentioned earlier is slated to cover a quarter of its seas, although not all will be highly protected. Some areas will permit less damaging kinds of fishing to continue, like lobster trapping. France has declared it will protect 20 per cent of its waters by 2020 and half will exclude all fishing. In many countries, renewable energy may add to overall protection, since wind farms often prohibit fishing to protect cabling. Given the enthusiastic embrace of wind energy, the combined effect could secure 30–40 per cent of coastal waters, which is where we need to be.

Beyond the borders of these future protected areas, we need to scale back the intensity and destructiveness of fishing. This doesn't mean forgoing catches in the long run, although there may be short-term costs. Re-inflating fish stocks from their present nadir will bring increased catches within five to ten years, taken at lower cost with less fuel burned. It does, however, mean phasing out or greatly limiting the footprint of the most damaging fishing gears, like trawls and dredges, and unselective longlines. That move is long overdue. The fourteenth-century English peasants who pleaded with their king to ban the newly invented trawl that 'runs so heavily and hardly over the ground when fishing that it destroys the flowers of the land below water'[13] perfectly understood the dangers the method entailed. In the long run, those who decide to protect and treasure their natural assets are the ones who benefit most. Many of the gains that come from establishing protected areas and banning trawling are most apparent locally. Reinvigorating the sea isn't an altruistic act of self-sacrifice; it makes sense on the grounds of self-interest too.

There has been huge progress in the last fifteen years on some fronts and retreat in others. New high-level commitments to establish protected areas are encouraging and invigorating and I see incredible energy being thrown behind implementing these initiatives. Protection is at last being extended to some places in international waters far from coasts where the foundations of a high seas reserve network are being laid. In Europe, it was finally acknowledged in 2011 that throwing away dead fish after you have caught them is madness. It is better still to let the ones you don't want live on in the sea by fishing less in the first place and using more selective fishing methods. But, against these gains, the tide of human pressure continues to swell. In October 2011, as I was completing this book, the human population passed the 7 billion mark. And, in this globalized world, the time between discovery and depletion of a new type of fish or shellfish, or new fishing ground, has shrunk from decades to the space of a few years. The urgency for decisive change is greater than ever before.

Epilogue: The Sea Ahead

A few years ago, Inupiat hunters from the North Slope of Alaska caught a bowhead whale. When they cut up its carcass they found an iron harpoon point buried within its shoulder of a kind that had not been used for more than a hundred years. It turned out that this whale was 130 years old. Other older bowheads have been caught with stone harpoon points in them, which indicate that these animals can live for 200 years. I wonder what a 200-year-old whale would make of the changes it has experienced in its lifetime.

In the early nineteenth century, there were perhaps as many as 100,000 bowheads around the Bering Strait that separates North America from Asia. The sea would have resounded to their calls as they went about the noisy business of life. For more than half the year, much of their world was frozen. Then in 1848, a whaling boat penetrated these waters and slaughtered them like sheep. The bowhead is a docile creature and killing was easy. A year later, their southern grounds exhausted, more than 150 other whaling vessels did the same. In the space of a couple of decades, bowhead numbers were slashed. For every hundred whales before, only one or two were left and the whale chorus fell silent. A hundred years on, those seas began to fill with the noise of ships, and then in 1984 the world called off its commercial hunt. By then the bowheads' seas were pierced with a new sound – the deafening thump of seismic exploration for oil and gas. Sea ice retreated and unfamiliar fish and plankton began to move north. Over long years, ever so slowly, the numbers of bowhead whales began to recover. Now there are over 10,000 in Alaskan waters and their lowing calls once again fill the seas, although this time in competition with the roar of ship engines.

Within the space of a single bowhead lifetime, the world of whales changed for ever. We are not quite the Methuselahs they are, but the oldest among us have experienced transformations of similar magnitude. Our world is changing faster than at any time in human history. While our relationship with the oceans has hit a low, there is still time for us to change course. The oceans and life within them will endure, albeit in forms that may be unfamiliar today. The largest hurdle on the road ahead is perhaps not amenable to any technical fix – our humanity, shaped by nearly 200,000 years of struggle to survive. It is essential for ocean life and our own that we transform ourselves from being a species that uses up its resources to one that cherishes and nurtures them.

People have a deep emotional connection to the sea. The oceans inspire, thrill and soothe us. Some think we owe our clever brains and the success they brought to our ancestors' close link to the sea. But our relationship with the sea stretches back through time much further than this: all the way to the origins of life itself. We are creatures of the ocean.

In this book, I have made the case for a wholesale reversal of present trends of wildlife decline and environmental degradation. We are inseparable from the living fabric of this world. We depend on other creatures to fix energy for us, so in the most fundamental sense their lives make our own possible. Our present and future prosperity and happiness requires that we stop taking nature for granted, recognize what it does for us, and accord it the highest importance in human affairs. It is not enough to preserve bits and pieces of nature here and there simply to remind ourselves of what we once had.

I am not suggesting that we stop fishing altogether, but it is clear that we can't afford to go on fishing the way we do. The value of the fishing industry, large as it might seem, is dwarfed by its environmental costs of lost function and resilience in marine ecosystems. Many of us are only slowly beginning to recognize that the hurricanes and tornadoes, droughts and floods that have been ravaging the world are connected to our degradation of the environment. We are losing our natural defences, and nature unleashed is a terrible thing. We need to refurbish the oceans so they once again heave with life. Doing this will give the world more food and would reduce or eliminate many other problems of reduced wildlife abundance that now plague us. All of the countless things that the natural world does for us and we take for granted would be delivered at higher levels, with greater security, at less cost. Flourishing ecosystems are more resilient to shocks and change, and better placed to cope with chronic stress. Fewer species would require expensive protection. Our quality of life will rise with the abundance of wildlife and the seas can continue to inspire us as well as provide for us.

The changes I have chronicled in this book are those that most affect the oceans today. But new challenges continue to emerge. The economics of growing resource scarcity will soon make it worthwhile to mine metals from the deep sea. In many places the seabed thousands of metres below the reach of light is studded with fist-size nodules rich in scarce metals, a tempting prize for those who can devise a way to bring them to the surface. Already oil companies have pushed their wells into water more than a thousand metres deep. These activities bring with them new impacts and risks to marine life, as the horrific Deepwater Horizon blowout showed.

The pressure is on for aquaculture to use the latest genetic technologies to boost production from the sea. But we must be careful because our engineered hybrids will escape into the wild, and, once they have done so, they will be almost impossible to control. Melting ice sheets may soon open up the poles for use and exploitation, just as acidification begins to rip a hole in the polar carbonate system that could transform life there. And, just as we sort out how to neutralize our toxic contaminants, we will continue to invent new problems for ourselves, such as microplastic particles in facial scrubs and pharmaceutical pollution. Fixing ocean problems is not a one-time effort. It will be a continuous process of revision and adjustment.

Some of the backstory in this book is disheartening, and the picture gets worse looking forward if we blithely continue on our present course. But I am hugely encouraged by efforts in the last ten years. People have really noticed the spread of human influence across and beneath the sea, and there are countless efforts underway to redress the harm. I have never seen so much energy or commitment to tackle problems, from the humblest village to the debating halls of the United Nations. This is why I remain an optimist. We can change. We can turn around our impacts on the biosphere. We can live alongside wild nature. The alternative is self-destruction.

Appendix 1: Seafood with a Clear Conscience

In developed countries, there has never before been such a great variety of seafood on offer. Globalization serves up the delicacies of shoreline, reef and open sea from the furthest reaches of the planet to anybody willing to pay. Supermarket counters groan under a cornucopia of shellfish and finfish that please the eye and ravish the palate. I am often asked if it is possible to be a seafood lover and a lover of life in the sea. As an enthusiast of both, I have to answer 'yes', but it is a qualified yes. Buyers with a conscience must choose carefully, because, as I have showed in this book, for many species on offer the price paid by the environment for their capture is far greater than the ticket price.

There are four main questions to ask when choosing seafood:

- Is the species in trouble in the wild where the animals were caught?
- Does fishing for the species damage ocean habitats?
- Is there a large amount of unwanted bycatch taken with the target species?
- Does the fishery have a problem with discards – generally undersized animals caught and thrown away because their market value is low?

It is difficult to give blanket advice on what to buy and what to avoid, since fisheries for the same species can have very different impacts depending on how and where the animals are caught. But I will attempt a few rules of thumb here. Avoid large, long-lived species that mature late in life, as they are easily overfished. Examples include skates, sharks, swordfish, marlin, some tunas (such as bluefin and bigeye), wolffish, halibut and sturgeon (caviar). Sharkfin soup is a big no, not just because sharks are in steep decline everywhere, but because the fins are often cut from living animals that are then thrown back overboard to suffer lingering death. Avoid all deep-sea species because they are very easily overfished and are slow to recover. They also tend to be long-lived, slow growing and mature late in life. However, added to these problems, deep-sea fishing causes immense habitat damage and has severe bycatch problems. Almost none of the bycatch survives because of extreme pressure and temperature changes experienced on being brought to the surface. Examples include orange roughy, Patagonian toothfish (also known as Chilean seabass), oreo dories, scabbardfish (espado), grenadiers (hoki), black halibut, redfish and deep-water prawns.

Many fishing methods have terrible impacts on habitats and other marine life, as I have said throughout this book. Bottom trawls, scallop dredges and hydraulic clam dredges tear up or bury fragile marine life growing on the bottom like corals, seafans and sponges. Gill nets hang like ghostly walls in the water and drown thousands of marine mammals and seabirds, as well as catching many economically worthless species that are thrown away. Prawn and shrimp fisheries have the worst bycatch records, with five to ten times the amount of usable catch tossed overboard, sometimes more. Think about how much of a pile those dead and discarded animals would make on your plate the next time you tuck into a meal of wild-caught shrimps or prawns! Choose animals caught with minimal damage to the environment. 'Hand-picked', 'diver-caught', 'hook and line-caught' (although beware as this can mean longlines; handlines are best), 'pole and line', 'creel' or 'trap-caught' species are usually good bets. Avoid trawls, dredges, gill nets, longlines and drift nets. Schooling species like herring, pilchard and anchovy are caught using nets but they are very clean fisheries with almost no bycatch, and are an excellent choice on health grounds (high oil content and little in the way of toxins since they eat low in the food chain).

'Dolphin-friendly' tuna brands are not always what they seem. As I've explained, much of the tuna caught in the eastern Pacific still involves surrounding dolphins with purse-seine nets. Even though few dolphins are killed outright, animals are stressed and mothers get separated from their young. Pole and line-caught tuna is the best choice; look for it on the tin. The same caveat goes for fresh tuna steaks in the supermarket, most of which will have been caught with longlines or purse seines. Many of those purse seines are now set around fish aggregating devices that have been deployed for days or weeks by the catching vessel to attract tuna. The trouble is they also gather together turtles, sharks, whales and dozens of other species that may also be caught and killed.

I covered the many problems of aquaculture in Chapter 16 so I will only give a brief recap here. Suffice to say that many farm-raised fish are certainly not guilt-free, and they may come loaded with contaminant chemicals used to prevent disease in overcrowded pens. Some fish farms are stocked with fry or young caught from the wild, like some prawns and bluefin tuna. Aquaculture also has direct impacts on the environment, including pollution from chemicals, excess feed and wastes produced, alien species' introductions and habitat destruction to make way for ponds or cages, not to mention the human rights abuses associated with unscrupulous businesses in some parts of the world. Shrimp farming has caused vast areas of mangrove forest to be cleared along tropical coasts. And you probably wouldn't want to eat anything that has come out of particularly filthy places, like China's Bohai Sea. However, not all aquaculture is bad. Choose fish that feed low in the food chain, such as tilapia and carp. Avoid predators like salmon, groupers and tuna, because they will usually have consumed far more wild-caught fish over their lives than the weight of flesh produced. However, some (usually high-end) suppliers do produce ethical varieties of these fish fed on things like fish trimmings or purpose-grown worms. Choose shellfish such as mussels and oysters. Farming shellfish can do favours for coastal water quality, as these animals filter plankton and other organic matter from the water. While organic farms cause less contamination of the environment from the use of chemicals, the animals still produce polluting faeces and are often still fed on wild-caught fish. Organic is a better choice, however.

Knowing all this is one thing, but applying it is another matter entirely. A problem we all face at the fish counter is a dearth of key information – namely, where and how the fish was caught (and sometimes what the fish is in the first place, as DNA testing shows a significant fraction are mislabelled, even in reputable stores). Don't be afraid to grill your fishmonger on the provenance of their fish. If this all seems too much to remember when buying fish or ordering dinner, you can download wallet-sized cards with advice on which fish to buy and which to avoid, or use mobile phone apps from the list below.

One of the pioneering and still best efforts to rate the sustainability of different fish species is the Monterey Bay Aquarium Seafood Watch Program, based in California (www.montereybayaquarium. org/cr/seafoodwatch.aspx). It produces wallet cards which list species to avoid and good choices. A US national card and six regional guides are available, several in English and Spanish. There is a sushi guide too (www.montereybayaquarium.org/cr/cr_seafoodwatch/download. aspx). There is also a Monterey Bay Aquarium Seafood Watch Android app that brings you up-to-date recommendations for ocean-friendly seafood and sushi. The newest version, with Project FishMap, lets you share the locations of restaurants and markets where you've found sustainable seafood. As the map grows, you'll also be able to see what others have found near you.

Another US-based organization that recommends sustainable seafood is the Blue Ocean Institute (www.blueocean.org/seafood/ seafood-guide). They also produce mobile phone and iphone applications (www.blueocean.org/fishphone).

For best seafood choices within Europe, the UK Marine Conservation Society's Fishonline (www.fishonline.org) is a great resource.

There are many other guides produced for other parts of the world in local languages. The following website provides links to guides from the Netherlands, Belgium, Sweden, Finland, Germany, Switzerland, Canada, Australia, Indonesia, Hong Kong and South Africa: overfishing.org/pages/guide_to_good_fish.php.

Finally, Fish2Fork (www.fish2fork.com) is a campaigning guide for those who want to eat fish in restaurants sustainably, and was founded by Charles Clover, who wrote *The End of the Line*. Currently it rates restaurants in the UK, USA, Spain, France and Belgium.

A WORD ON ECO-LABELLING

Some fish are provided with eco-labels that are designed to alert consumers to sustainable fish. The most recognizable (and, so far, only independent) label is that of the Marine Stewardship Council (www. msc.org). As I've explained, fish given the MSC label are supposed to come from fisheries that are exploited in accord with the following principles:

- Principle 1: Is the stock at levels considered to be sustainable over the long term, and is it well managed and monitored?
- Principle 2: Does the fishery damage habitats or kill other species, including marine mammals and sea birds?
- Principle 3: Are good management procedures in place to regulate fishing and ensure rules are followed?

Although the MSC has been criticized (by myself included), especially with respect to not fully implementing Principle 2, it is still a good guide to the sustainability of the labelled fish itself, and is almost always a better choice than fish without the MSC label.

IS SEAFOOD SAFE?

Fortunately, you can avoid the worst of the problems from contaminant chemicals in seafood the same way you can avoid some of the worst overfishing problems: avoid big predators like tuna, swordfish and marlin, and eat low in the food web by choosing species like herring, anchovy, pilchards, tilapia and mussels. If you are worried about how to balance the health benefits of fish with the contaminant dark side of seafood, there is an outstanding source of information: the SeaWeb KidSafe Seafood programme (www.kidsafeseafood.org). This website contains a host of information on the best choices to avoid contamination and promote sustainable fishing, as well as lots of background on the pollutants involved.

Appendix 2: Conservation Charities Working to Protect Ocean Life

If, having read my book, you would like to support efforts to safeguard life in the sea, or just want to find out more, I can recommend some excellent places to begin. The following is a list of organizations that I have worked with over the years and can personally recommend for their dedication, innovation and effectiveness. Each has its own approach and priorities. Hopefully, you will find among them a match to your own interests.

SEAWEB (WWW.SEAWEB.ORG)

SeaWeb's motto is 'Leading voices for a healthy ocean'. High-quality, accurate scientific knowledge of the state of the sea forms the bedrock for all SeaWeb's programmes and campaigns. They take a unique approach to conservation, putting the latest understanding from communications science to work in their efforts to raise awareness of problems at sea, change mindsets and put in place measures to protect marine life. Its commitment to basic science has made SeaWeb a trusted source of information by those looking for the facts without the spin that is often imparted by those with agendas to pursue. I mentioned one of those information portals in Appendix 1 – KidSafe Seafood (www.kidsafeseafood.org) – which helps parents gain the health benefits of fish for their children while avoiding the toxic downside posed by compounds like mercury. Over the years I have benefited greatly from one of SeaWeb's services to ocean professionals: a regular email update on the latest scientific research (Marine Science Review: www.seaweb.org/science/MSRnewsletters/msr_archives.php).

For those who want to see some of the issues I have covered in a more graphic way, SeaWeb runs a Marine Photobank (www.marinephotobank. org) that contains thousands of incredible photos from all over the world, donated by the photographers and available for use by the media.

Rather than considering the corporate world an enemy, as some conservation bodies do, SeaWeb has sought to engage with them, with considerable success. Its flagship programme, Seafood Choices, has for years been building better relationships between the seafood industry and conservationists. To achieve this it has enlisted the help of celebrity chefs, restaurateurs and captains of industry to speak out and encourage others to follow the path towards more sustainable fisheries. Its hugely popular annual 'Seafood Summit' provides neutral ground at which conservation and consumption can meet and exchange ideas. SeaWeb's work has broadened over the years from a focus on campaigns to protect species threatened by overfishing, such as swordfish and sturgeon (caviar), to tackling the threats posed by pollution and climate change. As a Board Member of SeaWeb since 2007, I know at first hand that the SeaWeb team works incredibly hard to foster greater understanding of the importance of the sea to us.

FAUNA AND FLORA INTERNATIONAL
(WWW.FAUNA-FLORA.ORG)

Fauna and Flora International (FFI) was founded in the UK in 1903. Early on it focused on the protection of big game in Africa, where hunting was decimating iconic animals like lions, cheetahs and leopards. FFI has since expanded worldwide and today has satellite organizations in America, Australia and Singapore that guide extensive regional programmes. With over a century of practical conservation behind it, FFI is one of the most experienced organizations around. However, it was only in 2010 that they founded a fully fledged marine programme, in recognition of the growing peril faced by life at sea, supported by generous donations from Lisbet Rausing, Peter Baldwin and Arcadia. The shape of this programme is still evolving, but its core emphasis is to support the creation and effective management of marine protected areas. As a Council Member of FFI since 2007, I have been lucky enough to help FFI launch its efforts seawards.

BLUE MARINE FOUNDATION
(WWW.BLUEMARINEFOUNDATION.COM)

Blue Marine Foundation grew out of the success of the 2009 movie *The End of the Line*. It was founded by George Duffield and Chris Gorrell Barnes, who produced the film, and is chaired by Charles Clover, who wrote the book of the same name. Blue notched up its first major success in 2010, raising funds to finance management of the newly created Chagos Marine Reserve in the Indian Ocean. They believe that there are many businesses and individuals who would willingly support greater protection for the oceans, if only they could find a simple and reliable way to donate to worthy causes. Blue is dedicated to finding excellent projects in need of support, and raising the funds to help them. I have been on the Board of Directors of Blue since it was founded in 2010.

WWF (WWW.WORLDWILDLIFE.ORG (US); WWW.WWF.ORG.UK (UK); WWW.PANDA.ORG (INTERNATIONAL) AND MANY OTHER NATIONAL OFFICES)

WWF (World Wildlife Fund in the US and World Wide Fund for Nature everywhere else) celebrated its fiftieth year of conservation effort in 2011. It is a vast organization that now claims millions of members in over a hundred countries of the world, and has projects in many more. WWF works with partners all the way from village level in places like Africa, through national governments, to international bodies. It has long taken the view that it must work directly with governments and businesses to promote environmental protection, drawing criticism from those who consider some of these relationships too toxic to be pursued. However, through these initiatives WWF has often gained privileged access to high-level meetings and fora that has enabled it to score many landmark conservation successes. I served two terms on the National Council of WWF-US and have been an Ambassador for WWF-UK since 2009.

RARE (WWW.RARECONSERVATION.ORG)

Rare is an organization that works internationally from the bottom up, by promoting a stronger cultural and emotional connection between people and nature. I first came across them years ago when I was studying the effectiveness of marine reserves in St Lucia. This Caribbean island was where it all began for Rare, with a passionate campaign to save the native parrot, then on the brink of extinction. The parrot was uniquely St Lucian and the successful campaign to save it tapped into a growing sense of national identity and pride. Since then, Rare has rolled out this approach in more than fifty countries, training local conservation fellows to apply lessons learned from these efforts to achieve rapid and lasting benefits for the environment, and thereby enhance quality of life for their communities. In recent years they have begun vigorous efforts to increase the numbers and effectiveness of marine reserves in places that desperately need them, such as the Philippines.

CONSERVATION INTERNATIONAL (WWW.CONSERVATION.ORG)

Conservation International (CI) was founded in the US over twenty years ago around the notion that rather than protect a smattering of places as 'relics of the past', we should move towards a future with a healthy environment as the green beating heart of a prosperous society. As such, it has become a champion for indigenous communities as well as for the wildlife and habitats their homelands support. Like FFI, its emphasis was initially land-based, but CI soon developed an active marine programme. Like its terrestrial counterpart, CI's efforts in the sea have sought protection for some of the richest habitats on the planet, especially coral reefs like those of Southeast Asia's Coral Triangle, the heartland of marine biodiversity. It also co-ordinates regional initiatives in places like the eastern Pacific.

GREENPEACE (WWW.GREENPEACE.ORG)

Greenpeace has its origins in the environmental movement of the late 1960s, and from the beginning campaigned on ocean issues from its vessel *Rainbow Warrior*. In recent years the organization has been a passionate advocate for high seas and deep-sea protection in the global commons beyond the limits of national waters. My research group did a report for Greenpeace a few years ago in which we asked the question, what might a network of high seas marine protected areas look like if we were to create it today. It was a flag-waving exercise to highlight the almost complete lack of protection these areas had. I am glad to say that, through Greenpeace's efforts, some areas we proposed have now been placed off limits to tuna fishing in the Pacific, and OSPAR has further developed suggestions for North Atlantic MPAs and implemented them in 2010. I always enjoy working with Greenpeace – they have a refreshing idealism coupled with steely determination.

PEW ENVIRONMENT GROUP (WWW.PEWENVIRONMENT.ORG)

The Pew Environment Group was set up with support from The Pew Charitable Trusts, which have long had an intense interest in the sea. They run the well-known Pew Fellows Program in Marine Conservation which supports people in mid-career who are dedicated to promoting greater protection for the oceans (I am lucky enough to be one). Under the tireless leadership of Josh Reichert, Pew has been a powerful advocate for the oceans, establishing the Pew Oceans Commission that reported on the state of the sea in the US in 2003. Although its focus has traditionally been concentrated in America, increasingly its reach extends outwards as it turns to solving problems that respect no borders. Efforts include campaigns to establish very large and highly protected areas, like the north-west Hawaiian Islands, British Indian Ocean Territory (Chagos Islands) and Australia's Coral Sea. Among many other things, Pew is working hard to put in place strategies to protect sharks. The Pew Environment Group has given strong support to basic scientific research that can inform and underpin conservation.

WILDAID (WWW.WILDAID.ORG)

Wildaid focuses its efforts on reducing demand for wildlife products ('when the buying stops, the killing can too'). It is one of the few conservation groups to have made serious inroads into the Chinese centre of Asian seafood consumption. It is best known for its quirky and beautifully made infomercials that urge people to change their eating habits (such as shunning sharkfin soup in China or turtle eggs in Mexico). Wildaid enlists celebrities to carry these messages to a wide audience, including Leonardo DiCaprio, Richard Branson and Chinese basketball ace Yao Ming.

THE OCEAN CONSERVANCY (WWW.OCEANCONSERVANCY.ORG)

The Ocean Conservancy is based in the US and has historically concentrated on protecting national waters. It has dedicated itself to causes such as reducing plastics and rubbish that despoil oceans and coasts, and to establishing more and better Marine Protected Areas (MPAs). The Ocean Conservancy was among a small number of players that were instrumental in getting California's Marine Life Protection Act on to the statute books and using it to build a network of MPAs that will, in my view, become a source of national pride over the course of time. It is also campaigning for a fresh approach to environmental protection in the Gulf of Mexico following BP's Deepwater Horizon blowout of 2010. The organization sees opportunity in the aftermath of disaster, and proposes that some of the damages fund be put towards ending years of environmental neglect and degradation in the Gulf of Mexico (remember those seas of trash, I mean artificial reefs, I mentioned!).

OCEANA (NA.OCEANA.ORG)

Oceana claims to be the world's biggest organization dedicated wholly to ocean conservation. It has over half a million members and is active all over the world, especially in the US and Europe. Oceana runs campaigns across a wide spectrum of ocean problems, from overfishing to ocean acidification to mercury pollution. Its successes include efforts to protect the oceans from bottom trawling in Alaska, banning shark finning in Chile and banning the trade in shark fins in California (and as the saying goes, where California leads, the nation follows). Oceana is one of the few organizations to tackle the pernicious issue of harmful government subsidies that prop up overfishing. It has waged a long campaign to remove subsidies that is not yet won, but the rewards are potentially huge. Withdrawal of subsidies from high seas and deep-sea fisheries could eliminate overnight much of the harm these industries perpetrate, since they are only economic with generous taxpayer handouts.

MARINE CONSERVATION INSTITUTE (WWW.MARINE-CONSERVATION.ORG)

The Marine Conservation Institute (MCI) is a highly respected outfit that is headquartered in the US but whose reach extends far beyond US borders. In no small part that is due to the enormous energy and boundless vision of its founding President, Elliott Norse. Elliott is a master strategist and a tireless campaigner. And, although too modest to admit it widely, he and his team have notched up a string of successes, notably in campaigns to establish huge marine protected areas during the Clinton and George W. Bush presidencies. The Institute has dedicated enormous effort to seabed protection from bottom trawling, especially in the deep sea. It was a core partner of the Deep Sea Conservation Coalition that gained UN support for deep sea protection in 2008. Like SeaWeb, MCI is committed to generating and using excellent science in the cause of greater protection for marine life.

SYLVIA EARLE ALLIANCE – SEA (OCEAN. NATIONALGEOGRAPHIC.COM/OCEAN/MISSIONBLUE/ SYLVIA-EARLE-ALLIANCE)

Sylvia Earle is the most eminent advocate for ocean conservation in the world. She has had a long and starred career as scientist, oceanographer, explorer, government appointee and author. Although now in her seventies, she relishes hard work and dedicates greater energy to conservation than most people half her age. She was awarded the TED Prize in 2009 which gives recipients the chance to make a wish. Hers was that we 'use all means at your disposal – Films! Expeditions! The web! More! – to ignite public support for a global network of marine protected areas, "hope spots" large enough to save and restore the ocean, the blue heart of the planet'. The upshot has been dramatic. Sylvia has done a huge amount to raise awareness of the perils facing life in the sea among people who have significant resources or means to do something about them. She founded SEA as part of her TED prize and has since raised millions of dollars for marine conservation.

THE NATURE CONSERVANCY (WWW.NATURE.ORG)

The Nature Conservancy was founded in the US in the 1950s but has since spread its wings internationally. For many years, it devoted most of its funds to buy land to create refuges for threatened habitats and species. The oceans are publicly owned and so this approach cannot be translated directly except in coastal areas, where the organization has used it to great effect to protect wetlands and shores. Further seawards, the Conservancy has been working to develop ways to promote resilience in natural habitats so they can better deal with the slew of stresses I have described in this book. Like Conservation International, it has an especially active programme of reef conservation in the Coral Triangle.

CLIENT EARTH (WWW.CLIENTEARTH.ORG)

Client Earth is a group of idealistic lawyers (they do exist!) who have committed themselves to fighting environmental abuses with the law. Their starting point is in human rights: the right that everyone has to live in a healthy environment. But, as they point out, that right is in its legal infancy and is still fragile. As their name suggests, they consider the Earth to be their client and their brief is to defend it. Sometimes that means the creative application of rather arcane legislation, but often it just means ensuring that good laws are used to their full force.

And, where no suitable laws exist, their aim is to nurture them into being. In the marine realm, Client Earth is working to ensure Europe's Common Fisheries Policy is reformed so that overfishing is ended, to halt the reckless slide to oblivion for bluefin tuna, and to challenge supermarkets over shaky environmental claims made on fish packaging, among other initiatives.

SEA SHEPHERD (WWW.SEASHEPHERD.ORG)

This is the only organization in this list that I have not worked with or experienced personally in some capacity. But, for those of you who prefer your conservation to be more confrontational, Sea Shepherd can probably not be bettered. Its mission is to protect wildlife at sea, and there is not much its activists won't do to get between those intent on killing animals and their quarry. Their most high-profile campaign has been against Japanese whaling in Antarctica, but they have also been active in many other places. Among dozens of campaigns, they have helped patrol the Galapagos Marine Reserve against shark poachers. They have tried to protect Taiji's migrating dolphins from slaughter in Japan (whose plight was brought to international attention by the movie *The Cove*). And for many years they have fought against Canada's harp seal hunt.

Forgive me if I haven't mentioned your favourite organization here. There are literally hundreds worldwide. I have concentrated on those working internationally in the above list so my apologies to all those excellent groups working to make things better nationally, like Canada's Living Oceans Society, the Australian Conservation Foundation, the UK's Marine Conservation Society, Mexico's Communidad y Biodiversidad, all of whom I can vouch for personally, as well as countless more.

Acknowledgements

This is a book that I hadn't planned to write. My first, *The Unnatural History of the Sea*, took in a thousand years of history and had the world as its stage. It took five years to research and write, so by the time it was complete I thought I would try something a little less daunting. My agent, Patrick Walsh, and Will Goodlad, my soon-to-be editor at Penguin UK, had other ideas. Over a splendid lunch, they persuaded me to go even bigger: the story of the oceans from the beginning of the world to their possible futures a hundred years from now. I am not sure how or why I agreed, but I am glad I did, even though it has taken another five years! I am grateful to both of you for the boldness of your vision and your faith in me to take it on.

I had the great good fortune to have Joy de Menil from Viking as my editor. Joy has a gift of seeing what is essential and showing an author how to say it with elegance and clarity. She suggested I cut many things from my draft, some of which I thought at first were indispensable. If you have made it this far, you will be glad she wielded her red pen so skilfully.

In researching this book I visited subjects I scarcely even knew existed at the outset. I have been privileged to have so many excellent guides who have been generous with their advice and time. I am indebted to the following for their comments on draft chapters, suggestions for improvements, and for aiding my understanding: Rebecca Atkins, Geoff Bailey, Andrew Bakun, Bryce Beukers-Stewart, Dee Boersma, Alistair Boxall, Ian Boyd, Sally Brown, John Bruno, Ken Caldeira, Jim Carlton, William Cheung, Daniel Conley, Sarika Cullis-Suzuki, Curtis Ebbesmeyer, Jon Erlandson, Rainer Froese, Jason Hall-Spencer, Julie Hawkins, Mark Hixon, Leigh Howarth, Joanie

Kleypas, Dane Klinger, Andrew Knoll, Kevin Lafferty, James McCarthy, Carol Milner, Charles Moore, Roz Naylor, Maggy Nugues, Beth O'Leary, Daniel Pauly, Howard Peters, Jurgenne Primavera, Yasmin Primavera, Nancy Rabalais, Judith Sealey, John Shepherd, Steven Simpson, Albert Tacon, Ruth Thurstan, Carol Turley, Peter Tyack, Charlie Veron and Oliver Wurl. These folks spared me from many errors. Those that remain are my own. I have been lucky to be able to try out many of the ideas in this book on dozens of enthusiastic and very bright students. They tested the limits of my understanding and forced me to improve my explanations when they were found wanting. I am also grateful for the kindness of the following colleagues who let me use their images in this book: Marco Octavio Aburto-Oropeza, Bryce Beukers-Stewart, Nancy Boucha, Victor Hugo Casillas Romo, Aaron J. Cavosie, Sarika Cullis-Suzuki, Katharina Fabricius, Dave Harasti (who is a superb underwater photographer and whose pictures can be seen at: www.daveharasti.com), Elizabeth Gates, Ryan Goehrung, Karin Malmstrom, Guy Marcovaldi, Curtis Marean, Loren McClenachan, Stephen McGowan, Tim Pusack, Link Roberts, William Rodriguez Schepis (Instituto EcoFaxina), Steve Spring, Guy Stevens (The Manta Trust: www.mantatrust.org), Silke Stukenbrock, Bob Talbot, Glen Tepke, John Valley, Kyle Van Houten, Cynthia Vanderlip and Reuven Walder. Thanks also to Devin Harvey of Sea-Web for his great help in sourcing images.

I have my wife Julie and two wonderful daughters to thank for patience, understanding, encouragement and love throughout the course of writing this book. I am fortunate indeed.

Notes

PROLOGUE

1. McClenachan, L. (2009) Documenting loss of large trophy fish from the Florida Keys with historical photographs. *Conservation Biology*, 23: 636–43.

I. FOUR AND A HALF BILLION YEARS

1. Valley, J. W. (2005) A cool early Earth? *Scientific American*, October: 58–65.
2. We can date the timing of Earth formation very precisely to 4.567 billion years ago based on the age of the earliest meteorites. Valley, J. W. (2006) Early Earth. *Elements*, 2: 201–4.
3. Sleep, N. H. et al. (2001) Initiation of clement surface conditions on the earliest Earth. *Proceedings of the National Academy of Sciences*, 98: 3666–72.
4. Fedonkin, M. A. et al. (2007) *The Rise of Animals: Evolution and Diversification of the Kingdom* Animalia. The Johns Hopkins University Press, Baltimore.
5. Strangely, it has been very hard to find evidence for higher and more energetic tides in the geological record. As Harvard professor Andrew Knoll put it to me, perhaps we don't know what to look for.
6. Wilde, S. A. et al. (2001) Evidence from detrital zircons for the existence of continental crust and oceans on the Earth 4.4 Gyr ago. *Nature*, 409: 175–8.
7. Drake, M. J. (2005) Origin of water in the terrestrial planets. *Meteoritics and Planetary Science*, 40: 519–27.
8. Isotopes are variants of elements that have different atomic weight because they contain different numbers of neutrons. They help us to

distinguish the origins of ocean water, since the isotopic composition of the source must be similar to that of the oceans.

9. Hartogh, P. et al. (2011) Ocean-like water in the Jupiter-family comet 103P/Hartley 2. *Nature*, 478: 218–20.

10. Schopf, T. J. M. (1980) *Paleoceanography*. Harvard University Press, Cambridge, Mass.

11. Koeberl, C. (2006) Impact processes on the early Earth. *Elements*, 2: 211–16.

12. Charette, M. A. and Smith, W. H. F. (2010) The volume of Earth's ocean. *Oceanography*, 23: 112–14.

13. Snelgrove, P. V. R. (2010) *Discoveries of the Census of Marine Life*. Cambridge University Press, Cambridge.

14. Hawkesworth, C. J. et al. (2010) The generation and evolution of continental crust. *Journal of the Geological Society*, 167: 229–48.

15. Nutman, A. P. (2006) Antiquity of the oceans and continents. *Elements*, 2: 223–7.

16. Hessen, D. O. (2008) Solar radiation and the evolution of life, in E. Bjertness (ed.), *Solar Radiation and Human Health*. The Norwegian Academy of Science and Letters, Oslo, pp. 123–36.

17. Battistuzzi, F. U. et al. (2004) A genomic timescale of prokaryote evolution: insights into the origin of methanogenesis, phototrophy, and the colonization of land. *BMC Evolutionary Biology*, 4:44, doi:10.1186/1471-2148-4-44.

18. Some of this warming probably came about through splitting of methane in the upper atmosphere and conversion of ethane, another powerful greenhouse gas. Haqq-Misra, J. D. et al. (2008) A revised, hazy methane greenhouse for the Archean Earth. *Astrobiology*, 8: 1127–37.

19. This lack of physical evidence means that the study of earliest life is highly controversial and is constantly under review as new evidence emerges.

20. Our best estimates suggest the oceans of the deep past were tens to hundreds of times less productive than today.

21. Canfield, D. E. et al. (2006) Early anaerobic metabolisms. *Philosophical Transactions of the Royal Society B*, 361: 1819–36.

22. Canfield, D. E. (2005) The early history of atmospheric oxygen: homage to Robert M. Garrels. *Annual Reviews of Earth and Planetary Science*, 33: 1–36.

23. Buick, R. (2008) When did oxygenic photosynthesis evolve? *Philosophical Transactions of the Royal Society B*, 363: 2731–43.

24. Anbar, A. D. et al. (2007) A whiff of oxygen before the great oxidation event? *Science*, 317: 1903–6.

25. Kopp, R. E. et al. (2005) The Paleoproterozoic snowball Earth: a climate disaster triggered by the evolution of oxygenic photosynthesis. *Proceedings of the National Academy of Sciences*, 102: 11131–6.

26. Williams, D. M. et al. (1998) Low-latitude glaciation and rapid changes in the Earth's obliquity explained by obliquity-oblateness feedback. *Nature*, 396: 453–5.

27. Kappler, A. et al. (2005) Deposition of banded iron formations by anoxygenic phototrophic Fe(II)-oxidising bacteria. *Geology*, 33: 865–8.

28. Anbar, A. D. and Knoll, A. H. (2002) Proterozoic ocean chemistry and evolution: a bioinorganic bridge? *Science*, 297: 1137–42.

29. Sulphide is toxic to oxygen-producing cells, so they were excluded from deeper layers.

30. The earliest oxygen users might have gained their oxygen from the breakdown of hydrogen peroxide. Dismukes, G. C. et al. (2001) The origin of atmospheric oxygen on Earth: the innovation of oxygenic photosynthesis. *Proceedings of the National Academy of Sciences*, 98: 2170–75.

31. Gill, B. C. et al. (2011) Geochemical evidence for widespread euxinia in the Later Cambrian ocean. *Nature*, 469: 80–83.

32. Dahl, T. et al. (2010) Devonian rise in atmospheric oxygen correlated to the radiations in terrestrial plants and large predatory fish. *Proceedings of the National Academy of Sciences*, 107: 17911–15; Lenton, T. M. (2001) The role of land plants, phosphorus weathering and fire in the rise and regulation of atmospheric oxygen. *Global Change Biology*, 7: 613–29.

33. Lambert, O. et al. (2010) The giant bite of a new raptorial sperm whale from the Miocene epoch of Peru. *Nature*, 466: 105–8.

34. Gill et al. (2011), op. cit.

35. Benton, M. J. and Twitchett, R. J. (2003) How to kill (almost) all life: the end-Permian extinction event. *Trends in Ecology and Evolution*, 18: 358–65.

36. Kidder, D. L. and Worsley, T. R. (2004) Causes and consequences of extreme Permo-Triassic warming to globally equable climate and relation to the Permo-Triassic extinction and recovery. *Palaeo*, 203: 207–37.

37. Knoll, A. H. et al. (2007) Paleophysiology and end-Permian mass extinction. *Earth and Planetary Science Letters*, 256: 295–313.

2. FOOD FROM THE SEA

1. Landau, M. (1984) Human evolution as narrative. *American Scientist*, 72: 262–7.

2. Verhaegen, M. et al. (2007) The original econiche of the Genus *Homo*: open plain or waterside?, in S. I. Munoz (ed.), *Ecology Research Progress*. Nova Science Publishers, New York, pp. 155–86.

3. Recently the term Hominin has replaced the previous use of Hominid. The term Hominin includes humans and all their ancestors, while Hominid refers to all of the great apes (including us) and their ancestors.

4. Tishkoff, S. A. et al. (2009) The genetic structure and history of Africans and African Americans. *Science*, 324: 1035–44.

5. Marean, C. W. et al. (2007) Early human use of marine resources and pigment in South Africa during the Middle Pleistocene. *Nature*, 449: 905–8.

6. Although Blombos was occupied for longer, it wasn't inhabited as early as Pinnacle Point.

7. Brown, K. S. et al. (2009) Fire as an engineering tool of early modern humans. *Science*, 325: 859–62.

8. Braun, D. R. et al. (2010) Early hominin diet included diverse terrestrial and aquatic animals 1.95 Ma in East Turkana, Kenya. *Proceedings of the National Academy of Sciences*, 107: 10002–7.

9. For a long time it was thought that Neanderthals and modern humans overlapped in their habitation of Europe. But new dates for deposits seem to contradict that, pushing back the dates of the last Neanderthals to before our arrival. However, researchers from Spain suggest that their dates of 32,000 years ago are robust. No doubt the debate will continue. http://news.sciencemag.org/sciencenow/2011/05/were-neandertals-and-modern-huma.html?ref=hp; accessed 18 May 2011. Stringer, C. B. et al. (2008) Neanderthal exploitation of marine mammals in Gibraltar. *Proceedings of the National Academy of Sciences*, 105: 14319–24.

10. Clottes, J. and Courtin, J. (1996) *The Cave Beneath the Sea: Paleolithic Images at Cosquer*. H. N. Abrams, New York.

11. Erlandson, J. (2001) The archaeology of aquatic adaptations: paradigms for a new millennium. *Journal of Archaeological Research*, 9: 287–348; Erlandson, J. M. et al. (2007) The kelp highway hypothesis: marine ecology, the coastal migration theory, and the peopling of the Americas. *Journal of Island and Coastal Archaeology*, 2: 161–74.

12. There is evidence for earlier sea journeys by other hominins. For example, *Homo erectus* made it to the island of Flores in Indonesia between 800,000 and 700,000 years ago, involving a couple of short sea journeys. Those journeys would never have required them to be out of sight of land, however. Similarly, Neanderthals probably crossed short stretches of sea in the Mediterranean before 22,000 years ago (Erlandson (2001), op. cit.).

13. Hine, P. et al. (2010) Antiquity of stone-walled tidal fish traps on the Cape Coast, South Africa. *South African Archaeological Bulletin*, 65: 35–44.

14. O'Connor, S. et al. (2011) Pelagic fishing at 42,000 years before the present and the maritime skills of modern humans. *Science*, 334: 1117–21.

15. Erlandson, J. M. et al. (2009) Fishing up the food web? 12,000 years of maritime subsistence and adaptive adjustments on California's Channel Islands. *Pacific Science*, 63: 711–24.

16. Geoff Bailey, University of York, personal communication.

17. Johannes, R. E. (1981) *Words of the Lagoon*. University of California Press, Berkeley.

18. O'Connor et al. (2011), op. cit.

19. Kvavadze, E. et al. (2009) 30,000-year-old wild flax fibers. *Science*, 325: 1359.

20. Soffer, O. (2004) Recovering perishable technologies through use-wear on tools: preliminary evidence for Upper Palaeolithic weaving and net-making. *Current Anthropology*, 45: 407–13.

21. Erlandson, J. M. and Rick, T.C. (2008) Archaeology, historical ecology, and the future of ocean ecosystems, in T. C. Rick and J. M. Erlandson (eds.), *Human Impacts on Ancient Marine Ecosystems*. University of California Press, Berkeley, pp. 297–308.

22. Walters, I. (1988) Fish hooks: evidence for dual social systems in south-eastern Australia? *Australian Archaeology*, 27: 98–114.

23. Radcliffe, W. (1921) *Fishing from Earliest Times*. John Murray, London.

24. Mair, A. W. (1928) *Oppian, Colluthus and Tryphiodorus, with an English translation*. William Heinemann, London.

25. It is a testament to the popularity of Oppian's poem that 58 contemporary copies survive. Bartley, A. N. (2003) *Stories from the Mountains, Stories from the Sea. The Digressions and Similes of Oppian's Halieutica and Cynegetica*. Vandenhoeck and Ruprecht, Gottingen.

26. Trakadas, A. (2003) The archaeological evidence for fish processing in the western Mediterranean, in T. Bekker-Nielsen, *Ancient Fishing and Fish Processing in the Black Sea Region*. Aarhus University Press, Aarhus, pp. 47–82.

27. It might also be effective for catching large bottom dwellers like skates and rays.

28. Gosnell, M. (2007) *Ice: The Nature, the History, and the Uses of an Astonishing Substance*. Chicago University Press, Chicago.

29. Trakadas (2003), op. cit.

30. Corcoran, T. H. (1963) Roman fish sauces. *The Classical Journal*, 58: 204–10.

31. Ibid.
32. Aquerreta, Y. et al. (2002) Use of exogenous enzymes to elaborate the Roman fish sauce 'garum'. *Journal of the Science of Food and Agriculture*, 82: 107–12.
33. Curtis, R. I. (2003) Source for production and trade of Greek and Roman processed fish, in T. Bekker-Nielsen, *Ancient Fishing and Fish Processing in the Black Sea Region*. Aarhus University Press, Aarhus, pp. 31–46.
34. Ejstrud, B. (2003) Size matters: estimating trade of wine, oil and fish-sauce from amphorae in the first century AD, in T. Bekker-Nielsen, *Ancient Fishing and Fish Processing in the Black Sea Region*. Aarhus University Press, Aarhus, pp. 171–82.
35. Barrett, J. H. et al. (2004) The origins of intensive marine fishing in Medieval Europe: the English Evidence. *Proceedings of the Royal Society B*, 271: 2417–21; Barrett, J. H. et al. (2004) Dark Age economics revisited: the English fish bone evidence AD 600–1600. *Antiquity*, 78: 618–36.
36. Barrett, J. H. et al. (2008) Detecting the medieval cod trade: a new method and first results. *Journal of Archaeological Science*, 35: 850–61.
37. Merwe, van der, P. (ed.) (1986) *Hooking, Drifting and Trawling: 500 Years of British Deep Sea Fishing*. National Maritime Museum, Greenwich.
38. Roberts, C. M. (2007) *The Unnatural History of the Sea*. Island Press, Washington, DC.
39. Duhamel du Monceau, M. (1769) *Traité général des pesches et histoire des poissons*. Saillant and Nyon, Paris.
40. Johannes (1981), op. cit.
41. Webster, G. (2004) The invention of the kite. *The Kiteflyer*, 98: 9–14.

3. LESS FISH IN THE SEA

1. Smith, J. (1624) *The Generall Historie of Virginia, New-England, and the Summer Isles*. Reprinted 1907 by J. MacLehose, Glasgow.
2. Thurstan, R. H. et al. (2010) The effects of 118 years of industrial fishing on UK bottom trawl fisheries. *Nature Communications*, 1:15, doi: 10.1038/ncomms1013.
3. We are very grateful to G. H. Engelhard for making the calculations that made our analysis possible. Engelhard, G. H. (2008) One hundred and twenty years of change in fishing power of English North Sea trawlers, in A. Payne, J. Cotter and T. Potter (eds.), *Advances in Fisheries Science 50 years on from Beverton and Holt*, Blackwell Publishing, Oxford, pp. 1–25.

4. Fish like menhaden were often caught in such vast numbers they were used as fertilizer since most of the catch spoiled before it could be sold for human consumption and only a limited amount could be fed to available livestock. Brown Goode, G. (1884) *The Fisheries and Fishery Industries of the United States. Section I. Natural History of Useful Aquatic Animals.* Government Printing Office, Washington, DC.

5. Hoover, H. (1917) The food armies of liberty. The winning weapon: food. *National Geographic*, September, pp. 187–212.

6. Collins, J. W. (1887) *Report on the Investigation of Fishing Grounds in the Gulf of Mexico.* Government Printing Office, Washington, DC.

7. Lewis Anspach, quoted in *The Literary Gazette and Journal of Belles Lettres for the Year* 1819. William Pople, London.

8. Wallace, S. and Gisborne, B. (2006) *Basking Sharks. The Slaughter of BC's Gentle Giants.* New Star Books, Vancouver.

9. One vivid account from Ireland in 1744 describes the slaughter of a large group of porpoises: 'Yesterday being a great spring tide, a vast army of porpusses came up Lough Foyle in pursuit of salmon. As they rolled by Londonderry, the sailors pursued them in their boats, and killed them all the way, drove them six miles farther up the lough, to the flats about Mount Gavelling. There a new chase began by our fishermen and country people, who stretched a net across the lough, and drove them up to the narrow passages of the Great Island, which lies a mile below this town; there they fell on them with guns, swords, hatchets, and all kinds of weapons, and made a terrible slaughter. There were killed here above one hundred and sixty, besides as many mortally wounded and carried off by the flood. Including those the men of Londonderry killed, there have at least fallen in this battle five hundred porpoises, generally weighing from 1,000 to 1,500 [lbs] weight, and very good oil. Some of them were full of young ones as big as calves; and some had from six to ten salmon in their stomachs. But we hope that since these grand devourers are destroyed, our fishing will hereafter flourish, and we are pretty well repaid by this oil for the damage they have done.' From the *Post-Boy*, dated 12 November 1744, and quoted in T. de Voe (1866) *Market Assistant. Containing a brief description of every article of human food sold in the public markets of the cities of New York, Boston, Philadelphia and Brooklyn.* Hurd and Houghton, New York. The weights quoted seem too high for porpoises, suggesting they may have been dolphins.

10. NOAA National Marine Fisheries Service (2011) 2010 Report to Congress: Status of U.S. Fisheries. Washington, DC. http://www.nmfs.noaa

.gov/sfa/statusoffisheries/2010/2010_Report_to_Congress.pdf; accessed 18 November 2011.

11. Continental shelves are the shallow areas adjacent to landmasses and are typically less than about 200m deep. Watling, L. and Norse, E. A. (1998) Disturbance of the seabed by mobile fishing gear: A comparison to forest clearcutting. *Conservation Biology*, 12: 1180–97.

12. Bertram, J. G. (1873) *The Harvest of the Sea*. John Murray, London.

13. Bailey, D. M. et al. (2009) Long-term changes in deep-water fish populations in the northeast Atlantic: a deeper reaching effect of fisheries? *Proceedings of the Royal Society B*, 276: 1965–9, doi: 10.1098/rspb .2009.0098.

14. The study in which this figure first appeared (Myers, R. A. and Worm, B. (2003) Rapid worldwide depletion of predatory fish communities. *Nature*, 423: 280–83) has been hotly argued over since it came out and the figures questioned. Some critics say the wrong analysis was used and that all fish stocks will have collapsed by early in the twenty-second century rather than the mid-twenty-first. Others contend that catches are an inappropriate metric of the condition of fish stocks, given that they depend on fishing effort as well as how many fish are in the sea. In fact, the authors never suggested that all fish stocks would be gone by 2048, just that they might all have experienced collapse, defined as catches falling below 10 per cent of their maximum. It is perfectly possible for some collapsed fisheries to have recovered by then, and for others to still produce fish, but in much lower quantities than they could at higher abundance. The picture has been revised in recent years (Worm, B. et al. (2009) Rebuilding global fisheries. *Science*, 325: 578–85) to reflect efforts to recover fish stocks in the US and elsewhere, and to account for differences when biomass rather than catches are used as measures of the condition of stocks (Branch, T. A. et al. (2011) Contrasting global trends in marine fishery status obtained from catches and from stock assessments. *Conservation Biology*, doi: 10.1111/j.1523-1739.2011.01687.x), but the picture looks bleak at a global scale.

15. Rick, T. C. and Erlandson, J. M. (eds.) (2008) *Human Impacts on Ancient Marine Ecosystems*. University of California Press, Berkeley.

16. Swain, D. P. et al. (2007) Evolutionary response to size-selective mortality in an exploited fish population. *Proceedings of the Royal Society B*, 274: 1015–22; Mollet, F. M. et al. (2007) Fisheries-induced evolutionary changes in maturation reaction norms in North Sea sole *Solea solea*. *Marine Ecology Progress Series*, 351: 189–99.

17. Hsieh, C. et al. (2006) Fishing elevates variability in the abundance of exploited species. *Nature*, 443: 859–62.

18. Daniel Pauly coined the term 'fishing down the food web' (Pauly, D. et al. (1998) Fishing down marine foodwebs. *Science*, 279: 860–63). The phenomenon has been questioned by some scientists who suggest what is happening is fishing through the food web (Essington, T. E. et al. (2006) Fishing through marine foodwebs. *Proceedings of the National Academy of Sciences*, 103: 3171–5) and that species are simply added to the fishery over time as the big ones are still caught, or that the effect doesn't exist (Branch, T. A. et al. (2010) The trophic fingerprint of marine fisheries. *Nature*, 468: 431–5). However you slice it, serial depletion of big fish is easy to see wherever there is a gradient of fishing pressure. Intensively exploited places lack the biggest species and big individuals that less fished places have. Although the phenomenon is most often cited in relation to trophic level in the food web, in reality trophic level is just a correlate of what drives fishermen, which is profit. So fishers are really fishing down the value chain (Sethi, S. A. et al. (2010) Global fishery development patterns are driven by profit but not trophic level. *Proceedings of the National Academy of Sciences*, 107: 12163–7).

19. Discovery of a new cod depot. Quoted in *Friends' Intelligencer*, Volume XVIII, T. Ellwood Zell, Philadelphia, pp. 618–19.

20. Beaufoy, H. (1785) Third report from the Committee appointed to inquire into the state of the British fisheries, and into the most effectual means for their improvement and extension. Sixteenth Parliament of Great Britain: Second session (25 January 1785–2 August 1785). In Reports from the Committees of the House of Commons 1715–1801 (1803). *Miscellaneous Subjects Vol. 10: 1785–1801*, pp. 18–189.

21. Thurstan, R. H. and Roberts, C. M. (2010) Ecological meltdown in the Firth of Clyde, Scotland: two centuries of change in a coastal marine ecosystem. *PLoS ONE*, 5: e11767. doi:10.1371/journal.pone.0011767.

22. http://www.dunoon-observer.com/index.php/news/past-stories-covered-in-cowal-and-argyll/549-fishermens-leaders-slam-clyde-report; accessed 27 December 2011.

23. O'Leary, B. et al. (2011) Fisheries mismanagement. *Marine Pollution Bulletin*, 62: 2642–8.

24. Fromentin, J. M. (2009) Lessons from the past: investigating historical data from bluefin tuna fisheries. *Fish and Fisheries*, 10: 197–216.

4. WINDS AND CURRENTS

1. Lozier, M. S. (2010) Deconstructing the conveyor belt. *Science*, 328: 1507–11.
2. Had they been able to measure deeper, they would have found the temperature drop to just 3° or 4°C.
3. Benjamin, Count of Rumford (1757) An account of the manner in which heat is propagated in fluids, and its general consequences in the economy of the universe. *A Journal of Natural Philosophy, Chemistry and the Arts*, Volume 1. William Nicholson, London.
4. Denny, M. W. (2008) *How the Ocean Works. An Introduction to Oceanography*. Princeton University Press, Princeton.
5. The total flow of world rivers is approximately 40,700km^3 per year. The Amazon accounts for 15 per cent of this flow. The flow of the Atlantic Meridional Overturning Circulation, as the Atlantic arm of the global ocean conveyor current is known, is about 490,000 km^3 per year, or just one third of 1 per cent of the volume of the sea. Willis, J. K. (2010) Can in situ floats and satellite altimeters detect long-term changes in Atlantic Ocean overturning? *Geophysical Research Letters* 37: art. Lo6602; Postel, S. L. et al. (1996) Human appropriation of renewable fresh water. *Science*, 271: 785–8.
6. Lenton, T. M. et al. (2008) Tipping elements in the Earth's climate system. *Proceedings of the National Academy of Sciences*, 105: 1786–93.
7. Arrhenius won the Nobel Prize for Chemistry in 1903, not for the spectacular leap of logic he made in predicting global warming, but for his theory of the dissociation of ions in solution.
8. Planetary wobbles, called Milankovich cycles after their discoverer, are also critical to glaciations (http://en.wikipedia.org/wiki/Milankovitch _cycles; accessed 22 May 2011).
9. Arrhenius, S. (1908) *Worlds in the Making. The Evolution of the Universe*. Harper & Brothers, London and New York.
10. http://www.asi.org/adb/02/05/01/surface-temperature.html.
11. 388ppm of CO_2 was the value for 2007. IPCC (2007) *Climate Change 2007: Synthesis Report. Contribution of Working Groups I, II and III to the Fourth Assessment Report of the Intergovernmental Panel on Climate Change* (Core Writing Team, Pachauri, R. K and Reisinger, A. (eds.)). IPCC, Geneva, Switzerland.
12. Calculated over a time horizon of 100 years; since methane reacts with other compounds over time, its global warming potential is high to begin with and declines as time passes: http://en.wikipedia.org/wiki/Global _warming_potential.

13. There are also vast quantities of methane hydrates deep in the oceans; past episodes of intense global warming, such as that at the great end Permian extinction, led to their release.

14. 1kg of seawater contains about 4.13 times the heat of 1kg of air at the same temperature and pressure. A kilogram of air has a volume of 856.2 litres at a pressure of one atmosphere and 25°C, compared to a volume of 1 litre for seawater under the same conditions. So the heat capacity of water is 3,536 times higher than air under these conditions.

15. Schubert, R. et al. (2006) *The Future Oceans – Warming Up, Rising High, Turning Sour*. German Advisory Council on Global Change, Berlin.

16. Hoegh-Guldberg, O. and Bruno, J. F. (2010) The impact of climate change on the world's marine ecosystems. *Science*, 328: 1523–8.

17. Schofield, O. et al. (2010) How do polar marine ecosystems respond to rapid climate change? *Science*, 328: 1520–23.

18. http://www.timesonline.co.uk/tol/news/uk/article767459.ece; accessed 26 September 2011.

19. Kerr, R. A. (2007) Is battered Arctic sea ice down for the count? *Science*, 318: 33–4.

20. This notion of the Gulf Stream as the North Atlantic's winter warmer has recently been questioned (Seager, R. et al. (2002) Is the Gulf Stream responsible for Europe's mild winters? *Quarterly Journal of the Royal Meteorological Society*, 128: 2563–86). The Rocky Mountains squeeze air into a thinner layer as it flows from the west, and twists winds to the north in the process. As air flows east beyond the Rockies, the twist unwinds to the south, bending winds above the warm Sargasso Sea that surrounds Bermuda. There they pick up heat, which is carried northeast. Europe's mild climate is a result of ocean warmth, but the Gulf Stream carries little of that heat, or so the authors argue. However, this view is at odds with sea surface temperature maps built up from satellite measurements, which show warm ocean waters pulled far into the north-eastern Atlantic by the global ocean conveyor. Because of the much greater heat capacity of water than air, far more heat would be required to warm those waters than mere winds could transfer blowing from the tropics.

21. Rahmstorf, S. (2002) Ocean circulation and climate during the past 120,000 years. *Nature*, 419: 207–14.

22. Nesje, A., Svein, O. D. and Bakke, J. (2004) Were abrupt late-glacial and early-Holocene climatic changes in northwest Europe linked to freshwater outbursts to the North Atlantic and Arctic Oceans? *The Holocene*, 14: 299–310.

23. Sachs, J. P. and Lehman, S. J. (1999) Subtropical North Atlantic temperatures 60,000 to 30,000 years ago. *Science*, 286: 756–9.

24. Rahmstorf (2002), op. cit.

25. Water at the surface veers at about 45° to the wind direction. Friction transfers wind energy from surface to deeper layers. If you take a pack of cards, place your finger in one corner and fan them using a finger in the opposite corner, friction moves each card a little less than the one above until the cards do not move at all. In the same way, friction moves water less as you go deeper. However, a peculiarity of the Coriolis force is that as depth increases, water twists a little further, so the angle of flow relative to the wind increases with depth, rather than lessens. The net flow, taking into account the whole water column, is at right angles to the wind direction.

26. Thermoclines are well-developed and virtually permanent features of tropical seas, but they are poorly developed or absent from polar seas where warming is less. In between, they tend to be seasonal, forming in spring and breaking up in autumn when storms stir the sea.

27. Bakun, A. et al. (2010) Greenhouse gas, upwelling-favorable winds, and the future of coastal ocean upwelling ecosystems. *Global Change Biology*, 16: 1213–28.

28. Bakun, A. and Weeks, S. J. (2004) Greenhouse gas buildup, sardines, submarine eruptions and the possibility of abrupt degradation of intense marine upwelling ecosystems. *Ecology Letters* 7: 1015–23.

29. Ibid.

30. Bakun, A. (1990) Global climate change and intensification of coastal ocean upwelling. *Science*, 247: 198–201.

31. Ekau, W. et al. (2010) Impacts of hypoxia on the structure and processes in pelagic communities (zooplankton, macro-invertebrates and fish). *Biogeosciences*, 7: 1669–99.

32. Daniel Pauly, whose breadth of understanding always amazes me, has come up with an elegant theory to explain much about why fish are built the way they are and do the things they do. He articulates the case for oxygen as a limiting constraint in Pauly, D. (2010) *Gasping Fish and Panting Squids: Oxygen, Temperature and the Growth of Water-Breathing Animals.* International Ecology Institute, Germany.

33. Bakun, A. (2010) Afterword, in D. Pauly, *Gasping Fish and Panting Squids: Oxygen, Temperature and the Growth of Water-Breathing Animals.* International Ecology Institute, Germany, pp. 137–40.

34. Field, J. (2008) Jumbo squid (*Dosidicus gigas*) invasions in the Eastern Pacific Ocean. *CalCOFI Report*, 49: 79–81.

35. Stramma, L. et al. (2008) Expanding oxygen-minimum zones in the tropical oceans. *Science*, 320: 655–8.

36. Chan, F. et al. (2008) Emergence of anoxia in the California Current Large Marine Ecosystem. *Science*, 319: 920.

37. http://www.sciencedaily.com/releases/2006/08/060812155855.htm; accessed 22 May 2011.

38. Zeidberg, L. D. and Robinson, B. H. (2008) Invasive range expansion by the Humboldt squid, *Dosidicus gigas*, in the eastern North Pacific. *Proceedings of the National Academy of Sciences*, 104: 12948–50.

5. LIFE ON THE MOVE

1. Roberts, C. M., Hawkins, J. P. et al. (2002) Marine biodiversity hotspots and conservation priorities for tropical reefs. *Science*, 295: 1280–84; Hawkins, J. P. et al. (2000) The threatened status of restricted range coral reef fish species. *Animal Conservation*, 3: 81–8.

2. Ushatinskaya, G. T. (2008) Origin and dispersal of the earliest brachiopods. *Paleontological Journal*, 42: 533–9.

3. Sepkoski, J. J. (1997) Biodiversity: past, present and future. *Journal of Paleontology*, 71: 533–9. Microbes, by contrast, seem almost immutable over vast stretches of geologic time.

4. Perry, A. et al. (2005) Climate change and distribution shifts in marine fishes. *Science*, 308: 1912–15.

5. Brander, K. (2009) Impacts of climate change on marine ecosystems and fisheries. *Journal of the Marine Biological Association of India*, 51: 1–13.

6. Beaugrand, G. et al. (2003) Plankton effect on cod recruitment in the North Sea. *Nature*, 426: 661–4; Beaugrand, G. et al. (2002) Reorganisation of North Atlantic marine copepod biodiversity and climate. *Science*, 296: 1692–4.

7. Personal communication from John Pitchford, University of York.

8. Cheung, W. W. L. et al. (2009) Large-scale redistribution of maximum fisheries catch potential in the global ocean under climate change. *Global Change Biology*, 16: 24–35.

9. Sheppard, C. R. C. (2003) Predicted recurrences of mass coral mortality in the Indian Ocean. *Nature*, 425: 294–7.

10. Cheung, W. W. L. et al. (2008) Application of macroecological theory to predict effects of climate change on global fisheries potential. *Marine Ecology Progress Series*, 365: 187–97.

11. Hamilton, L. et al. (2000) Social change, ecology and climate in 20th century Greenland. *Climatic Change*, 47: 193–211.

6. RISING TIDES

1. Countering this spurt in wetland creation, coastal construction and reclamation has reduced wetland area in some places.

2. With the caveat that solids such as ice can take up more space than their warmer liquid phases.

3. Lambeck, K. et al. (2004) Sea level in Roman time in the central Mediterranean and implications for recent change. *Earth and Planetary Science Letters*, 224: 563–75.

4. Pilkey, O. and Young, R. (2009) *The Rising Sea*. Island Press, Washington, DC.

5. Nicholls, R. J. and Cazenave, A. (2010) Sea-level rise and its impact on coastal zones. *Science*, 328: 1517–20.

6. Jevrejeva, S., Grinsted, A. and Moore, J. C. (2009) Anthropogenic forcing dominates sea level rise since 1850. *Geophysical Research Letters*, 36: L20706. doi:10.1029/2009GL040216.

7. Fanos, A. M. (1995) The impact of human activities on the erosion and accretion of the Nile delta coast. *Journal of Coastal Research*, 11: 821–33.

8. Bamber, J. L. et al. (2009) Reassessment of the potential sea-level rise from a collapse of the West Antarctica ice sheet. *Science*, 324: 901–3. doi 10.1126/science.1169335. This study revises downwards the widely reported six-metre rise in sea levels that the melting of the West Antarctic ice shelf was expected to produce.

9. Weiss, J. L. et al. (2011) Implications of recent sea level rise science for low-elevation areas in coastal cities of the coterminous U.S.A. *Climatic Change*, 105: 635–45.

10. Blanchon, P. et al. (2009) Rapid sea-level rise and reef back-stepping at the close of the last interglacial highstand. *Nature*, 458: 881–4.

11. Kopp, R. E. et al. (2009) Probabilistic assessment of sea level during the last interglacial stage. *Nature*, 462: 863–8.

12. Nicholls, R. J. et al. (2011) Sea-level rise and its possible impacts given a 'beyond 4°C world' in the twenty-first century. *Philosophical Transactions of the Royal Society A*, 369: 161–81.

13. Jenkins, A. et al. (2010) Observations beneath Pine Island Glacier in West Antarctica and implications for its retreat. *Nature Geoscience*, 3: 468–72. doi 10.1038/ngeo890.

14. Connor, S. (2011) Vast methane 'plumes' seen in Arctic ocean as sea ice retreats. *Independent*, 13 December, www.independent.co.uk/news/science/vast-methane-plumes-seen-in-arctic-ocean-as-sea-ice-retreats-6276278.html.

15. UN-HABITAT (2008) *State of the World's Cities 2008/2009 – Harmonious Cities*. UN-HABITAT (United Nations Human Settlement Programme), Nairobi, Kenya: http://www.unhabitat.org/pmss/getPage .asp?page=bookView&book=2562; accessed 18 August 2011.

16. Kabat, P. et al. (2009) Dutch coasts in transition. *Nature Geoscience*, 2: 450–52.

17. Subsidence rates in the Maldives are around 0.15mm per year, which is very slow compared to subsidence rates in many deltas. Fürstenau, J. et al. (2010) Submerged reef terraces of the Maldives, Indian Ocean. *Geo-Marine Letters*, 30: 511–15.

18. http://www.parliament.uk/documents/post/postpn342.pdf; accessed 22 May 2011. Two other factors explain the difference: Scotland has fewer people and more hard coastal rock than England, so it has less to defend and less need for coastal defence.

19. Large dams are defined as those with walls higher than 15m. Syvitski, J. P. M. et al. (2005) Impact of humans on the flux of terrestrial sediment to the global coastal ocean. *Science*, 308: 376–80.

20. Engelkemeir, R. et al. (2010) Surface deformation in Houston, Texas using GPS. *Tectonophysics*, 490: 47. doi: 10.1016/j.tecto.2010.04.016.

21. Extremes will rise faster than the average if the extent of the variation around an average value is proportional to the value of the average.

22. Kerr, R. A. (2010) Models foresee more-intense hurricanes in the greenhouse. *Science*, 327: 399.

23. Ericson, J. P. et al. (2006) Effective sea-level rise and deltas: causes of change and human dimension implications. *Global and Planetary Change*, 50: 63–82.

24. Ibid.

25. Other effects of climate change, including higher temperatures and altered rainfall patterns, will also affect agricultural production.

26. Tanaka, N. et al. (2007) Coastal vegetation structures and their function in tsunami protection: experience of the recent Indian Ocean tsunami. *Landscape and Ecological Engineering*, 3: 33–45.

27. Analyses of satellite images show that 14 of the world's largest deltas lost 3 per cent of their wetlands in the last 14 years of the twentieth century. Coleman, J. M., Huh, O. K. and Braud, D. (2008) Wetland loss in world deltas. *Journal of Coastal Research*, 24: 1–14.

28. Alongi, D. M. (2002) Present state and future of the world's mangrove forests. *Environmental Conservation*, 29: 331–49.

29. Waycott, M. et al. (2009) Accelerating loss of seagrasses across the globe threatens coastal ecosystems. *Proceedings of the National Academy of Sciences*, 106: 12377–81.

30. Pilkey and Young (2009), op. cit.

31. Chatenoux, B. and Peduzzi, P. (2007) Impacts from the 2004 Indian Ocean Tsunami: analysing the potential protecting role of environmental features. *Natural Hazards*, 40: 289–304.

32. King, S. E. and Lester, J. N. (1995) The value of salt marsh as a sea defence. *Marine Pollution Bulletin*, 30: 180–89.

7. CORROSIVE SEAS

1. W. Russell, quoted in Littell, E. (1846) *Littell's Living Age*. Volume X. Waite, Pearce & Company, Boston.

2. Hall-Spencer, J. and Rauer, E. (2009) Champagne seas – foretelling the ocean's future? *Journal of Marine Education*, 25: 11–12; Hall-Spencer, J. M. et al. (2008) Volcanic carbon dioxide vents show ecosystem effects of ocean acidification. *Nature*, 454: 96–9.

3. If you are familiar with the language of chemistry, the equations that represent the dissolution of carbon dioxide in the sea are:

CO_2 (carbon dioxide) + H_2O (water) $\leftrightarrow$ H_2CO_3 (carbonic acid)

H_2CO_3 (carbonic acid) $\leftrightarrow$ HCO_3^- (bicarbonate ion) + H^+ (hydrogen ion)

$H^+ + CO_3^{2-}$ (carbonate ion) $\leftrightarrow$ HCO_3^-

4. Forest clearance and wetland drainage liberate large quantities of stored carbon, most of which ends up in the atmosphere as carbon dioxide.

5. Secretariat of the Convention on Biological Diversity (2009) *Scientific Synthesis of the Impacts of Ocean Acidification on Marine Biodiversity*. Montreal, Technical Series No. 46.

6. Ridgwell, A. and Schmidt, D. N. (2010) Past constraints on the vulnerability of marine calcifiers to massive carbon dioxide release. *Nature Geoscience*, 3: 196–200. doi: 10.1038/NGEO755.

7. This assumes business-as-usual in growth of carbon dioxide emissions.

8. Caldeira, K. and Wickett, M. E. (2003) Anthropogenic carbon and ocean pH. *Nature*, 425: 365.

9. De'ath, G. et al. (2009) Declining coral calcification on the Great Barrier Reef. *Science*, 323: 116–19.

10. Maier, C. et al. (2011) Calcification rates and the effect of ocean acidification on Mediterranean cold-water corals. *Proceedings of the Royal Society B*, doi:10.1098/rspb.2011.1763.

11. The depth below which carbonate dissolves is shallower in the Atlantic than in the Pacific, because deep water in the Pacific is 'older' (as you will remember, deep bottom water forms in the North and South Atlantic and

flows around the world from there) so it has had more time to accumulate carbon dioxide from the breakdown of organic matter.

12. Veron, J. E. N. et al. (2009) The coral reef crisis: the critical importance of 350 ppm CO_2. *Marine Pollution Bulletin*, 58: 1428–36. Some of my colleagues were pleased there was no agreement on emissions reduction in Copenhagen in 2010 because it left negotiating space for a tougher deal that could save coral reefs.

13. Fabricius, K. E. et al. (2011) Losers and winners in coral reefs acclimatized to elevated carbon dioxide concentrations. *Nature Climate Change*, 1: 165–9. doi: 10.1038/nclimate1122.

14. Deep waters are more acidic because they contain higher levels of carbon dioxide as a result of respiration by the organisms that live there, and less oxygen because there is no light for photosynthesis. Manzello, D. P. et al. (2008) Poorly cemented coral reefs of the eastern tropical Pacific: possible insights into reef development in a high-CO_2 world. *Proceedings of the National Academy of Sciences*, 105: 10450–55.

15. Yamamoto-Kawai, M. et al. (2009) Aragonite undersaturation in the Arctic Ocean: effects of ocean acidification and sea ice melt. *Science*, 326: 1098–100.

16. Tropical seas warmed about 5°C while those at the poles warmed up to 9°C. Deep bottom waters also warmed by 4–5°C. Zachos, J. et al. (2005) Rapid acidification of the ocean during the Paleocene–Eocene thermal maximum. *Science*, 308: 1611–15.

17. Kerr, R. A. (2010) Ocean acidification. Unprecedented, unsettling. *Science*, 328: 1500–501.

18. Carbon dioxide was removed by weathering of silicate rocks above sea level, as well as dissolution of carbonates in the sea.

19. Scheibner, C. and Speijer, R. P. (2008) Decline of coral reefs during late Paleocene to early Eocene global warming. www.electronic-earth.net/3/19/2008/. doi:10.5194/ee-3-19-2008. Charlie Veron thinks that there was a major reef coral extinction event as well, but the research has yet to be done to document it.

20. Zachos et al. (2005), op. cit.

21. There is a possibility that lack of oxygen in bottom waters killed these species, rather than being a direct effect of acidification. Zachos, J. C. et al. (2008) An early Cenozoic perspective on greenhouse warming and carbon cycle dynamics. *Nature*, 451: 279–83.

22. Gibbs, S. J. et al. (2006) Nannoplankton extinction and origination across the Paleocene-Eocene Thermal Maximum. *Science*, 314: 1770–73.

23. Although oxygen is fundamental to life, there is little chance we will run out of it quickly. Even if all photosynthesis were to cease tomorrow, it would take over a thousand years to draw down atmospheric oxygen to levels that would threaten human existence.

24. Fernandez, E. et al. (1993) Production of organic and inorganic carbon within a large-scale coccolithophore bloom in the northeast Atlantic Ocean. *Marine Ecology Progress Series*, 97: 271–85.

25. Despite these encouraging results from lab experiments, a recent study has shown that calcification by coccolithophores in the sea is reduced by higher dissolved carbon dioxide levels. Whether this affects other aspects of their growth and oxygen production is not known. Hutchins, D. A. (2011) Forecasting the rain ratio. *Nature*, 476: 41–2.

26. Suttle, C. A. (2005) Viruses in the sea. *Nature*, 437: 356–61. By weight, all the viruses in the sea are equivalent to 75 million adult blue whales (of which there are about 10,000 left today).

27. Shi, D. et al. (2010) Effect of ocean acidification on iron availability to marine phytoplankton. *Science*, 327: 676–9.

28. Dixson, D. et al. (2010) Ocean acidification disrupts the innate ability of fish to detect predator olfactory cues. *Ecology Letters*, 13: 68–75.

29. Munday, P. et al. (2009) Ocean acidification impairs olfactory discrimination and homing ability of a marine fish. *Proceedings of the National Academy of Sciences*, 106: 1848–52.

30. Cigliano, M. et al. (2010) Effects of ocean acidification on invertebrate settlement at volcanic CO_2 vents. *Marine Biology*, 157: 2489–502.

31. Kerr (2010), op. cit.

8. DEAD ZONES AND THE WORLD'S GREAT RIVERS

1. Actually the flush toilet was a reinvention – the Romans had them and there are claims in the archaeological literature for even more ancient examples.

2. Another important source of nutrients to the sea is atmospheric deposition, especially of nitrogen.

3. Beman, J. M. et al. (2005) Agricultural runoff fuels large phytoplankton blooms in vulnerable areas of the ocean. *Nature*, 434: 211–14.

4. Diaz, R. J. and Rosenberg, R. (2008) Spreading dead zones and consequences for marine ecosystems. *Science*, 321: 926–9.

5. Rabalais, N. N. et al. (2007) Sediments tell the history of eutrophication and hypoxia in the northern Gulf of Mexico. *Ecological Applications*, 17, Supplement: S129–S143.

6. http://edition.cnn.com/2008/TECH/science/08/18/dead.zone/; accessed 19 November 2011. Dead zones are actually far from lifeless. In these places we have reproduced the conditions of ancient oceans and awakened the sulphur-loving microbial communities of old.

7. Rabalais et al. (2007), op. cit.

8. Fahlbusch, H. (2009) Early dams. *Proceedings of the Institution of Civil Engineers*, 162: 13–18. doi: 10.1680/ehh2009.162.1.13.

9. Chen, C.-T. A. (2002) The impact of dams on fisheries: case of the Three Gorges Dam, in W. Steffen et al. (eds.), *Challenges of a Changing Earth*. Springer, Berlin, Chapter 16.

10. Ibid.

11. Gwo-Ching, G. et al. (2006) Reduction of primary production and changing of nutrient ratio in the East China Sea: effect of the Three Gorges Dam? *Geophysical Research Letters*, 33, L07610, doi:10.1029/2006GL025800.

12. Ryan, W. B. F. et al. (1997) An abrupt drowning of the Black Sea shelf. *Marine Geology*, 138: 119–26. A recent study suggests the flood was not as severe as this, with only a 30m height difference between the Mediterranean and Black Sea lake, rather than 80m: Giosan, L. et al. (2009) Was the Black Sea catastrophically flooded in the early Holocene? *Quaternary Science Reviews*, 28: 1–6.

13. The Black Sea basin gets abundant freshwater inflow from rivers. This mixes with seawater to form a low density, low salinity layer, which flows out through the Bosporus into the eastern Mediterranean. Some of the outflowing water is replaced by a deeper return flow of high salinity water from the Mediterranean which sinks into the Black Sea basin. The two layer nature of the Black Sea is thus maintained by salinity and temperature contrasts between deep and shallow water, which mean that the surface water layer is much less dense and floats above the deep layer.

14. Savage, C. et al. (2010) Effects of land use, urbanization, and climate variability on coastal eutrophication in the Baltic Sea. *Limnology and Oceanography*, 55: 1033–46.

15. Conley, D. J. et al. (2007) Long-term changes and impacts of hypoxia in Danish coastal waters. *Ecological Applications*, 17, Supplement: S165–S184.

16. Jackson, J. B. C. et al. (2001) Historical overfishing and the recent collapse of coastal ecosystems. *Science*, 293: 629–38.

17. Boesch, D. F. et al. (2007) *Coastal Dead Zones and Global Climate Change: Ramifications of Climate Change for Chesapeake Bay Hypoxia*.

Pew Center on Global Climate Change and University of Maryland Center for Environmental Science.

18. The causes of Florida red tides are less straightforward than a simple response to increased nutrient pollution, but this doubtless plays an important role in triggering and sustaining them. Alcock, F. (2007) *An Assessment of Florida Red Tide: Causes, Consequences and Management Strategies*. Technical Report 1190, Mote Marine Laboratory, Sarasota.

19. Kirkpatrick, B. et al. (2006) Environmental exposures to Florida red tides: effects on emergency room respiratory diagnoses admissions. *Harmful Algae*, 5: 526–33; Kirkpatrick, B. et al. (2010) Gastrointestinal emergency room admissions and Florida red tide blooms. *Harmful Algae*, 9: 82–6. See also www.topcancernews.com/news/1797/1/Algal-toxin-commonly-inhaled-in-sea-spray-attacks-and-damages-DNA.

20. Knowler, D. J. et al. (2002) An open-access model of fisheries and nutrient enrichment in the Black Sea. *Marine Resource Economics*, 16: 195–217.

21. Richardson, A. J. et al. (2009) The jellyfish joyride: causes, consequences and management responses to a more gelatinous future. *Trends in Ecology and Evolution*, 24: 312–22.

9. UNWHOLESOME WATERS

1. Safina, C. (2011) The 2010 Gulf of Mexico oil well blowout: a little hindsight. *PLoS Biology*, 9: e1001049. doi:10.1371/journal.pbio.1001049.

2. Crone, T. J. and Tolstoy, M. (2010) Magnitude of the 2010 Gulf of Mexico oil leak. *Science*, 330: 634.

3. This figure took me by surprise when I first saw it; the wellhead was gushing so fast at 68,000 barrels a day. But tankers are monumental these days, and it would take a long time to fill one at that rate. The rate and total amount of oil loss was estimated by experts in fluid dynamics who used footage of the leaking wellhead to measure the speed at which the oil gushed from it.

4. I am grateful to Nigel Haggan of the University of British Columbia for sharing this witticism!

5. Schrope, M. (2010) The lost legacy of the last great oil spill. *Nature*, 466: 305–6.

6. http://www.birdlife.org/news/news/2010/06/seabird-petition.html.

7. At the time of writing, October 2011, an area of 1,041 square miles around the Deepwater Horizon wellhead remains closed: http://sero.nmfs.noaa.gov/media/pdfs/2010/Area%206%20and%207%20Press%20Release_FINAL.pdf; accessed 29 October 2011.

8. Safina (2011), op. cit.

9. Jernelöv, A. (2010) How to defend against future oil spills. *Nature*, 466: 182–3.

10. National Research Council (2003) *Oil in the Sea III: Inputs, Fates, and Effects*. Washington, DC.

11. Wurl, O. and Obbard, J. P. (2004) A review of pollutants in the sea-surface microlayer (SML): a unique habitat for marine organisms. *Marine Pollution Bulletin*, 48: 1016–30.

12. Quite a lot of pollutants are also carried long distances by wind alone to be deposited into the sea far from the sources.

13. Wurl and Obbard (2004), op. cit.

14. Kessler, J. D. et al. (2011) A persistent oxygen anomaly reveals the fate of spilled methane in the deep Gulf of Mexico. *Science*, 331: 312–15.

15. Wells, R. S. et al. (2005) Integrating life-history and reproductive success data to examine potential relationships with organochlorine compounds for bottlenose dolphins (*Tursiops truncatus*) in Sarasota Bay, Florida. *Science of the Total Environment*, 349: 106–19.

16. Reddy, M. L. et al. (2001) Opportunities for using Navy marine mammals to explore associations between organochlorine contaminants and unfavourable effects on reproduction. *Science of the Total Environment*, 274: 171–82.

17. Hoover, S. M. (1999) Exposure to persistent organochlorines in Canadian breast milk: a probabilistic assessment. *Risk Analysis*, 19: 527–45.

18. AMAP (2009) *Arctic Pollution 2009*. Arctic Monitoring and Assessment Programme, Oslo.

19. Porterfield, S. P. (1994) Vulnerability of the developing brain to thyroid abnormalities: environmental insults to the thyroid system. *Environmental Health Perspectives*, 102: 125–30.

20. AMAP (2009), op. cit.

21. Sunderland, E. M. et al. (2009) Mercury sources, distribution, and bioavailability in the North Pacific Ocean: insights from data and models. *Global Biogeochemical Cycles*, 23, doi: 10.1029/2008GB003425.

22. Anh-Thu, E. V. et al. (2011) Temporal increase in organic mercury in an endangered pelagic seabird assessed by century-old museum specimens. *Proceedings of the National Academy of Sciences*, 108: 7466–71.

23. Frederick, P. and Jayasema, N. (2010) Altered pairing behaviour and reproductive success in white ibises exposed to environmentally relevant concentrations of methylmercury. *Proceedings of the Royal Society B*, doi: 10.1098/rspb.2010.2189.

24. Sunderland et al. (2009), op. cit.

25. Lowenstein, J. H. (2010) DNA barcodes reveal species-specific mercury levels in tuna sushi that pose a health risk to consumers. *Biology Letters*, doi:10.1098/rsbl.2010.0156.

26. Tarbox, B. M. (2010) *Toxic Fish Counter*. www.gotmercury.org.

27. Muir, D. C. G. and Howard, P. H. (2006) Are there other persistent organic pollutants? A challenge for environmental chemists. *Environmental Science and Technology*, 40: 7157–66.

28. Law, R. J. et al. (2006) Levels and trends of brominated flame retardants in the European environment. *Chemosphere*, 64: 187–208; Darnerud, P. O. (2003) Toxic effects of brominated flame retardants in man and in wildlife. *Environment International*, 29: 841–53.

29. Roze, E. et al. (2009) Prenatal exposure to organohalogens, including brominated flame retardants, influences motor, cognitive, and behavioral performance at school age. *Environmental Health Perspectives*, 117: 1953–8.

30. Arnold, K. et al. (in press) Medicating the environment: impacts on individuals and populations. *Trends in Ecology and Evolution*.

31. Heckmann, L.-H. et al. (2007) Chronic toxicity of ibuprofen to *Daphnia magna*: effects on life history traits and population dynamics. *Toxicology Letters*, 172: 137–45.

32. Daigle, J. K. (2010) Acute Responses of Freshwater and Marine Species to Ethinyl Estradiol and Fluoxetine. MSc Thesis, Louisiana State University.

33. Hannah, W. and Thompson, P. B. (2008) Nanotechnology, risk and the environment: a review. *Journal of Environmental Monitoring*, 10: 291–300.

34. Koehler, A. et al. (2008) Effects of nanoparticles in *Mytilus edulis* gills and hepatopancreas – a new threat to marine life? *Marine Environmental Research*, 66: 12–14.

35. Ylitalo, G. M. et al. (2009) High levels of persistent organic pollutants measured in blubber of island-associated false killer whales (*Pseudorca crassidens*) around the main Hawaiian Islands. *Marine Pollution Bulletin*, 58: 1922–52.

10. THE AGE OF PLASTIC

1. Ebbesmeyer, C. C. and Scigliano, E. (2009) *Flotsametrics and the Floating World*. HarperCollins, New York.

2. http://www.epsomandewellhistoryexplorer.org.uk/Yarsley.html; accessed 3 November 2011.

3. Yarsley, V. E. and Couzens, E. G. (1941) *Plastics*. Penguin Books, Harmondsworth.

4. Thompson, R. C. et al. (2009) Plastics, the environment and human health: current consensus and future trends. *Philosophical Transactions of the Royal Society B*, 364: 2153–66.

5. Hohn, D. (2011) *Moby Duck*. Viking, New York.

6. Maury, M. F. (1883) *The Physical Geography of the Sea*. Thomas Nelson and Sons, Edinburgh and New York.

7. Purdy, J. (1845) *Memoir Descriptive and Explanatory, to Accompany the Charts of the Northern Atlantic Ocean*. R. H. Laurie, London.

8. Ebbesmeyer and Scigliano (2009), op cit.

9. Moore, S. L. et al. (2001) Composition and Distribution of Beach Debris in Orange County, California. ftp://www.sccwrp.org/pub/download/DOCUMENTS/AnnualReports/1999AnnualReport/09_ar10.pdf; accessed 20 May 2011.

10. Barnes, D. K. A. (2005) Remote islands reveal rapid rise of southern hemisphere sea debris. *The Scientific World Journal*, 5: 915–21.

11. Barnes, D. K. A. et al. (2009) Accumulation and fragmentation of plastic debris in global environments. *Philosophical Transactions of the Royal Society B*, 364: 1985–98.

12. Yamashita, R. and Tanimura, A. (2007) Floating plastic in the Kuroshio Current area, western North Pacific Ocean. *Marine Pollution Bulletin*, 54: 464–88.

13. Gregory, M. R. (2009) Environmental implications of plastic debris in marine settings – entanglement, ingestion, smothering, hangers-on, hitch-hiking and alien invasions. *Philosophical Transactions of the Royal Society B*, 364: 2013–25.

14. Moore, C. J. et al. (2001) A comparison of plastic and plankton in the North Pacific central gyre. *Marine Pollution Bulletin*, 42: 1297–1300.

15. Lavender Law, K. (2010) Plastic accumulation in the north Atlantic subtropical gyre. *Science*, 329: 1185–8.

16. Young, L. C. et al. (2009) Bringing home the trash: do colony-based differences in foraging distribution lead to increased plastic ingestion in Laysan albatrosses? *PLoS ONE*, 4: e7623. doi:10.1371/journal.pone .0007623.

17. As any frustrated golfer knows, golf balls sink in fresh water, but they float in higher density salt water.

18. Ryan, P. G. et al. (2009) Monitoring the abundance of plastic debris in the marine environment. *Philosophical Transactions of the Royal Society B*, 364: 1999–2012.

19. Mrosovsky, N. (2009) Leatherback turtles: the menace of plastic. *Marine Pollution Bulletin*, 58: 287–9.
20. Plot, V. and Georges, J.-Y. (2010) Plastic debris in a nesting leatherback turtle in French Guiana. *Chelonian Conservation and Biology*, 9: 267–70.
21. Teuten, E. L. et al. (2009) Transport and release of chemicals from plastics to the environment and wildlife. *Philosophical Transactions of the Royal Society B*, 364: 2027–45.
22. Eriksson, C. and Burton, H. (2003) Origins and biological accumulation of small plastic particles in fur seals from Macquarie Island. *Ambio*, 32: 380–84.
23. Stamper, M. A. et al. (2006) Case study: morbidity in a pygmy sperm whale (*Kogia breviceps*) due to ocean-borne plastic. *Marine Mammal Science*, 22: 719–22.
24. Tarpley, R. J. and Marwitz, S. (2003) Plastic debris ingestion by cetaceans along the Texas coast: two case reports. *Aquatic Mammals*, 19: 93–8.
25. Jacobsen, J. K. et al. (2010) Fatal ingestion of floating net debris by two sperm whales (*Physeter macrocephalus*). *Marine Pollution Bulletin*, 60: 765–7.
26. Ryan et al. (2009), op. cit.
27. American Chemical Society (2010) Hard plastics decompose in oceans, releasing endocrine disruptor BPA. ScienceDaily, 24 March, http://www.sciencedaily.com-/releases/2010/03/100323184607.htm; accessed 20 May 2011.
28. Mato, Y. et al. (2001) Plastic resin pellets as a transport medium for toxic chemicals in the marine environment. *Environmental Science and Technology*, 35: 318–24.
29. According to Captain Charles Moore, plastic debris from Central and North American coasts probably also arrives there via ocean highways joining the gyre from coastal inputs without having to go around.
30. Boerger, C. M. et al. (2010) Plastic ingestion by planktivorous fishes in the North Pacific Central Gyre. *Marine Pollution Bulletin*, 60: 2275–8.
31. Teuten et al. (2009), op. cit.
32. Ebbesmeyer and Scigliano (2009), op. cit.
33. Ryan et al. (2009), op. cit.

11. THE NOT SO SILENT WORLD

1. Frisk, G. V. (undated) Noiseonomics: the relationship between ambient noise levels and global economic trends. http://pruac.apl.washington.edu/abstracts/Frisk.pdf; accessed 14 February 2011.

2. Denny, M. W. (1993) *Air and Water: The Biology and Physics of Life's Media*. Princeton University Press, Princeton.

3. Doug Anderson's favourite marine mammal call is that of the Weddell Seal of Antarctica. You can hear a clip of this and many other marine mammals at Discovery of sound in the sea: www.dosits.org/audio/marinemammals/pinnipeds/weddellseal/; accessed 1 January 2012.

4. http://news.bbc.co.uk/1/hi/sci/tech/7641537.stm; accessed 20 May 2011.

5. Frantzis, A. (1998) Does acoustic testing strand whales? *Nature*, 329: 29.

6. Not all military sonars produce sound within the hearing range of whales. Some use lower frequencies that may be perfectly safe. Cook, M. L. H. et al. (2006) Beaked whale auditory evoked potential hearing measurements. *Journal of Comparative Physiology A*, 192: 489–95.

7. Fernández, A. et al. (2005) Gas and fat embolic syndrome involving a mass stranding of beaked whales (family *Ziphiidae*) exposed to anthropogenic sonar signals. *Veterinary Pathology*, 42: 446. doi: 10.1354/vp.42-4-446.

8. Tyack, P. L. et al. (2006) Extreme diving of beaked whales. *Journal of Experimental Biology*, 209: 4238–53.

9. Nosengo, N. (2009) The neutrino and the whale. *Nature*, 462: 560–61.

10. Frankel, A. S. and Clark, C. W. (2002) ATOC and other factors affecting the distribution and abundance of humpback whales (*Megaptera novaeangliae*) off the north shore of Kauai. *Marine Mammal Science*, 18: 644–62.

11. Kroll, D. A. et al. (2002) Only male fin whales sing loud songs. *Nature*, 417: 809.

12. Researchers played dolphins the whistle sounds of others that had been synthesized electronically to ensure dolphins responded to the call rather than the sound of the voice of the caller. They still recognized one another as individuals. Janik, V. N. et al. (2006) Signature whistle shape conveys identity information to bottlenose dolphins. *Proceedings of the National Academy of Sciences*, 103: 8293–7.

13. Buckstaff, K. C. (2004) Effects of watercraft noise on the acoustic behaviour of bottlenose dolphins, *Tursiops truncatus*, in Sarasota Bay, Florida. *Marine Mammal Science*, 20: 709–25.

14. Wysocki, L. E. et al. (2006) Ship noise and cortisol secretion in European freshwater fishes. *Biological Conservation*, 128: 501–8.

15. Papoutsoglou, S. E. et al. (2007) Effect of Mozart's music (Romanze-Andante of 'Eine Kleine Nacht Musik', sol major, K525) stimulus on common carp (*Cyprinus carpio* L.) physiology under different light conditions. *Aquacultural Engineering*, 36: 61–72.

16. Fay, R. (2009) Soundscapes and the sense of hearing of fishes. *Integrative Zoology*, 4: 26–32.

17. Simpson, S. D. et al. (2005) Homeward sound. *Science*, 308: 221.

18. Vermeij, M. J. A. et al. (2010) Coral larvae move towards reef sounds. *PLoS One*, 5: e10660.

19. Lohse, D., Schmitz, B. and Versluis, M. (2001) Snapping shrimp make flashing bubbles. *Nature*, 413: 477–8.

20. Rowell, T. et al. (2010) Presentation given at Gulf and Caribbean Fisheries Institute Meeting, Puerto Rico, titled 'Use of passive acoustics to map grouper spawning aggregations, with emphasis on Red hind, *Epinephelus guttatus*, off Western Puerto Rico': http://www.gcfi.org/Conferences/63rd/GCFIBook_Of_AbstractsEngPDF.pdf; accessed 22 May 2011.

21. Vascocelos, R. O. et al. (2007) Effects of ship noise on the detectability of communication signals in the Lusitanian toadfish. *Journal of Experimental Biology*, 210: 2104–12.

22. Bass, A. H. et al. (2008) Evolutionary origins for social vocalization in a vertebrate hindbrain-spinal compartment. *Science*, 321: 417–21.

23. Anderson, A. (1985) Humming fish disturb the peace. *New Scientist*, 12 September, pp. 64–5.

24. Parks, S. E. et al. (2010) Individual right whales call louder in increased environmental noise. *Biology Letters*, doi: 10.1098/rsbl.2010.0451. See also http://www.whoi.edu/oceanus/viewArticle.do?id=84868; accessed 14 February 2011.

25. Foote, A. D. et al. (2004) Whale-call response to masking boat noise. *Nature*, 428: 910.

26. Slabbekoorn, H. and Den Boer-Visser, A. (2006) Cities change the song of birds. *Current Biology*, 16: 2326–31.

27. Aguilar Soto, N. A. et al. (2006) Does intense ship noise disrupt foraging in deep-diving Cuvier's beaked whales (*Ziphius cavirostris*)? *Marine Mammal Science*, 22: 690–99.

28. Mooney, T. A. et al. (2009) Sonar-induced temporary hearing loss in dolphins. *Biology Letters*, 5: 565–7.

29. A short-fin pilot whale also had profound hearing loss, while twelve animals from other species had normal hearing. Mann, D. et al. (2010) Hearing loss in stranded Odontocete dolphins and whales. *PLoS ONE*, 5: e13824. doi:10.1371/journal.pone.0013824.

30. Frisk (undated), op. cit.

12. ALIENS, INVADERS AND THE
HOMOGENIZATION OF LIFE

1. Coates, A. G. et al. (1992) Closure of the Isthmus of Panama: the near-shore marine record of Costa Rica and western Panama. *Geological Society of America Bulletin*, 104: 814–28.
2. Brawley, S. H. et al. (2009) Historical invasions of the intertidal zone of Atlantic North America associated with distinctive patterns of trade and emigration. *Proceedings of the National Academy of Sciences*, 106: 8239–44.
3. There is a widely recounted tale, which turns out to be apocryphal, that lionfish were released into the Atlantic during Hurricane Andrew in 1992, when the storm surge overwhelmed Biscayne Bay aquarium. The source of the story has since retracted it: http://news.sciencemag.org/scienceinsider/2010/04/mystery-of-the-lionfish-dont-bla.html#more; accessed 17 November 2011.
4. Albins, M. A. and Hixon, M. A. (2011) Worst case scenario: potential long-term effects of invasive predatory lionfish (*Pterois volitans*) on Atlantic and Caribbean coral-reef communities. *Environmental Biology of Fishes*, doi: 10.1007/s10641-011-9795-1.
5. Mark Hixon was quoted in *The Times*, 20 October 2008; http://www.timesonline.co.uk/tol/news/environment/article4974396.ece; accessed 22 May 2011.
6. Bax, N. et al. (2003) Marine invasive alien species: a threat to global biodiversity. *Marine Policy*, 27: 313–23.
7. Carlton, J. T. (1999) The scale and ecological consequences of biological invasions in the world's oceans, in O. T. Sandlund et al. (eds.), *Invasive Species and Biodiversity Management*. Kluwer Academic Publishers, Dordrecht, pp. 195–212.
8. Shiganova, T. A. and Bulgakova, Y. V. (2000) Effects of gelatinous plankton on Black Sea and Sea of Azov fish and their food resources. *ICES Journal of Marine Science*, 57: 641–8. doi:10.1006/jmsc.2000.0736.
9. Saltonstall, K. (2002) Cryptic invasion by a non-native genotype of *Phragmites australis* into North America. *Proceedings of the National Academy of Sciences*, 99: 2445–9.
10. Cohen, A. N. and Carlton, J. T. (1998) Accelerating invasion rate in a highly invaded estuary. *Science*, 279: 555–8.
11. Personal communication from James Carlton, Williams College, Mystic, Conn.

12. Ray, G. L. (2005) Invasive marine and estuarine animals of California. ERDC/TN ANSRP-05-2, US Army Engineer Research and Development Center, Vicksburg, Miss.

13. Meinesz, A. (1999) *Killer Algae*. University of Chicago Press, Chicago.

14. Musée Océanographique de Monaco.

15. Meinesz (1999), op cit.

16. Carlton, J. T. and Eldredge, L. (2009) Marine bioinvasions of Hawai'i: the introduced and cryptogenic marine and estuarine animals and plants of the Hawaiian archipelago. *Bishop Museum Bulletin of Cultural and Environmental Studies*, 4: 1–202.

17. Tan, K. S. and Morton, B. (2006) The invasive Caribbean bivalve *Mytilopsis sallei* (Dreissenidae) introduced to Singapore and Johor Bahru, Malaysia. *The Raffles Bulletin of Zoology*, 54: 429–34.

18. Reid, P. C. et al. (2007) A biological consequence of reducing Arctic ice cover: arrival of the Pacific diatom *Neodenticula seminae* in the North Atlantic for the first time in 800 000 years. *Global Change Biology*, 13: 1910–21.

19. Scheinin, A. P. et al. (2011) Gray whale (*Eschrichtius robustus*) in the Mediterranean Sea: anomalous event or early sign of climate-driven distribution change? *Marine Biodiversity Records*, doi:10.1017/S1755267211000042.

20. Lewis, P. N. et al. (2005) Assisted passage or passive drift: a comparison of alternative transport mechanisms for non-indigenous coastal species into the Southern Ocean. *Antarctic Science*, 17: 183–91.

21. Ricciardi, A. et al. (2011) Should biological invasions be managed as natural disasters? *BioScience*, 61: 312–17.

22. Galil, B. S. (2007) Loss or gain? Invasive aliens and biodiversity in the Mediterranean Sea. *Marine Pollution Bulletin*, 55: 314–22.

23. One experimental test of this idea proved the case at a very small scale: Stachowicz, J. J. et al. (1999) Species diversity and invasion resistance in a marine ecosystem. *Science*, 286: 1577–9.

24. Some scientists, like Williams College's James Carlton, think this conclusion has more to do with our ignorance of marine life than a real absence of extinctions.

25. Myers, J. H. et al. (2000) Eradication revisited: dealing with exotic species. *Trends in Ecology and Evolution*, 15: 316–20.

26. Ecospeed, for example. See Hydrex Underwater Technology: www.hydrex.be/ecospeed_hull_coating_system; accessed 2 January 2012.

13. PESTILENCE AND PLAGUE

1. Nugues, M. M. and Nagelkerken, I. (2006) Status of aspergillosis and sea fan populations in Curaçao ten years after the 1995 Caribbean epizootic. *Revista de Biologia Tropical*, 54: 153–60.

2. Darwin, C. (1886) *A Naturalist's Voyage*. John Murray, London.

3. Garrison, V. H. et al. (2003) African and Asian dust: from desert soils to coral reefs. *BioScience*, 53: 469–80. Others, like the University of North Carolina's John Bruno, doubt the dust explanation, pointing out that African dust has been falling for thousands of years without obvious harm. On the other hand, it may only have been recently that populations of animals like seafans became sufficiently stressed to be susceptible to epidemics.

4. Patterson Sutherland, K. et al. (2010) Human sewage identified as likely source of white pox disease of the threatened Caribbean elkhorn coral, *Acropora palmata*. *Environmental Microbiology*, 12: 122–31.

5. Gardner, T. A. et al. (2003) Long-term region-wide declines in Caribbean corals. *Science*, 301: 958–60.

6. Arthur, K. et al. (2008) The exposure of green turtles (*Chelonia mydas*) to tumour promoting compounds produced by the cyanobacterium *Lyngbya majuscula* and their potential role in the aetiology of fibropapillomatosis. *Harmful Algae*, 7: 114–25.

7. Ward, J. R. and Lafferty, K. D. (2004) The elusive baseline of marine disease: are diseases in ocean ecosystems increasing? *PLoS Biology*, 2: 542–7.

8. Lafferty, K. D. et al. (2008) Reef fishes have higher parasite richness at unfished Palmyra Atoll compared to fished Kiritimati Island. *EcoHealth*, 5: 338–45.

9. Thornton, R. (1990) *American Indian Holocaust and Survival: A Population History Since 1492*. University of Oklahoma Press, Norman, Okla., pp. 26–32.

10. Lessios, H. (1988) Mass mortality of *Diadema antillarum* in the Caribbean: what have we learned? *Annual Reviews in Ecology and Systematics*, 19: 371–93.

11. Gulland, F. M. D. and Hall, A. J. (2007) Is marine mammal health deteriorating? Trends in global reporting of marine mammal disease. *EcoHealth*, 4: 135–50.

12. Harvell, D. et al. (2004) The rising tide of ocean diseases: unsolved problems and research priorities. *Frontiers in Ecology and Environment*, 2: 375–82. See also http://news.bbc.co.uk/1/hi/4729810.stm; accessed 16 February 2011.

13. Swart, R. L. de et al. (1996) Impaired immunity in harbour seals (*Phoca vitulina*) exposed to bioaccumulated environmental contaminants: review of a long-term feeding study. *Environmental Health Perspectives*, 104: 823–8.

14. Carey, C. (2000) Infectious disease and worldwide declines of amphibian populations, with comments on emerging diseases in coral reef organisms and in humans. *Environmental Health Perspectives*, 108: 143–50.

15. Nugues, M. M. (2002) Impact of a coral disease outbreak on coral communities in St. Lucia: what and how much has been lost? *Marine Ecology Progress Series*, 229: 61–71.

16. Watling, L. and Norse, E. A. (1998) Disturbance of the seabed by mobile fishing gear: a comparison to forest clearcutting. *Conservation Biology*, 12: 1180–97.

17. Ford, S. E. (1996) Range extension by the oyster parasite *Perkinsus marinus* into the Northeastern United States: response to climate change? *Journal of Shellfish Research*, 15: 45–56.

18. Sokolow, S. (2009) Effects of a changing climate on the dynamics of coral infectious disease: a review of the evidence. *Diseases of Aquatic Organisms*, 87: 5–18.

19. Extreme warmth of 31–35°C seems to damp down growth of pathogens. See Sokolow (2009), op. cit.

20. Webster, N. S. (2007) Sponge disease: a global threat? *Environmental Microbiology*, 9: 1363–75.

21. Bruno, J. F. et al. (2003) Nutrient enrichment can increase the severity of coral diseases. *Ecology Letters*, 6: 1056–61.

22. Wood, C. L. et al. (2010) Fishing out marine parasites? Impacts of fishing on rates of parasitism in the ocean. *Ecology Letters*, 13: 761–75.

23. Raymundo, L. J. et al. (2009) Functionally diverse reef-fish communities ameliorate coral disease. *Proceedings of the National Academy of Sciences*, 106: 17067–70; Lafferty, K. D. (2004) Fishing for lobsters indirectly increases epidemics in sea urchins. *Ecological Applications*, 14: 1566–73.

14. MARE INCOGNITUM

1. Cohen, J. E. (2003) Human population: the next half century. *Science*, 302: 1172–5.

2. FAO (1998) *State of Fisheries and Aquaculture*. UN Food and Agriculture Organization, Rome.

3. Clarke, S. (2004) *Shark Product Trade in Hong Kong and Mainland China and Implementation of Cites Shark Listings.* TRAFFIC East Asia, Hong Kong, China.

4. Montgomery, D. R. (2007) *Dirt: The Erosion of Civilizations.* University of California Press, Berkeley.

5. Richardson, A. J. et al. (2009) The jellyfish joyride: causes, consequences and management responses to a more gelatinous future. *Trends in Ecology and Evolution*, 24: 312–22.

6. Arai, M. N. (2005) Predation on pelagic coelenterates: a review. *Proceedings of the Marine Biological Association of the UK*, 85: 523–36.

7. Jackson, J. B. C. (2008) Ecological extinction and evolution in the brave new ocean. *Proceedings of the National Academy of Sciences*, 105: 11458–65.

8. Boyce, D. G. et al. (2010) Global phytoplankton decline over the past century. *Nature*, 466: 591–6. The 40 per cent decline in phytoplankton productivity since 1950 claimed in this study has been challenged as an artefact of differences in sampling methods over time. The authors have rejected this criticism although it seems well-founded. This is an area where more research is needed for clarity to emerge. See also Stramma, L. et al. (2008) Expanding oxygen-minimum zones in the tropical oceans. *Science*, 320: 655–8.

9. Roopnarine, P. et al. (2010) The whale pump: marine mammals enhance primary productivity in a coastal basin. *PLoS ONE*, e13255 doi: 10.1371/journal.pone.0013255; http://harvardmagazine.com/2011/05/why-whales; accessed 23 May 2011.

10. Iglesias-Rodriguez, M. D. et al. (2008) Phytoplankton calcification in a high CO_2 world. *Science*, 320: 336–40.

11. Riebesell, U. et al. (2000) Reduced calcification of marine plankton in response to increased atmospheric CO_2. *Nature*, 407: 364–7.

12. Moy, A. D. et al. (2009) Reduced calcification in modern Southern Ocean planktonic foraminifera. *Nature Geoscience*, doi: 10.1038/NGEO460.

13. Kolbert, E. (2011) Acid oceans. *National Geographic*, April, pp. 100–121.

14. Muir, J. (1911) *My First Summer in the Sierra.* Houghton Mifflin, Boston and New York.

15. Ward, P. and Myers, R. A. (2005) Shifts in open ocean fish communities coinciding with the commencement of commercial fishing. *Ecology*, 86: 835–47.

16. Myers, R. A. et al. (2007) Cascading effects of the loss of apex predatory sharks from a coastal ocean. *Science*, 315: 1846–50.

17. Airoldi, L. et al. (2008) The gray zone: relationships between habitat loss and marine biodiversity and their implications for conservation. *Journal of Experimental Marine Biology and Ecology,* 366: 8–15.

18. Nugues, M. M. (2004) Algal contact as a trigger for coral disease. *Ecology Letters,* 7: 919–23.

19. McIntyre, P. B. et al. (2007) Fish extinctions alter nutrient recycling in tropical freshwaters. *Proceedings of the National Academy of Sciences,* 104: 4461–6; Schindler, D. E. (2007) Fish extinctions and ecosystem functioning in tropical ecosystems. *Proceedings of the National Academy of Sciences,* 104: 5707–8.

15. ECOSYSTEMS AT YOUR SERVICE

1. Seaman, O. (2011) The uses of the ocean, in H. Watson, ed., *Ode to the Sea. Poems to Celebrate Britain's Maritime History.* National Trust Books, London, pp. 162–3.

2. Bellamy, J. C. (1843) *The Housekeeper's Guide to the Fish-market for Each Month of the Year, and an Account of the Fish and Fisheries of Devon and Cornwall.* Longman, Brown, Green and Longmans, London.

3. Shephard, S. et al. (2011) Benthivorous fish may go hungry on trawled seabed. *Proceedings of the Royal Society B,* doi: 10.1098/rspb.2010.2713.

4. Thurstan, R. H. et al. (2010) Effects of 118 years of industrial fishing on UK bottom trawl fisheries. *Nature Communications,* 1: 15, doi: 10.1038/ncomms1013.

5. US Department of Health (2008) *The Burden of Asthma in Washington State: 2008 Update.* DOH Pub 345-240 Rev 05/2009.

6. Millennium Ecosystem Assessment (2005) *Ecosystems and Human Wellbeing: Synthesis.* Island Press, Washington, DC.

7. Wackernagel, M. (2002) Tracking the ecological overshoot of the human economy. *Proceedings of the National Academy of Sciences,* 99: 9266–71.

8. Stern, N. (2007) *The Economics of Climate Change* (The Stern Review). The Cabinet Office, London. http://www.webcitation.org/5nCeyEYJr.

9. Myers, N. (1997) Environmental Refugees. *Population and Environment,* 19: 167–82.

10. Diamond, J. (2005) *Collapse. How Societies Choose to Fail or Survive.* Penguin Books, London.

11. Ibid.

16. FARMING THE SEA

1. Herubel, M. A. (1912) *Sea Fisheries. Their Treasures and Toilers.* T. Fisher Unwin, London.

2. To see this trend, we have to first correct for systematic over-reporting of Chinese catches. In a centrally planned economy, it seems it is better to cook the books than not meet production targets. We also have to take out Peruvian anchovy catches, which make up the largest single species landed in good years. Anchovy abundance jumps up and down in response to the El Niño-Southern Oscillation climate swings that alter the strength of upwelling in the eastern Pacific, and therefore dictate the productivity of these seas. Watson, R. and Pauly, D. (2001) Systematic distortions in world fisheries catch trends. *Nature*, 414: 534–6.

3. Thurstan, R. and Roberts, C. M. (in press) Health recommendations and global fish availability: are there enough fish to go around? The figures represent fish available after processing losses (inedible bits like heads, tails, bones and shells). Of course, these calculations are only a mind-game because fish are not distributed evenly, and not all of the global fish catch is eaten directly. For countries such as the Philippines and Sierra Leone, fish provide most of the dietary protein. For others, fish are hardly eaten at all; Brazil springs to mind as a place where restaurants seem to sell almost nothing that isn't red and meaty.

4. FAO (2010) *The State of World Fisheries and Aquaculture 2010.* UN Food and Agriculture Organization, Rome.

5. Costa-Pierce, B. A. (1987) Aquaculture in ancient Hawaii. *BioScience*, 37: 320–31.

6. Higginbotham, J. (1997) *Piscinae. Artificial Fishponds in Roman Italy.* University of North Carolina Press, Chapel Hill and London.

7. Pliny the Elder, *Naturalis Historiae*, Book 9, Chapter LIV.

8. Higginbotham (1997), op. cit.

9. Costa-Pierce (1987), op. cit.

10. http://www.teara.govt.nz/en/1966/fish-introduced-freshwater/1; accessed 8 March 2011.

11. Herubel (1912), op. cit.

12. There has recently been renewed interest in restocking and enhancing wild fisheries using hatchery-reared animals, but evidence for success is still limited. Bell, J. D. et al. (2006) Restocking and stock enhancement of coastal fisheries: potential, problems and progress. *Fisheries Research*, 80: 1–8.

13. Williams, J. (2006) *Clam Gardens: Aboriginal Mariculture on Canada's West Coast*. New Star Books, Vancouver.

14. Naylor, R. N. et al. (2009) Feeding aquaculture in an era of finite resources. *Proceedings of the National Academy of Science*, 106: 15103–10.

15. Ibid.

16. Lobo, A. S et al. (2010) Commercializing bycatch can push a fishery beyond economic extinction. *Conservation Letters*, 3: 277–85.

17. Tacon, A. G. J. and Metian, M. (2009) Fishing for feed or fishing for food: increasing global competition for small pelagic forage fish. *Ambio*, 38: 294–302.

18. Brashares, J. S. et al. (2004) Bushmeat hunting, wildlife declines, and fish supply in West Africa. *Science*, 306: 1180–83.

19. Schiermeier, Q. (2010) Ecologists fear Antarctic krill crisis. *Nature*, 467: 15.

20. Krkošek, M. et al. (2006) Epizootics of wild fish induced by farm fish. *Proceedings of the National Academy of Sciences*, 103: 15506–10.

21. Murray, A. G. (2002) Shipping and the spread of infectious salmon anemia in Scottish aquaculture. *Emerging Infectious Diseases*, 8: 1–5.

22. Cabello, F. C. (2006) Heavy use of prophylactic antibiotics in aquaculture: a growing problem for human and animal health and for the environment. *Environmental Microbiology*, 8: 1137–44.

23. Sapkota, A. et al. (2008) Aquaculture practices and potential human health risks: current knowledge and future priorities. *Environment International*, 34: 1215–26.

24. MacGarvin, M. (2000) *Scotland's Secret? Aquaculture, nutrient pollution, eutrophication and toxic algal blooms*. WWF Scotland, Aberfeldy, www.wwf.org.uk/fileLibrary/pdf/secret.pdf; accessed 24 May 2011.

25. Biao, X. and Kaijin, Y. (2007) Shrimp farming in China: operating characteristics, environmental impact and perspectives. *Ocean and Coastal Management*, 50: 538–50.

26. Primavera, J. H. (2006) Overcoming the impacts of aquaculture on the coastal zone. *Ocean and Coastal Management*, 49: 531–45.

27. *Caixin Times*, 9 September 2011. http://english.caixin.cn/2011-09-14/100304938.html; accessed 28 October 2011. See also Wang, B. et al. (2011) Water quality in marginal seas off China in the last two decades. *International Journal of Oceanography*, doi:10.1155/2011/731828.

28. Marris, E. (2010) Transgenic fish go large. *Nature*, 467: 259.

29. Cao, L. et al. (2007) Environmental impact of aquaculture and countermeasures to aquaculture pollution in China. *Environmental Science and Pollution Research*, 14: 452–62.

30. Azad, A. K. (2009) Coastal aquaculture development in Bangladesh: unsustainable and sustainable experiences. *Environmental Management*, 44: 800–809.

31. Cressey, D. (2009) Future fish. *Nature*, 458: 398–400.

17. THE GREAT CLEAN-UP

1. Piskaln, C. H. et al. (1998) Resuspension of bottom sediment by bottom trawling in the Gulf of Maine and potential geochemical consequences. *Conservation Biology*, 12: 1223–29.

2. Stockholm Convention: chm.pops.int/Countries/StatusofRatification/tabid/252/language/en-US/Default.aspx; accessed 2 January 2012. The number of countries is a rough total since it often changes.

3. Jones, O. A. H. et al. (2007) Questioning the excessive use of advanced treatment to remove organic micropollutants from wastewater. *Environmental Science and Technology*, 41: 5085–9.

4. http://www.guardian.co.uk/environment/2011/may/04/eu-fishermen-catch-plastic; accessed 24 May 2011.

5. Thompson, R. C. et al. (2009) Plastics, the environment and human health: current consensus and future trends. *Philosophical Transactions of the Royal Society B*, 364: 2153–66.

6. O'Brine, T. and Thompson, R. C. (2010) Degradation of plastic carrier bags in the marine environment. *Marine Pollution Bulletin*, 60: 2279–83.

7. http://www.mcsuk.org/downloads/pollution/beachwatch/latest2011/Methods%20&%20Results%20BW10.pdf; accessed 28 December 2011.

8. Page, B. et al. (2004) Entanglement of Australian sea lions and New Zealand fur seals in lost fishing gear and other marine debris before and after Government and industry attempts to reduce the problem. *Marine Pollution Bulletin*, 49: 33–42; Arnould, J. P. Y. and Croxall, J. P. (1995) Trends in entanglement of Antarctic fur seals (*Arctocephalus gazella*) in man-made debris at South Georgia. *Marine Pollution Bulletin*, 30: 707–12.

9. Mee, L. D. et al. (2005) Restoring the Black Sea in times of uncertainty. *Oceanography*, 18: 100–111.

10. Conley, D. J. et al. (2009) Tackling hypoxia in the Baltic Sea: is engineering a solution? *Environmental Science and Technology*, 43: 3407–11.

11. Weaver, D. (2008) Environmental impacts of bottom trawling suspended solids generation. Report for the United Anglers of Southern California, http://web.me.com/deweaver/bottom_trawling/Links_to_Docs_files/Bottom_Trawling3.pdf; accessed 24 May 2011. See also Palanques, A.

et al. (2001) Impact of bottom trawling on water turbidity and muddy sediment of an unfished continental shelf. *Limnology and Oceanography*, 46: 1100–1110.

12. Kunzig, R. (2011) Seven billion. *National Geographic*, January, pp. 42–63. In 1900, there were an estimated 1.6 billion people on planet Earth. In 2011, the total reached 7 billion.

18. CAN WE COOL OUR WARMING WORLD?

1. Keith, D. W. (2009) Why capture CO_2 from the atmosphere? *Science*, 325: 1654–5.
2. Bollmann, M. et al. (2010) *World Ocean Review. Living With the Ocean.* Maribus, GmbH, Hamburg.
3. Lindeboom, H. J. et al. (2011) Short-term ecological effects of an offshore wind farm in the Dutch coastal zone: a compilation. *Environmental Research Letters*, doi:10.1088/1748-9326/6/3/035101.
4. http://www.eurekalert.org/pub_releases/2008-11/ci-wnc110708.php; accessed 21 March 2011.
5. We will also have to deal with sources of methane. However, as methane has a much shorter life in the atmosphere than carbon dioxide (a half life of about seven years as compared to about a century for carbon dioxide), direct methane removal will not be necessary – only reduced emissions.
6. Jones, I. S. F. and Young, H. E. (2009) The potential of the ocean for the management of global warming. *International Journal of Global Warming*, 1: 43–56.
7. Strong, A. L. et al. (2009) Ocean fertilization. Science, policy and commerce. *Oceanography*, 22: 236–61.
8. Anon. (2009) Carbon sequestration. *Science*, 325: 1644–5.
9. Haszeldine, R. S. (2009) Carbon capture and storage: how green can black be? *Science*, 325: 1647–52.
10. Rasch, P. J. et al. (2009) Geoengineering by cloud seeding: influence on sea ice and climate system. *Environmental Research Letters*, 4: 1–8.
11. Keith, D. W. et al. (2010) Research on global sunblock needed now. *Nature*, 463: 426–7.
12. Turner, W. R. et al. (2009) A force to fight global warming. *Nature*, 462: 278–9.
13. Nellemann, C. et al. (eds.) (2009) *Blue Carbon. A Rapid Response Assessment.* United Nations Environment Programme, GRID-Arendal, www.grida.no.

14. These figures seem high to me and may not stand up to long-term scrutiny. But they are the best we have at the moment.

19. A NEW DEAL FOR THE OCEANS

1. Hueber, A. (2011) Fake seagrass could boost fish numbers. *New Zealand Herald*, 30 January.
2. http://www.outdooralabama.com/fishing/saltwater/where/artificial -reefs/reefhist.cfm; accessed 11 April 2011.
3. All US States have regulations that require contaminant chemicals like engine oil to be removed before dumping.
4. Community of Arran Seabed Trust: www.arrancoast.com.
5. Howarth, L. M. et al. (2011) Complex habitat boosts scallop recruitment in a fully protected marine reserve. *Marine Biology*, doi: 10.1007/s00227-011-1690-y.
6. Alcala, A. C. and Russ, G. R. (1990) A direct test of the effects of protective management on abundance and yield of tropical marine resources. *Journal du Conseil Internationale pour L'Exploration de la Mer*, 46: 40–47; Russ, G. R. and Alcala, A. C. (1996) Marine reserves: rates and patterns of recovery and decline of large predatory fish. *Ecological Applications*, 6: 947–61; Russ, G. R. and Alcala, A. C. (2011) Enhanced biodiversity beyond marine reserve boundaries: The cup spillith over. *Ecological Applications*, 21: 241–50.
7. García-Charton, J. et al. (2008) Effectiveness of European Atlanto-Mediterranean MPAs: do they accomplish the expected effects on populations, communities and ecosystems? *Journal for Nature Conservation*, 16: 193–221; Ault, J. S. et al. (2006) Building sustainable fisheries in Florida's coral reef ecosystem: positive signs in the Dry Tortugas. *Bulletin of Marine Science*, 78: 633–54.
8. Here is a way you can visualize how marine reserves help to steady the variability of fish stocks. Think of habitats in the sea as being like a sponge that is capable of holding only so much life. Imagine that sponge is held beneath a tap that rains new life upon it to replace the life that dribbles away through mortality by natural and human causes. The tap sometimes flows faster, sometimes more slowly depending on how favourable conditions are for survival of young animals and plants. Fishing reduces reproduction so the tap merely trickles and the sponge is only occasionally saturated. Creating marine reserves turns the tap on more forcefully. Although its flow still varies, the sponge is kept full.

9. Thurstan, R. et al. (2010). The effects of 118 years of industrial fishing on UK bottom trawl fisheries. *Nature Communications*, 1: 15. doi: 10.1038/ncomms1013.

10. *Fishing News*, 15 April 2011.

11. Mascia, M. B. et al. (2010) Impacts of marine protected areas on fishing communities. *Conservation Biology*, 24: 1424–9.

12. Goñi, R. et al. (2008) Spillover from six western Mediterranean marine protected areas: evidence from artisanal fisheries. *Marine Ecology Progress Series*, 366: 159–74.

13. Murawski, S. A. et al. (2005) Effort distribution and catch patterns adjacent to temperate MPAs. *ICES Journal of Marine Science*, 62: 1150–67.

14. Hoskin, M. G. et al. (2011) Variable population responses by large decapod crustaceans to the establishment of a temperate marine no-take zone. *Canadian Journal of Fisheries and Aquatic Sciences*, 68: 185–200.

15. PISCO (2011) The science of marine reserves (European edition), www.piscoweb.org. Lester, S. E. et al. (2009) Biological effects within no-take marine reserves: a global synthesis. *Marine Ecology Progress Series*, 384: 33–46.

16. Svedäng, H. (2010) Long-term impact of different fishing methods on the ecosystem in the Kattegat and Öresund. European Parliament, IP/B/PECH/IC/2010_24.

17. Schindler, D. W. et al. (2010) Population diversity and the portfolio effect in an exploited species. *Nature*, 465: 609–12.

18. Kean, S. (2010) The secret lives of ocean fish. *Science*, 327: 264.

19. Unpublished research by myself and colleagues Julie Hawkins, Gemma Aitken and John Bainbridge, Environment Department, University of York.

20. LIFE REFURBISHED

1. The World Bank (2009) *The Sunken Billions. The Economic Justification for Fisheries Reform*. The World Bank, Washington, DC and the UN Food and Agriculture Organization, Rome.

2. Froese, R. and Proelß, A. (2010) Rebuilding fish stocks no later than 2015: will Europe meet the deadline? *Fish and Fisheries*, doi: 10.1111/j.1467-2979.2009.00349.x

3. Worm, B. et al. (2009) Rebuilding global fisheries. *Science*, 325: 578–85.

4. O'Leary, B. et al. (2011) Fisheries mismanagement. *Marine Pollution Bulletin*, 62: 2642–8.

5. Prince Charles made this comment in a speech at St James's Palace, London in March 2011.

6. Bromley, D. W. (2009) Abdicating responsibility: the deceits of fisheries policy. *Fisheries*, 34: 280–90.

7. Essington, T. E. (2010) Ecological indicators display reduced variation in North American catch share fisheries. *Proceedings of the National Academy of Sciences*, 107: 754–9.

8. The Seth Macinko quote is from a talk I heard him give in Arran, Scotland in 2010.

9. Schrope, M. (2010) What's the catch? *Nature*, 465: 540–42.

10. Fox, C. J. (2010) West coast fishery trials of a twin rigged *Nephrops* trawl incorporating a large mesh topsheet for reducing commercial gadoid species bycatch. Scottish Industry Science Partnership Report Number 03/10. Marine Scotland.

11. I offer more detailed advice in Appendix 1 of this book.

12. Wooldridge, S. A. and Done, T. J. (2009) Improved water quality can ameliorate effects of climate change on corals. *Ecological Applications*, 19: 1492–9.

13. Gibson, L. and Sodhi, N. S. (2010) Habitats at risk: a step forward, a step back. *Science*, 331: 1137.

14. Stern, N. (2007) *The Economics of Climate Change* (The Stern Review). The Cabinet Office, London.

15. Mörner, N. et al. (2004) New perspectives for the future of the Maldives. *Global and Planetary Change*, 40: 177–82.

21. SAVING THE GIANTS OF THE SEA

1. Swimmer, S. et al. (2010) Sustainable fishing gear: the case of modified circle hooks in a Costa Rican longline fishery. *Marine Biology*, 158: 757–67. The authors tested a modified hook that would reduce these horrific bycatch rates. The new hooks reduced the number of turtles caught per mahi-mahi from 2.3 to 'just' 1.7, but the authors concluded they would not be acceptable to the fishing industry because they also reduced capture rates of mahi-mahi by 15 per cent.

2. Edwards, E. F. (2007) Fishery effects on dolphins targeted by tuna purse-seiners in the eastern tropical Pacific Ocean. *International Journal of Comparative Psychology*, 20: 217–27. See also Cramer, K. L. et al. (2008) Declines in reproductive output in two dolphin populations depleted by the yellowfin tuna purse-seine fishery. *Marine Ecology Progress Series*, 369: 273–85.

3. The best option for guilt-free tuna is 'pole and line' caught. Look out for this on the label. Anything that doesn't mention pole and line will probably have been caught with longlines or purse seines, so many other animals will have been killed to put your fish in the can.

4. See www.topp.org for exciting examples of how tagging can reveal the enigmatic lives of ocean wanderers.

5. Fujiwara, M. and Caswell, H. (2001) Demography of the endangered North Atlantic right whale. *Nature*, 414: 537–41.

6. Personal communication from Guy Stevens, Maldivian Manta Ray Project.

7. This effort involved many people. My team comprised Beth O'Leary, Rachel Brown, Melanie O'Rourke, Andrew Davies and Tina Molodtsova. Of the rest, particular credit should go to the German delegation to OSPAR under the direction of Henning von Nordheim, together with Jeff Ardron and Tim Packeiser, WWF International, and the delegations of the Netherlands and Portugal, all ably assisted by David Johnson and his team within the OSPAR secretariat.

8. O'Leary, B. C. et al. (2012) The first network of marine protected areas (MPAs) in the high seas: the process, the challenges and where next. *Marine Policy*, 36: 598–605.

9. Iceland tried to scupper the North Atlantic marine protected areas, but fortunately were persuaded against it this time. I find it ironic that a country that knows more about environmental degradation than almost any other should wish the same upon the world oceans. In truth, I think the majority of Icelanders don't want that. From what I have heard it comes down more to the prejudices of their international negotiators whose worldview was forged in the 1960s' fog of cod wars and unilateral enlargement of sovereign waters.

10. Haeckel, E. (1899) *Kunstformen der Natur*. Bibliographisches Institut, Leipzig.

22. PREPARING FOR THE WORST

1. Barnosky, A. D. et al. (2011) Has the world's sixth mass extinction already arrived? *Nature*, 471: 51–7.

2. Lutz, W. et al. (2001) The end of world population growth. *Nature*, 412: 543–5.

3. Butchart, S. H. M. et al. (2010) Global biodiversity: indicators of recent declines. *Science*, 328: 1164–8.

4. The Chagossian people were forcibly removed from their homeland by the UK Government in the 1960s and 1970s to make way for the US military base.

5. Sheppard, C. R. C. et al. (2008) Archipelago-wide coral recovery patterns since 1998 in the Chagos Archipelago, central Indian Ocean. *Marine Ecology Progress Series*, 362: 109–17.

6. In late 2011, the Australian government launched a consultation on a proposal to turn the whole Coral Sea into a Marine Protected Area, half of which would be protected from all exploitation. Australian Government Department of Sustainability, Environment, Water, Population and Communities: http://www.environment.gov.au/coasts/mbp/coralsea/consultation/index.html; accessed 3 January 2012.

7. http://www.unep-wcmc.org/medialibrary/2011/06/23/b3b09e87/Gough%20and%20Inaccessible%20Islands.pdf; accessed 2 January 2012.

8. Roberts, C. M. et al. (2002) Marine biodiversity hotspots and conservation priorities for tropical reefs. *Science*, 295: 1280–84; Tittensor, D. P. et al. (2010) Global patterns and predictors of marine biodiversity across taxa. *Nature*, 466: 1098–101; Trebilco, R. et al. (2011) Mapping species richness and human impact drivers to inform global pelagic conservation prioritisation. *Biological Conservation*, 144: 1758–66.

9. Fuller, R. A. et al. (2010) Replacing underperforming protected areas achieves better conservation outcomes. *Nature*, 466: 365–7. To be fair to the authors they did mention in passing that their cost–benefit metric might be replaced by another that measured amenity value, but this seemed very much an afterthought.

10. In other words, to levels required to achieve the maximum long-term yield from a stock, i.e. maximum sustainable yield.

11. Roberts, C. M. et al. (2008) Guidance on the size and spacing of Marine Protected Areas in England. Natural England Commissioned Report NECR037, Peterborough.

12. Munday, P. L. et al. (2008) Climate change and coral reef connectivity. *Coral Reefs*, 28: 379–95.

13. Alward, G. L. (1932) *The Sea Fisheries of Great Britain and Ireland*. Albert Gait, Grimsby.

Index

diagrams, figures and maps are given in italics